The Postcolonial

Science and Technology

Studies Reader

T0261764

# The Postcolonial Science and Technology Studies Reader

EDITED BY

Sandra Harding

Duke University Press  Durham and London  2011

© 2011 Duke University Press
All rights reserved
Printed in the United States of America on acid-free paper ∞
Designed by C. H. Westmoreland
Typeset in Minion with Magma Compact by Achorn International, Inc.
Library of Congress Cataloging-in-Publication Data appear on the
last printed page of this book.

# Contents

PREFACE ix

ACKNOWLEDGMENTS XV

INTRODUCTION  Beyond Postcolonial Theory:
Two Undertheorized Perspectives on Science and Technology 1


I.  Counterhistories 33

1. Discovering the Oriental West
*John M. Hobson* 39

2. Long-Distance Corporations, Big Sciences, and
the Geography of Knowledge
*Steven J. Harris* 61

3. Heroic Narratives of Quest and Discovery
*Mary Terrall* 84

4. Maria Sibylla Merian: A Woman of Art and Science
*Ella Reitsma* 103

5. Prospecting for Drugs: European Naturalists in the West Indies
*Londa Schiebinger* 110

6. Science and Colonial Expansion:
The Role of the British Royal Botanic Gardens
*Lucile H. Brockway* 127

7. Out of Africa: Colonial Rice History in the Black Atlantic
*Judith Carney* 140


II.  Other Cultures' Sciences 151

8. Navigation in the Western Carolines: A Traditional Science
*Ward H. Goodenough* 159

9. Science for the West, Myth for the Rest? The Case of James Bay
Cree Knowledge Construction
*Colin Scott* 175

10. Ecolinguistics, Linguistic Diversity, Ecological Diversity
*Peter Mühlhäusler* 198

11. Gender and Indigenous Knowledge
*Helen Appleton, Maria E. Fernandez, Catherine L. M. Hill,
and Consuelo Quiroz* 211

12. Whose Knowledge, Whose Genes, Whose Rights?
*Stephen B. Brush* 225

13. The Role of the Global Network of Indigenous
Knowledge Resource Centers in the Conservation
of Cultural and Biological Diversity
*D. Michael Warren* 247

III. Residues and Reinventions 263

14. Development and the Anthropology of Modernity
*Arturo Escobar* 269

15. Tradition and Gender in Modernization Theory
*Catherine V. Scott* 290

16. Security and Survival:
Why Do Poor People Have Many Children?
*Betsy Hartmann* 310

17. Call for a New Approach
*Committee on Women, Population, and the Environment* 318

18. The Human Genome Diversity Project: What Went Wrong?
*Jenny Reardon* 321

19. Bioprospecting's Representational Dilemma
*Cori Hayden* 343

IV. Moving Forward: Possible Pathways 365

20. Islamic Science: The Contemporary Debate
*Ziauddin Sardar* 373

21. Mining Civilizational Knowledge
*Susantha Goonatilake* 380

22. Towards the Integration of Knowledge Systems:
Challenges to Thought and Practice
*Catherine A. Odora Hoppers* 388

23. Human Well-Being and Federal Science:
What's the Connection?
*Daniel Sarewitz* 403

24. Science in an Era of Globalization: Alternative Pathways
*David J. Hess* 419

25. Civic Science for Sustainability: Reframing the Role of Experts,
Policy-Makers, and Citizens in Environmental Governance
*Karin Bäckstrand* 439

COPYRIGHT ACKNOWLEDGMENTS 459

INDEX 463

# Preface

Critical discussions of Western colonialism and imperialism and of what the term *postcolonialism* could mean, require, and enable first became acceptable in literature and cultural studies departments in the United States some three decades ago. Yet it has been much harder to create such discussions of sciences and technologies. Especially resistant are those departments where the West's scientific rationality and technical expertise have long been lovingly explained and "served up" for use in corporate and nationalist policies: sociology, philosophy, economics, and international relations, as well as the natural sciences themselves. Now such discussions are beginning to arise here and there in government departments, international agencies, and research disciplines. The introduction of postcolonial issues in Western universities is a powerful political and intellectual move, contrary to the view of some skeptics. As the historian Robert Young points out in the epigraph that begins this volume's introduction, the institutionalization of postcolonialism in higher education marks "the first time . . . [that] the power of western academic institutions has been deployed against the west." To focus these kinds of critical perspectives on sciences, technologies, and their philosophies is to direct them to the most fundamental rationale invoked by the West in support of its claims to unique, universally valid achievements and, consequently, political entitlement.

A number of circumstances have conjoined to increase the possibility of such discussions. The West's so-called development programs for the Third World have been recognized as failures for some time now, even in the West. Those programs largely created de-development and maldevelopment for the majority of the world's most vulnerable citizens while further enriching already advantaged elites in the West and their allies around the globe. Western scientific rationality and technical expertise have been perceived as the motor of progress for these development projects from the beginning, so the failures of development raise questions about the value of Western rationality and expertise. Another influence has been the recent outpouring of empirically and theoretically well-grounded postcolonial analyses in English (as well as in other languages) on scientific and technological issues. Much of this work has arisen from the intellectual arms of activist groups seeking to serve local needs and interests rather than those of transnational corporations and foreign governments and investors. Two central focuses have been the development of anti-Eurocentric primary, secondary, and university curricula, and of appropriate and sustainable forms of development.

Equally important, in the last three decades, many U.S. campuses have been transformed into sites of global education in several senses of the term. They have experienced a huge increase in the numbers of students who either come from around the globe or are U.S. citizens interested in the cultural legacies and current situations of the homelands of their foreign-born parents or partners. Moreover, researchers and scholars born or with roots in the Third World have joined the most prestigious faculties. They have produced flourishing literatures on the nature and effects on science agendas of colonialism, postcoloniality, postcolonialism, Western neocolonialism, the new global empire, nationalisms, and indigenous knowledge, and also on the effects of Western sciences on such phenomena. Furthermore, U.S.-born students and faculty travel in the Third World more often than was the case even one generation ago. They often return with more realistic understandings of how global economic, political, and social relations work than can easily be gathered at home. They also bring back networks of friends and intellectual colleagues from other countries. Obviously these trends are more visible on some campuses than on others. Yet across the country and around the globe there is increasing interest in providing the kinds of education that young people will need to make intelligent decisions in their lives and work. This collection intends to contribute to such projects.

Additionally, the United Nations Educational, Scientific, and Cultural Organization (UNESCO), as well as other international and national agencies and institutions, has sponsored conferences around the world on the topics of this volume. By now a number of special issues of journals in the fields of science studies and postcolonial studies have attempted to integrate the literary-cultural, political, and science-technology sets of concerns, and several journals devoted to such topics have appeared. They join an older array of journals and books on topics such as ethnobotanies and ethnomathematics produced by anthropologists, botanists, and historians. Gone are the days when it could appear uncontroversial to assume that Western sciences are or ever have been autonomous from society, value free, and maximally objective, or that their standard for rationality is universally valid. To be sure, there is much in this Enlightenment legacy to be cherished today, as many authors here will argue in different ways. Yet it remains as much a challenge to separate the still valuable and useful from the archaic and dangerous in these Western traditions as it does in other cultures' traditions. There are still many scientists and citizens who have never heard of postcolonial issues, or at least not ones focused on sciences and technologies. And there are others who have heard of them but cling to the illusion that conventional Western views of our sciences and technologies might turn out after all to be viable. But their numbers have radically dwindled. Their articulation of such views is increasingly regarded as revealing more about them and the unfortunate limitations of their intellectual worlds than about the sciences they practice and value.

This new outpouring of research and scholarship, conferences and curricula, underlines the significance of older philosophical questions and generates new ones. As readers will see, it is not easy to provide satisfying answers to such questions. What are the best alternatives to the impossible choice of absolutism or relativism for our knowledge claims? Must the standard regulative ideals of Western scientific research, such as objectivity, rationality, and good method, be abandoned? Or can and should they be transformed and made useful in the environment of both these new historical and social studies of the sciences and activist attempts around the globe to transform our sciences and their philosophies in ways called for by postcolonial criticisms? Is it only a romantic view to claim that other cultures' knowledge systems offer serious challenges to how the West conceptualizes the value of its own sciences and technologies? Or could it be a realistic recognition of desirable—to "us" as well as to "them"—ongoing projects around the globe? What could and should be the relation between other cultures' still-useful knowledge systems and modern Western sciences? Who should determine the best answers to such questions? This collection is intended to respond to such concerns from many global standpoints.

For many years I have wanted to be able to assign students a book such as this one. Yet that splendidly flourishing literature on the topics in focus here has been scattered across many disciplines, languages, and regional publications. Important parts of it speak primarily to disciplinary specialists, especially in the history of science and medical anthropology. I kept hoping that someone else would produce a reader like this one.

I have worked in this field for two decades. In the early 1990s, I published a reader intended to make it easier to teach courses that addressed a number of issues about race and science that had at that point been located primarily in particular disciplinary literatures (*The "Racial" Economy of Science*). Later I attempted to provide for upper-level undergraduate and graduate students in the United States an account of how Western sciences and their philosophies were implicated in the ways in which the so-called voyages of discovery and the emergence of modern sciences in Europe (rather than elsewhere) each needed the success of the other for its own successes (*Is Science Multicultural?*). With Rob Figueroa I codirected a National Science Foundation grant to the American Philosophical Association that addressed issues of diversity in the philosophy of science. This resulted in a dozen or so programs at APA conferences and an edited collection (Figueroa and Harding, *Science and Other Cultures*). Because I am trained as a philosopher, my own contribution to such projects has been to ask what implications the new economic, political, historical, and social accounts of the origins, successes, and practices of modern Western sciences have for the powerful Western philosophies of science. And I had begun to try to perceive the links between postcolonial and gender issues—to bring each to bear on the other. With Uma Narayan I coedited a double issue of a philosophy journal on such topics (*Decentering the Center*). My horizons were broadened also by consultations on gender, science, and technology issues to several United Nations organizations, such as the Pan American Health Organization, the United Nations Development Fund for Women, the United Nations Committee on Science and Technology for Development, and UNESCO (Harding and McGregor, *Science and Technology: The Gender Dimension*). I also learned a lot from colleagues and students through visiting professorships, conferences, and lectures on six continents.

This collection draws together recent publications with materials I have been teaching for a decade or more, though the emphasis is on the new work. As is usual for such volumes, it has been painful to have to restrict the readings to such a small number. There are at least another dozen for each section that would add additional valuable perspectives. (Some of these are indicated in the introductions to each section.) And there are many more that are probably favorites of the increasing numbers of researchers and scholars working in these areas. But we must find ways to invite students and colleagues to start in on these issues in a manageable way. I hope that the particular kind of map of the field provided here will be supplemented by additional materials with which each teacher will be familiar—different materials for each of the relevant disciplines and parts of the world that are of interest. Many faculty have great depth of expertise in one or another area of postcolonial science and technology issues, but little in other relevant areas. This collection intends to emphasize the importance of each of those research, scholarly, and policy expertises by setting them in broader contexts. For example, in my own (quarter-system) classes, I add more material from science studies, epistemology, philosophy of science, gender studies, postcolonial theory, and science education. In graduate seminars, this added material amounts to about one-third of the course materials (readings and lectures). In undergraduate classes, such additions are made primarily through lectures.

In important respects, the content of this book takes readers beyond existing postcolonial theory. Readers can ask how the issues raised in these essays have implications for each and every nook and cranny in every research discipline. Moreover, I have tried to signal a greater multiplicity of sites around the globe where colonial and postcolonial issues arise. Yet I must admit that there is little here specifically on important parts of the world such as Eastern Europe, China, Japan, Australia, or the Middle East. Readers will recognize that the British and American colonialisms and imperialisms receive the most attention. This collection has focused on presenting as broad and deep as possible a range of issues, but at the cost of lacking equal representation to different areas of the globe, which is in itself an important issue with implications for the histories of science and technology outside missing areas.

Chapters have been selected for their accessibility to general readers while retaining the freshness, complexity, and sophistication in the ways these works pose relevant challenges. They have also been selected for the diversity of issues that they collectively address, and the interesting ways in which they raise the philosophical issues mentioned earlier. The epigraphs that begin each of the book's four parts alert readers to relevant influential earlier insights, as well as to significant issues that could not be represented by full texts. My introduction to the book identifies a number of issues and challenges that readers can pursue through the four parts, and the part introductions set each part in relevant contexts.

This is an especially auspicious time to be considering these fascinating but often perplexing issues. Yearnings for equitable social transformation have been arising with increasing force around the globe and within virtually every society. In each case they are met with resistance—sometimes well intentioned, alas. Recognition of the challenges that such issues raise for different groups, as well as of the array of strategies already in use to meet them, can contribute to more thoughtful discussions of where "we" are (whosoever you readers may be). They can also help us think more clearly about where we could and should be headed, and which actions we must pursue to see beyond the next few steps. I hope readers enjoy reflecting on these challenges as much as I have.

# Acknowledgments

Comments from many people have strengthened this collection. For reviews of earlier drafts of the introductions, I am grateful to Suman Seth, Sharon Traweek, Nancy Tuana, Marguerite Waller, and the anonymous reviewers for Duke University Press. For illuminating responses to the book's themes and many of the chapters, I thank the students in my graduate seminars and undergraduate classes on science and technology in a multicultural, postcolonial, and gendered world in the Departments of Education, Women's Studies, and Philosophy at the University of California, Los Angeles, from 2000 to 2009. I am indebted to seminar participants and lecture audiences on many university campuses for illuminating discussions of this project, especially at the University of Wisconsin and University of California, Riverside.

For provocative conversations and for their own pathbreaking work, I am grateful to many courageous and brilliant scholars and activists, especially to Warwick Anderson, Cynthia Enloe, Ulrich Beck, Susantha Goonatilake, David Hess, Françoise Lionnet, Jim Maffie, Ziauddin Sardar, Suman Seth, Shu-mei Shih, and Ann Tickner, as well as to all the contributors to this collection.

For his wise, patient, and gentle guidance, I thank Ken Wissoker at Duke University Press. Neva Pemberton provided superb assistance in the permissions process and in preparing the manuscript.

For nourishing my soul throughout this project, I thank Emily Abel, Françoise Lionnet, Gail Kligman, Sara Melzer, Marguerite Waller, Alice Wexler, and the many other friends in Los Angeles who have listened to my woes and pleasures on mountain hikes, over dinner, and in chance encounters. My deepest gratitude for the delights of daily life goes to my dear housemates Emily and Eva.

# Beyond Postcolonial Theory:

## Two Undertheorized Perspectives on

## Science and Technology

Orientalism depends for its strategy on this flexible positional superiority, which puts the Westerner in a whole series of possible relationships with the Orient without ever losing him the relative upper hand. And why should it have been otherwise, especially during the period of extraordinary European ascendancy from the late Renaissance to the present? . . . There emerged a complex Orient suitable for study in the Academy, for display in the museum, for reconstruction in the colonial office, for theoretical illustration in anthropological, biological, linguistic, racial, and historical theses about mankind and the universe, for instances of economic and sociological theories of development, revolution, cultural personality, national or religious character.—EDWARD SAID, *Orientalism*

Resistance to the critique of Eurocentrism is always extreme, for we are here entering the realm of the taboo. The calling into question of the Eurocentric dimension of the dominant ideology is more difficult to accept even than a critical challenge to its economic dimension. For the critique of Eurocentrism directly calls into question the position of the comfortable classes of this world. —SAMIR AMIN, *Eurocentrism*

Historically it was activists and intellectuals in or from the colonies and newly decolonized nations that most effectively articulated the opposition to colonialism, imperialism, and eurocentrism; these critiques were allied to those developed in the west. What is so striking in retrospect is the sheer energy, volume, and heroic commitment of the intellectual as well as political opposition to colonialism, and that productively continued into the postcolonial period. Postcolonial studies has developed that work to give it a disciplinary focus, and foregrounds its significance. For the first time, in a move that was the very reverse to that which Said describes in *Orientalism* (1978), the power of western academic institutions has been deployed against the west. For the first time, in the western academy, postcolonial subjects become subjects rather than the objects of knowledge. For the first time, tricontinental knowledge, cultural and political practices have asserted and achieved more or less equal institutional status with any other.—ROBERT J. YOUNG, *Postcolonialism*

ACCORDING TO WESTERN policymakers after World War II, the world peace that so many desired required greater investment in scientific and technical research.[1] World peace could not occur without democratic social relations, and this in turn required economic prosperity for all societies. Poverty drove desperate peoples to support irrational beliefs of the sort that had led to World War II. It was only Western scientific rationality and technical expertise that could boost economic prosperity for poor societies, thereby attracting people to rational forms of political participation. Consequently, it was the duty of Western societies to increase their scientific and technical research and to disseminate the results to poor societies. The newly established United Nations, joined by many Western countries, moved quickly to set up agencies to deliver economic development to poor societies around the globe. The green revolution in agriculture was just one of the results of such research projects.

This way of looking at science and social progress is grounded in modernization theory, which is itself rooted in the Enlightenment belief in the beneficial powers of scientific rationality. The West's sciences and technologies were supposed to be the jewels in the crown of modernity. To achieve social progress, value-neutral scientific rationality and technical expertise must replace traditional religious beliefs, myths, and superstitions about nature and social relations. To be sure, valuable aspects of this legacy endure. Some have said that we need much more rationality and modernity to engage in realistic and democratic ways with global challenges today (Harding 2008).

Yet the way this view is articulated in the preceding two paragraphs obscures perhaps as much as it reveals about science and society in history. Indeed, at the very moment that leaders of U.S. scientific institutions were proclaiming the autonomy of science from society as a reason to support increased funding for scientific and technical research, both the United States and its allies as well as the Soviet Union were vigorously directing research toward projects intended to win superiority in the Cold War arms race. Would this arms race bring world peace? Many believed it would. At any rate, at the time it seemed preferable to another "hot war," this time with nuclear weapons.

Moreover, the U.S. Congress noticed that it had been permitted no oversight over the huge expenditure of taxpayer funds in the Manhattan Project, which had created the atomic bomb. Nor would Congress get to have such oversight in how the newly established National Science Foundation would be distributing federal funds. As one historian notes, the "autonomy of science" rhetoric from leaders of the scientific community was specifically intended to forestall government "meddling" in the agendas and practices of the scientific community. Science was already a "little democracy," proclaimed spokesmen for the scientific community, so it needed no government oversight of the sort taxpayers usually expected (Hollinger 1996). Most of the time, this is widely regarded as a sensible precaution to protect research from the shifting winds of political whim. However, skeptics could well wonder if all this research directly spon- sored and directed by government interests, in addition to corporate in- terests, should still be regarded as economically, politically, socially, and culturally value neutral.

Soon the "unaligned nations," as they were named by the Cold War participants, transformed themselves into the Third World. Many Third World intellectuals began strategizing about what should be the science and technology policies of their own newly independent countries, since these countries were no longer under formal Western rule. As part of this project, they analyzed the contributions that Western sciences had pro- vided to colonialism and that colonialism, in turn, had provided to West- ern sciences. These new histories began to appear as early as 1959. In that year an influential essay by Frantz Fanon was published, which dem- onstrated that under colonialism, just as under Nazi rule, doctors were complicit with "state-sanctioned barbarism" (Fanon 2002). Philip Cur- tin's analysis of how Western medical achievements made it possible for Europeans to colonize the interior of Africa appeared two years later. One of his examples was the development of quinine for use against malaria (Curtin 1961). More counterhistories to the standard Western accounts of the history of science followed, alongside analyses by anthropologists and biologists of the strengths of traditional health, agriculture, and environ- mental practices. These produced additional reasons to question central assumptions of modernization theory and its Enlightenment-grounded philosophy of science.[2]

Now, five decades later, a good-sized literature has further developed a postcolonial framework for thinking about sciences and technologies.[3] It has produced startling insights about how sciences function in the everyday push and pull of local and global political, economic, social, and cultural relations. These intellectuals have argued from the begin- ning that modern Western sciences have been "epistemologically under- developed"; they lacked the resources necessary to recognize their own locations in social relations and history. It is the agendas of this work, including the policy debates, that lie behind most of the essays in this collection.[4]

It remains puzzling that the issues raised in this literature are only now beginning to attract the attention of broader audiences in the West. One can find relatively little engagement with these postcolonial science and technology writings in university curricula or in relevant research fields such as the sciences themselves or the philosophy and social studies of science. (Histories of non-Western sciences and the anthropology of medicine provide important exceptions here.) The postcolonial writings mostly seem to be over the horizon and out of view of other kinds of lively discussions of science and technology issues in our universities and research fields. However, in the last few years, promising signs of more robust encounters with issues raised in the postcolonial writings have begun to appear. A few journals in the field of science and technology studies have published special issues devoted to such topics.[5] The third edition of a prominent science and technology studies handbook included a provocative review article (Anderson and Adams 2007). A leading journal in postcolonial studies published its first issue devoted to the topic (Seth 2009a). Yet in 2009 one of the authors of that review article could still say that "most STS scholars have not seen the point of postcolonial theory. . . . and most postcolonial theorists . . . have flocked instead to the analysis of literary texts" (Anderson 2009, 390).

We could probably identify many causes of this history of disinterest in the West in postcolonial issues about science and technology. This is so even though these issues are about us in the West and our sciences and technologies and not just about distant others who are "out there" in the Third World. Western self-interest, Eurocentrism, racism, and a fascination with globalization theory have been mentioned as possible causes. Perhaps another cause worth considering is the preoccupation with Cold War agendas. From the end of World War II until the fall of the Soviet Union in 1989, such preoccupations made it difficult for Westerners to become interested in thinking about the tension between, on the one hand, assumptions that the value neutrality of modern sciences was both desirable and possible and, on the other hand, the clearly political and economic missions to which so much scientific research was dedicated.[6] Sympathetic attention to the science and technology concerns of these First World and Third World intellectuals would have seemed not only unrealistic but, more importantly, deeply unpatriotic. The Cold War was not a good time to articulate for Western ears skepticism about the empirical and theoretical adequacy or the political desirability of modern Western sciences and technologies. Perhaps today, two decades after the end of such relations between the First and Second Worlds, we are ready to move beyond a Cold War mentality.

To be sure, the antimilitarist and radical science movements that emerged in the United States and Europe during the Vietnam War did attract widespread attention, especially among the young in the 1960s. And ecology movements began to raise troubling questions about how Western sciences and technologies were affecting the environment. By the early 1970s, feminist movements were questioning the sexist biases of some of the most widely disseminated scientific theories, such as sociobiology, biological and medical theories about reproduction, and the assumptions about the environment that were the target of ecofeminists. Yet the message that most scientists, engineers, the educated public, and even university researchers and scholars took away from these claims and analyses was about the uses and abuses of scientific research. According to this view, these kinds of criticisms should focus on the politics in society, not on scientific and technical research itself. Such research could itself still be defended as value-free and committed to supposedly pure science and its basic research.

In the next section, I pave the way for many of the essays in this collection by briefly describing this science-focused kind of postcolonial theory, or "postcolonial science theory," as I refer to it. The third section takes up another challenge for those who would create more reliable sciences that have the resources to advance democratic social relations in today's world. That challenge is the continuing persistence of damaging gender stereotypes that guide science policies and practices around the globe, including those of many advocates of postcolonial science theory. These gender criticisms, too, first emerged during the Cold War.[7] The third section looks at both the important assumptions that postcolonial and gender science and technology studies share, and the conflicting assumptions that prevent them from making good use of each other's most valuable insights. Now to the two main focuses of postcolonial science theory.

## A Postcolonial Theory for Science and Technology Studies

As noted earlier, two issues were the focus of especially provocative questions from the beginnings of postcolonial science theory a half century ago. One was a historical question: what roles had Western sciences and technologies played in colonial histories, and what role had colonialism played in the histories of Western sciences and technologies? The other asked what the focus and character of science and technology policy should be in the newly independent Third World states. Attention to the second question had to focus also on aligning the policies and practices of the international and national aid agencies with the interests and desires of the poor people of the world rather than with only the interests and desires of the formerly colonial powers that now were the major funders of these programs. Many Western activists and researchers joined their Third World colleagues in working on this issue. As we shall see, the history and policy questions were linked. Their conjunction enables us to grasp the importance of thinking in terms of multiple modernities with their multiple sciences. Yet this recognition deeply challenges conventional Western epistemologies and philosophies of science, which have deep commitments to the existence of only one modernity and one real science.

## History

Third World theorists found especially problematic the exceptionalist and triumphalist assumptions of the conventional Western views of science and technology in history. Exceptionalism assumes that the West alone is capable of accurate understandings of the regularities of nature and social relations and their underlying causal tendencies. There is one world, and it has a single internal order. One and only one science is capable of understanding that order. And one and only one society is capable of producing that science: our Western society! This was the logic of the exceptionalist view. It has reigned in philosophy of science as the unity-of-science thesis.[8] Triumphalism assumes that the history of Western scientific and technological work consists only of a parade of admirable discoveries and inventions. Any harmful events or processes in which scientific or technical achievements are accused of playing a role—such as Hiroshima, environmental destruction, global warming, militarism, or colonialism itself—were said to be caused by the ignorance and bad politics of political leaders and the public that they court. That is, such events or processes cannot be attributed to any features of modern Western sciences and technologies themselves. Those who make these assumptions find it unintelligible to claim that other societies can and have produced competent sciences or that it is reasonable to think that certain attributes of modern sciences themselves have made contributions to natural and social disasters.

Of course, ignorance and bad politics have all too often left their marks on history. But critics of exceptionalism and triumphalism think that ignorance and bad politics cannot be the end of the story. They have argued that a consequence of such assumptions is that Western sciences and technologies have seemed legitimately to escape the kind of postcolonial analyses and criticisms that have been so insightful about other Western institutions and practices. The postcolonial writings that became familiar in Western universities in the 1980s had little effect on such attitudes about sciences and technologies. This is so even though Edward Said, members of the Indian Subaltern Studies group, and other early post-colonial theorists clearly pointed to the role of Western sciences, technologies, and their philosophies in colonial projects.[9] This work also did not address the Cold War politics that had helped to put the very nature of modern Western sciences outside the range of reasonable criticism in university classrooms as well as by media in the West.

To be sure, a few scholars in the early days of the new social histories of science did use a postcolonial lens to produce counterhistories of Western sciences and technologies and their interactions with colonized societies.[10] However, as the field of social studies of science and technology began to develop during the Cold War, its historians, sociologists, and ethnographers tended to focus on how scientific facts were socially constructed in laboratories, and on how knowledge travels from one place to another. As Warwick Anderson points out, this interest in how knowledge travels aligns with globalization theory, not postcolonial theory. It makes issues about the past and present of colonial relations no longer relevant or even comprehensible.[11] Consequently issues about relations between sciences and technologies, on the one hand, and colonialism, imperialism, and their recent residues and resurrections, on the other hand, have until recently remained largely unaddressed in science and technology studies, as well as in academic postcolonial studies.[12]

*History and Policy*

By the mid-1980s, UNESCO and other international agencies, as well as regional institutes in the Third World, were sponsoring large multinational conferences on the issues raised by postcolonial science theorists, mostly but not entirely from the Third World. From the perspective of North American science and technology theory writings, it is hard to get a sense of the huge number of scholars, policymakers, and activists who participated in such projects, the rich institutional networks and resources that supported them, or the thoughtful and provocative character of their concerns. In the United States at least, the occasional appearance of activists in these debates, such as Vandana Shiva or Ashis Nandy, could not convey the extensive global networks and institutional supports for this kind of postcolonial science theory.

One can get a quick grasp of the nature and range of these inquiries and debates by examining the proceedings of three international conferences that were published from the mid-1980s to the mid-1990s. Here I can only briefly describe them. *The Revenge of Athena: Science, Exploitation, and the Third World*, edited by Ziauddin Sardar, published twenty-one of the many dozens of conference presentations given in November 1986 at a seminar titled "The Crisis in Modern Science" sponsored by the

Consumer Association of Penang, Malaysia. The book is divided into three parts: "What's Wrong with Science?," "Science and Third World Domination," and "Third World Possibilities." Contributors to the collection include figures—now well known in the field—such as Vandana Shiva, Claude Alvares, Susantha Goonatilake, Seyyed Hossein Nasr, and Jerome Ravetz, as well as Sardar himself. In this volume as in the other two, many of the contributors are themselves scientists, engineers, or mathematicians. The conference's "Declaration on Science and Technology," subsequently republished as *The Crisis in Modern Science: A Third World Perspective*, by the Third World Network,[13] provides an extensive agenda for redirecting Third World science and technology projects to Third World needs and desires. As the authors argue, "Only when science and technology evolve from the ethos and cultural milieu of Third World societies will they become meaningful for our needs and requirements, and express our true creativity and genius. Third World science and technology can evolve only through a reliance on indigenous categories, idioms and traditions in all spheres of thought and action" (Third World Network 1993, 487).

A second example is the proceedings of an international colloquium titled "Science and Empires—a Comparative History of Scientific Exchanges: European Expansion and Scientific Development in Asian, African, American, and Oceanian Countries," which was held in 1990 in the UNESCO building in Paris. The colloquium was organized by the REHSEIS (Research on Epistemology and History of Exact Sciences and Scientific Institutions) group of the French National Center for Scientific Research (CNRS). The proceedings were published as *Science and Empires*, edited by Patrick Petitjean, Catherine Jami, and Anne Marie Moulin. The thirty-five essays are organized into two parts, "Problems about the Integration of Classical and Modern Science" and "European Scientific Expansion and Political Strategies." A number of First World scholars contribute papers, including the three editors, as well as Nancy Leys Stepan, Lewis Pyenson, and Michael A. Osborne.

Finally, another set of proceedings contains papers presented at a conference sponsored by ORSTOM, the French science institute for research outside France, and UNESCO. The conference took place in Paris in 1994. Its theme was "Twentieth Century Sciences: Beyond the Metropolis." It featured presentations by a large number (perhaps more than half) of the approximately two thousand participants from the Third World and Europe: researchers and scholars, scientists and engineers, policymakers and activists.[14] Seven volumes, edited by Roland Waast, of about 150 of the many conference papers were subsequently published. They address a wide range of topics. For example, volume 6, *Sciences in the South:*

*Current Issues*, has book parts titled "Sciences on the Periphery: Assessments," "Privatisation and Globalization," and "The Western Character of Science."

## A World of Sciences

The postcolonial science and technology perspectives provided by the essays in this collection provide distinctive arguments for recognizing the nature and value of "a world of sciences"—that is, multiple scientific and technological traditions, each relatively well adapted to regional needs and interests, though never perfectly so. They are joined in this project by work in modernity studies and by minority tendencies in Western science and technology studies itself, as indicated earlier.[15] Here the central argument is that modernization is not identical to Westernization, contrary to Western exceptionalist and triumphalist assumptions. Rather, most peoples around the world now live in societies that have separated from hunter-gatherer economic and political relations and from the feudal political economies from which the modern West slowly emerged. Moreover, the global reach of Western modernity's corporations, environmental destruction, and arms industries, not to mention its contributions to the production of pandemics, financial disasters, immigration, and refugees, permeates even societies that have received few or no benefits from Western modernity. Today every society lives in global modernity, even if only in the darkest corners of its most hideous effects. To be sure, dissemination from the West and from other societies also plays significant roles in creating all societies today, but so too do processes internal to each society, as was the case in the West. Moreover, the recipient society always changes what it borrows so that the new ideas, processes, or goods fit into the existing social order with minimum disruption.

Thus modernity is not only disseminated from the West to other societies. It is also produced independently within each and every society. Whether arriving from outside or inside a society—or, more likely, through negotiations between inside and outside—it must be "sutured" into existing economic, political, cultural, psychic, and material worlds. Thus modernity will always take on distinctive local features in its multiple regional appearances. Its epistemologies will be to some extent local.[16] And it always tends to appropriate and reshape to its own ends the social hierarchies that it finds. Feminist and postcolonial projects will always have to be multiple and distinctively local if they are to serve those escaping male-supremacist and Western-supremacist histories.

A number of the authors here think in terms of a world of sciences, each serving the economic, political, cultural, and psychic needs of its peoples. And all these sciences are in many kinds of interactions with each other: conflicts, negotiations, coalitions, appropriations, integrations of parts of one with the other, disseminations, and more.[17] The urge to integrate or assimilate other knowledge systems completely into modern Western sciences, leaving just one global knowledge system, should vigorously be resisted. This would continue the tragic destruction and suppression of fruitful cultural diversity in knowledge systems that has characterized Western colonialism and imperialism. Fortunately such tendencies toward a monological knowledge system are widely resisted at least in practice these days, as many of the essays will demonstrate. Developing epistemologies adequate to a world of sciences is at this point an uncompleted project.

Such theories of knowledge must confront the reality that the contrast between modernity and tradition that has been so important to modernization theorists is neither as clear nor as useful as modernization theorists imagined. In the postcolonial literatures, one can see the contrast blurred, undermined, or "worked"—manipulated and destabilized—in historical practices of Third World societies and in the West. For example, the modernization theorists argued that the policies they recommended would replace supposedly backward traditions with modern beliefs and practices. Yet modernization, whether in the hands of the neoliberal World Bank or of post-Marxian dependencia theory, simply appropriated and subjugated to its own ends traditional households, women's work, and traditional family relations in its nation-building practices (Catherine Scott). Essays on the importance of so-called indigenous knowledge to indigenous societies and to ours, as well as the complexity and sophistication of indigenous knowledge, undermine modernity's intellectual and pragmatic devaluation of such knowledge systems. The essays here about South Pacific navigation (Goodenough) and Cree hunting practices (Colin Scott) provide good examples of such knowledge systems. Today advocates for traditional knowledge systems defend them by using modern electronic technologies (Warren) and legal contracts (Brush, Hayden). Some do insist on the importance of further integrating such systems into modern Western sciences (Goonatilake), and others advocate integrating selected elements of Western sciences and technologies into them (Hoppers, Sardar). Readers can identify additional ways in which both supposedly traditional and supposedly modern societies work the boundaries between the two categories.

I have mentioned women here and there. Yet it would be a mistake to think that *gender* refers only to women. What is gender? How have gender issues shaped sciences and technologies in colonial, imperial, and postcolonial contexts? What resources can feminist (or gender) studies of science and technology provide for improving research and democratizing global social relations?[18]

## Gender and Postcolonial Science and Technology Studies: Separate, Conjoined, or Coconstituted Paths?

### *No Women, So No Gender Issues?*

The conquistadors, explorers, missionaries, merchants, indigenous rulers, scientists, historians, anthropologists and their informants, and theorists of modernity and development, as well as the leading scholars who contribute to postcolonial science and technology studies—these have been mostly men. Consequently some scholars seem to think there is little reason to raise gender issues in addressing topics in this field. Many assume that gender issues are relevant only if women are in sight, or perhaps even only if one is actually studying women. Yet the assumption that gender refers exclusively to women is false. It undermines the reliability as well as the legitimacy of accounts guided by it. In contrast, recent studies have found ways to ask how the very absence of women has influenced the selection of scientific problems, the methods of research and the regulatory ideals that guide them, what count as scientific communities, conceptions of natural processes, the interpretation of data, the results of research, and the dissemination of scientific applications and technologies, as well as at least some prevailing understandings of nature's order. In societies organized by male-supremacist gender hierarchies, men, their ideas and practices, cannot be unique models of the human. They can only mark historically specific masculine examples of the human.

The field of feminist science and technology studies has been developing in the global North and South since the 1970s. Yet this work and postcolonial science and technology studies too often ignore each other. The assumption that gender and postcolonial paths are separate damages the reliability and progressive promise of each.[19] Arguments for their "intersection" are preferable, and they have been useful in confronting the race and gender blindness of U.S. law (Crenshaw et al. 1995). However, this metaphor retains the false idea that somehow gender relations and colonial relations were at one time both functioning, and yet were separate from each other before they "intersected." That assumption could be made only by people privileged by their position in gender and colonial hierarchies. For colonized women, differences in their lives from the conditions of men, as well as from the conditions of their colonial rulers, are part of their everyday lived experiences. Here the argument will be that gender and colonial relations have coconstituted each other.

*Gender and Postcolonial Science and Technology Studies:*
*Weak and Strong Complementarity*

The agendas of feminist and postcolonial science and technology studies are similar in important respects and thus would seem to be complementary.[20] For example, both argue that the perspectives and interests of their particular constituencies are not well served by modern Western science and technology policies, practices, or philosophies. To be sure, modern Western policies, practices, and philosophies of science and technology have delivered some benefits to some women in the West. Yet these sciences and technologies were not designed to respond to any group of women's needs and desires in the West, let alone to the distinctive needs and desires of women in different classes, races, ethnicities, and cultural groups around the globe. They have been designed to respond primarily to the needs of states, militaries, and corporations, from the design and management of which women have systematically been excluded. Women (as well as most men) around the globe have borne disproportionate shares of the costs and received relatively fewer of the benefits of modern Western sciences and technologies.

Moreover, both offer alternatives that they claim are grounded in more realistic understandings of knowledge production processes, are more comprehensive, and can better serve the peoples for whom each speaks. Thus the agendas of each are always explicitly political as well as intellectual.

Additional reasons for each to be interested in the projects of the other can come from recognition that their constituencies are overlapping and their discourses are interlocked. More than half of the formerly colonized and those still under the control of neocolonialism and neoimperialism are women. Additionally, children and the elderly, disabled, and sick depend on women for their daily survival. To put the point the other way, a huge majority of the world's women and their dependents are among formerly colonized peoples and those now negatively impacted by residues and resurrections of colonialism and imperialism.

Furthermore, the dominant discourses that these social movements criticize, as well as the ones they themselves use, are deeply imbricated or locked into each other: colonialism, imperialism, and male supremacy have persistently represented gender in racial or colonial terms, and racial and colonial relations in gender terms (e.g., Stepan 1986). Women supposedly are not fully civilized, and non-Western men are supposedly not as manly as are men of European descent. Nor are women and non-Western men regarded as capable of managing their own lives as well as are men of European descent, according to such views.

Gender and racial-colonial categories still coconstitute each other today (Catherine Scott). Thus, because of their overlapping constituencies and interlocking discourses, each of these science and technology movements would seem to have to depend on the successes of the other to achieve its own professed goals. In this sense, they are *strongly complementary.*

Yet these two science and technology movements often seem committed to conflicting assumptions about the relevant social relations, the relevant sciences, and questions of who can and should be agents of the kinds of radical social and scientific change for which each calls. (There are important exceptions to this charge, as we will see.) Under such circumstances, neither social movement can deliver the benefits it envisions to the majority of those to whom it has professed accountability. So what are the contributions that gender and postcolonial studies of science and technology can make to each other's projects? Before we turn to this issue, it is worthwhile to recollect just what gender is and is not. In feminist work, the term *gender* is used in ways that may not be obvious.

## What Is Gender?

Gender is not another word for women, as noted earlier. Rather, like class and race, it designates particular kinds of social relations, here between men and women as well as between men and between women. These relations are "made, not born," to borrow from Simone de Beauvoir's famous observation. Moreover, gender relations are manifested not only by individuals but also in the structures and systematic practices of institutions (e.g., job classifications, legal regulations). They also appear in our symbolic systems, our meanings, as when nations are represented as women (Liberty, Columbia, Marianne) or when regulative ideals of research, such as objectivity and rationality, are represented as requiring a distinctively masculine character. Furthermore, gender relations organize hierarchical institutional structures of economic, political, and social power. However, gender never functions alone; it always interacts with other powerful social relations, such as race and class. Whether one conceptualizes such interactions as intersections or as processes of coconstitution, gender relations are always historically dynamic. Finally, like race and class, *gender* is both a descriptive and an analytic term. It designates both something "out there" in social relations and also a kind of analytic framework invoked to explain diverse manifestations of such social relations.[21]

*Gender and Science Studies*

This field is by now four decades old. I will not review that history here except to name five focuses of ongoing concern in the North, as well as everywhere that Northern sciences have found a home in the South (though there are additional gender issues in the South, as we shall see). Such projects have been initiated by groups with different kinds of disciplinary, political, and institutional interests in scientific and technological research. One such question is where women in the social structures of modern sciences are (and have been), and why there have been so few of them in the arenas of the design and management of research. Another is how and why "sexist sciences" have provided empirical support for the claimed inferiority of women. A third asks how technologies and the applications of the results of scientific research have been used against women's equality. Women's health, reproductive, and environmental concerns were among the earliest such focuses here. Fourth, how do scientific and technical education—pedagogy and curricula—restrict girls' and women's (and boys') development as scientists and engineers?[22] Finally, what is problematic about the epistemologies, methodologies, and philosophies of science that produce and support such sexist and androcentric practices?[23]

Such issues all remain important almost four decades after they were first posed—unfortunately. Some areas show significant progress—for example, in increasing access for women to scientific educations, publications, organizations, and lab and classroom jobs, and in establishing at least token presences of women in policy contexts. Moreover, significant changes in health and reproductive policies have occurred for women in already advantaged groups. As we will see, some feminist epistemological and methodological work has enabled new kinds of increasingly widespread debates about the relation of different human experiences to the production of knowledge. Yet women in Africa, Asia, and other places around the globe, as well as poor women in the West, have not much benefited from these kinds of progress.

However, neither postcolonial nor Northern feminist science and technology studies are likely to improve women's conditions as long as their fundamental assumptions conflict. From the perspective of Northern labs, science curricula, and federal policy, it is all too easy to be unaware of how Northern sciences and technologies function globally. In none of such contexts can one easily focus on postcolonial or gendered social relations, indigenous knowledge or feminist research innovations, or the possibility that Northern residents, men or women, will probably not be the most valuable agents of democratic social change in science and technology worlds. What are these conflicting assumptions made by feminist and postcolonial science and technology studies?

*Theoretical and Methodological Sites of Dissonance*

First, what are the relevant social relations to be examined for these two kinds of science and technology studies? Postcolonial science and technology scholars who are men rarely see gender relations as relevant either to the situations they observe or to their own theoretical or empirical concerns. Similarly, far too few Western feminists have focused on postcolonial science and technology relations, as indicated earlier. The exceptions here are to be found primarily in the long history of criticisms of science and technology aspects of development policies and practices and in environmental studies, as several essays in this collection show.

More than three decades ago, historians pointed out that recognizing women to be fully human—as fully human as their brothers—undermines traditional theoretical and methodological assumptions about social relations. This recognition raises provocative questions. For example, how should we account for the fact that women's conditions have tended to regress at precisely the moments marked in conventional histories as high points of human progress, such as the Renaissance, or the state formation resulting in Athenian democracy and, more than two millennia later, the United States? Even worse, it turns out that it was precisely because of the features identified as progressive that women's lives regressed. This is because whatever is extolled as progressive tends to be symbolized as virile and manly in societies structured by gender hierarchy. For example, it was not an accident that in the Renaissance women lost rights and opportunities that they had earlier possessed. Moreover, in state formation, women have invariably lost legal and political rights they had possessed in earlier periods, including the democratic revolutions of eighteenth-century Europe and the independence movements of newly postcolonial states after World War II (Kelly-Gadol 1976; Pateman 1988; Catherine Scott). Apparently conventional theories of social change have failed to account for the transformations they intend to chart insofar as they ignore women's role and fate in such processes. It has become clear that chronologies grounded only in what happens in men's lives, whether about the North or the South, leave no conceptual space for significant changes in women's lives or for examining the effects that the conditions of women's and men's lives have had on each other.

Yet postcolonial science and technology studies seem to assume that women and men benefit equally from men's progress, and that gender relations are irrelevant to the most adequate theories of social change.

With important exceptions, the relevant social relations in the accounts of postcolonial science and technology studies are those of presumably gender-free imperialism, colonialism, nation building, and the local, apparently gender-free acquiescences or resistances to such processes. Occasional references to "women's concerns" do not address gendered social structures or symbolic practices, let alone feminist epistemological, methodological, or philosophy of science issues. Consequently postcolonial theory cannot understand colonial, imperial, postcolonial, or today's neocolonial and neoimperial processes as long as its assumptions obscure women's realities and experiences, their standpoints on dominant social relations, and the gender relations that structure and give meaning to social institutions and the men's and women's lives lived within them. Important exceptions here that are focused on science and technology issues include the work of Anne Fausto-Sterling (1994, 2005), Donna Haraway (1989, 1991), and Vandana Shiva (1989).

It is encouraging to see that a few historians and ethnographers have begun to identify the gendered symbolic meanings and accompanying practices that have shaped Western sciences and technologies in colonial and imperial projects. For example, they have focused on scientists' claims that the greatest scientific value should go to the discoveries and inventions produced though the manly heroism of scientific quests (Terrall); on the gender, class, and colonial structure of Jesuit scientific communities in their overseas missions (Harris 2005; Rhodes 2005); on the masculine chivalric values of the knowledge gathering by Spanish conquistadors as well as British and French colonialists (Canizares-Esguerra 2005); and on the application of gender stereotypes to colonial relations in typical British representations of fitness and disease in the colonies (Harrison 2005). This is an area ripe for further exploration.

In a parallel way, much Western feminist work only rarely sees the social relations of colonialism and imperialism as having anything to do with women's or men's experiences of Western scientific and technological work. These scholars seem to think that as long as they are focused only on Western women and gender relations, social relations of colonialism and imperialism are irrelevant to the sciences and technologies they observe. Such assumptions leave us all ignorant both of the history and practices of sciences and technologies around the globe and of women's and men's variable participation in, and experiences of, such histories and practices. Thus similar arguments about treating non-Westerners as fully human reveal the limitations of traditional Eurocentric methodology in Western feminist science and technology studies—one that is shared, for the most part, with the larger field of science studies. Each field ignores powerful kinds of social relations that have shaped the content of sciences and technologies.

A second site of dissonance is the question of what the relevant sciences are for these two kinds of studies. For Western feminists (like Western science studies more generally), these have been almost entirely modern Western ones.[24] Courageous and brilliant work has been accomplished here in addressing the gender dimensions even of the sciences thought least susceptible to social fingerprints, such as physics and chemistry (Keller 1984; Potter 2001; Traweek 1988). Yet the history of modern Western sciences and analyses of their practices today are almost never set in the context of the history of Western appropriation of significant achievements of other cultures' sciences and technologies, or of Western destruction of them. Indigenous knowledge traditions, whether in the West or elsewhere, seem for the most part to be beyond the horizons of most of this work. Western feminist work, like much of the larger science studies movement in which it is embedded, is unaware of the counterhistories, the successes of indigenous knowledge, or the arguments for valuing multiple science traditions—a world of sciences. These kinds of studies call for radically rethinking conventional Western assumptions about scientific rationality and technical expertise. Consequently, the view of modern Western sciences and technologies from the standpoint of non-Western societies is also missing from Western feminist science and technology studies.[25]

Indigenous traditions, critical perspectives on modern Western sciences, and the design of science and technology policies and practices that integrate the best of both worlds are central projects for postcolonial scholars. They have produced diverse evaluations of these different traditions and accounts of possible future relations between indigenous and modern Western scientific knowledge systems.[26] Yet there has been little focus on women's domains of producing knowledge in these accounts of indigenous knowledge, and little awareness of the different kinds of experiences women have (different in different cultural contexts, but also different from men's in such contexts) that have informed Western feminism's innovative methodological and epistemological strategies. Nor have postcolonial science and technology scholars grasped the limitations of their own analyses and recommendations from the standpoint of women's interests. They have not treated women, their needs, interests, and insights, as fully human, nor have they considered them as equally crucial to social progress as they consider their own. Often feminism is perceived by men in formerly colonized societies as a Western import. In these cases, resistance to feminism is perceived to be an important part of resistance to Western imperialism. And this is so in spite of often vigorous and innovative feminist movements created locally by their female colleagues and compatriots. Evidently these otherwise brilliant intellectuals and activists take women to be more easily duped by the West than are men.[27]

Finally, the feminist and postcolonial accounts disagree on questions of who can and should be the agents of progressive transformations of societies and their sciences. Neither movement seems to think it necessary to center members of the other group in the envisioned design and management of its projects. Only a few women, such as Donna Haraway and Vandana Shiva, appear in the citations of contemporary postcolonial science studies scholars, and these are mostly the same few who appear occasionally in the feminist work.[28] The standpoint of women only rarely makes an appearance in this postcolonial work. Similarly, the standpoint of poor people in the Third World is missing from many Northern feminist analyses. Moreover, non-Western peoples do not appear as the designers or leaders of radical political and intellectual transformation in most Western feminist work. Other voices are hardly ever heard or reported except occasionally as special interests. That is, the others are never represented as being at the forefront in conceptualizing or leading social action toward goals and strategies that will produce widespread benefits, including but not limited to those purportedly special-interest groups themselves.[29]

Neither movement can deliver social progress to its professed constituencies without attending to the full range of issues addressed by both postcolonial and feminist science and technology studies. The existing separation of these two powerful conceptual frameworks must be ended.

The preceding section focused on one especially challenging contribution that postcolonial science theorists have made to global thinking about sciences and technologies. This is the conception of a world of sciences; that is, a world of multiple modern sciences, each with distinctive achievements, and each often in conflict with other scientific traditions. As noted earlier, Western science studies itself has recently produced a similar account focused entirely on modern Western sciences (Galison and Stump 1996; Kellert, Longino, and Waters 2006). Here we turn to just one of the compelling and yet provocative contributions to rethinking regulative ideals of scientific research made by feminist science studies.

*Standpoint Methodology*

This way of designing and conducting research projects has been theorized most extensively with respect to gender issues, though its logic is also usually invoked in postcolonial accounts, as it is in many other social

justice research projects (Harding 2004b). The concept of a methodological standpoint arose in Marxian writings about the importance of taking
the "standpoint of the proletariat" to understand how capitalism actually
worked, contrary to the bourgeoisie's continual justification of the necessity of exploiting manual laborers. So this geographical metaphor directs
attention to a location, a site in social relations, from which a disadvantaged group learns to observe and speak *for* itself and to the advantaged
group about how unjust and oppressive social relations affect their lives.
By starting off thought from the daily lives of workers, one could explain
the otherwise mysterious phenomenon of how wealth accumulated in
the lives of the already advantaged while misery accumulated in the lives
of the workers. One could do so without appealing to the typical biological, religious, social, or political justifications for such inequalities that
were promoted by the ruling groups of the day (and still in our neoliberal days).

Of course there are many problems with using such Marxian theory
today. Nevertheless the basic insight of this research methodology—its
logic—has remained useful to many disadvantaged groups around the
globe. In feminist hands, the standpoint strategy directed researchers
to begin thinking about any and every project from the standpoint of
women's lives instead of from the conceptual frameworks of research disciplines or of the social institutions that such disciplines serve. Women
had been excluded from the design and management of these disciplines
and institutions. Those frameworks had been designed to answer questions that were *for* the dominant social groups, not *for* women or other
exploited groups. The dominant institutions sponsored, funded, and
monitored research in the natural and social sciences; their policies were
grounded in gender stereotypes. They promoted the "conceptual practices of power," in the words of Dorothy Smith (1990).

Standpoint projects "studied up" (as the Marxists put it). They began
by thinking about the dominant institutions, their practices and cultures,
from the standpoint of the women's lives affected by them. Their goal was
not to produce ethnographies of women's worlds, valuable as those can
be. Rather, they intended to explain the high-level institutional decisions
and practices responsible for initiating and maintaining such situations.
In this respect, they differed from the ethnographies that were frequently
parts of such projects and with which they were often mistakenly conflated.[30] Standpoints are not to be conceptualized only as perspectives.
Everyone has perspectives on the world, but standpoints are intellectual
and political achievements in that a group has to work together to figure
out how to arrive at them. They require critical, scientific study to see
beneath the everyday social relations in which all have been forced to

live. They also require political struggles to gain access to the sites (the boardrooms, the command centers, the policy circles) where one could see how decisions have been made that directed and maintained sexist and androcentric social relations (Hartsock [1983] 2003).

Standpoint theory produced stronger standards for good method in the natural as well as the social sciences. Similarly, it produced revisions of other regulative ideals of the sciences, including "strong objectivity" and "robust reflexivity," and produced more rigorous and comprehensive standards for rationality (Haraway 1991; Harding 2004a, 2004b). Standpoint methodologies have by now explicitly been adopted across the social sciences, in some mixed social and natural sciences such as environmental and health studies, in several areas of biology, and in some technology studies. Moreover, the logic of such methodologies has an organic quality in that it seems to appear whenever a disadvantaged group tries to articulate the legitimacy of its own knowledge needs against the research practices that serve powerful groups. Thus the logic of standpoint epistemology and methodology is routinely evoked in postcolonial writings that start off from the lives of Third World peoples to think about Western assumptions, policies and practices, and indigenous knowledge systems; or about encounters with European voyagers, botanists, and physicians; or about modernization or development theory. In many of the essays in this collection, one can detect an appeal to a standpoint logic, whether or not it is explicitly articulated as such.

One can still ask, however, if standpoint methodology and epistemology are too Western to be fully useful elsewhere. Standpoint theory was initially formulated within the Marxian and Enlightenment philosophical and methodological traditions, even as it protests significant aspects of such legacies. Although it is positioned against both positivist regulatory ideals and practices in Western-origin natural and social sciences, positivism is not, to take just one case, one of the most problematic aspects of Indian society for women, as the philosopher Uma Narayan pointed out several decades ago. Moreover, standpoint theory's appeal to the value of women's experience can lose its critical edge in societies that conceptualize sex and gender differences as fundamentally complementary rather than hierarchical, and this is so regardless of whether such differences are in fact treated as hierarchical (Narayan 1989). There are other ways, with significant relations to standpoint theory, to articulate research methodologies that can distribute their benefits more effectively to the least advantaged groups.[31]

Yet standpoint theory remains a valuable strategy to articulate the logic of "a space of a different kind for polemics about the epistemological priority of the experience of various groups or collectivities," as Fredric

Jameson put the point. "The presupposition is that, owing to its structural situation in the social order and to the specific forms of oppression and exploitation unique to that situation, each group lives in the world in a phenomenologically specific way that allows it to see, or better still, that makes it unavoidable for that group to see and know, features of the world that remain obscure, invisible, or merely occasional and secondary for other groups" (Jameson 2004).[32] Standpoint approaches can recognize the positive scientific and political value of local knowledge without falling into claims either of its absolute, universal validity and applicability or of its legitimacy by only local standards. That is, standpoint approaches do not commit their users either to problematic older positivist regulative ideals or to a mere relativism of claims valid only in their local context. It is a symptom of the originality of this approach that so many readers can't resist interpreting it only as either absolutist or relativist in a damaging way. Yet, it is only from the perspective of the absolutist's exceptionalist position that these do appear to be the only choices.

## Provocations and Illuminations

This introduction has explored two theoretical frameworks that can illuminate some of the most puzzling and provocative intellectual and political challenges of the day. Sciences (and technologies) and their societies coconstitute each other. Each provides resources for the development of the other—and this can occur whether such development is politically and intellectually progressive or regressive. This insight supports postcolonial and feminist arguments that sciences and technologies are never completely value-free. How should we think about the virtues of modern Western sciences and technologies in light of these challenging views of them? How should we think about the knowledge systems of other cultures? How should we think about many non-Western sciences and technologies that today function effectively for cognitively valuable and politically admirable projects in their own world and yet still do not address women's needs where these differ from men's? What about those that do and must function in a modern world but find themselves in some of the most deprived locations in that world?

By now it should be clear that there cannot be a single recipe for science and technology research projects that are desirable from feminist and postcolonial standpoints. We can at least agree that we should not support one that conforms to the traditional Western conception of progress and how to achieve it. Women and men in different eras and places experience differently the nature and effects of colonialism, imperialism, postcolonialism, neocolonialism, and the sciences and technologies that these social relations create. Cultures have their own distinctive histories, legacies, resources, values, and interests, and it is the cultures themselves that must create discussions of how best to plot their own futures (with certain caveats about harming others and their own most vulnerable members, of course). Thus it would be arrogant and ineffective for any one culture to take it upon itself to determine what will be best for all, and especially for Western researchers and scholars to do so for non-Western societies. There already are and must be many different kinds of epistemological, scientific, and technological struggles over priorities, goals, and strategies.

Meanwhile our natural and social environments themselves constantly change. They continually produce unexpected phenomena such as retroviruses, ozone holes, and global warming, as well as deadly financial crises, hurricanes, fires, and mudslides. Our daily environments now seem crowded with risks to life and health that were not imagined even one generation ago (Beck 1992). Westerners have to learn how to live with not knowing how such relations between knowledge systems and with natural and social orders will turn out. The vitality of both nature and global tendencies toward democracy in all their local varieties depends on our learning to tolerate—even thrive—in the face of continually appearing uncertainties (Sarewitz). In this kind of world, postcolonial and feminist science and technology studies can help us locate innovative strategies for moving forward, not only by considering their illuminating but provocative challenges to conventional assumptions but also by exploring the alternatives that they are debating.

## Notes

1. The language I use here of West and non-West is problematic. It echoes the discredited Orientalism that makes the West the center of geography, history, and critical analyses and is one of the founding targets of postcolonial criticism. It obscures the fact, addressed in many essays here, that the West consistently appropriated scientific and technological insights and achievements of other societies for its own projects, to this day almost always without acknowledgment. It occludes the difficulty of fitting into this binary the science and technologies of many societies around the world that have developed their own forms of modernity. See, e.g., Eisenstadt 2000; Mignolo 2000; Rofel 1999.

Moreover, all the available alternative contrasts are also problematic: First World–Third World (an artifact of the Cold War), "developed-underdeveloped" (who defines this difference?), and more. Furthermore, any such contrast inaccurately homogenizes the two groups and obscures the more complex social relations that exist between and among various global groupings in the past and today, reifying a preoccupation with differences that hides shared interests and practices between peoples in very different social circumstances. Some authors prefer to discuss today's global social relations in supposedly more politically neutral language such as "globalization" or "transnationalism." Such terms can be useful in some contexts. They are not politically neutral, however, for they hide power relations that are the focus of this book's contributors. In light of such difficulties with alternatives, I continue here to use primarily "West–non-West" and, where appropriate, shift to "First World–Third World" or "North–South" when those terms better indicate the relevant context.

2. See Seth 2009b for these and many more citations to the rich history of anticolonial counterhistories of science to the standard Western ones.

3. The term *postcolonial* is highly contested, even within the field of postcolonial studies. Who and what is, and should be, included or left out of its domain? Is it by now archaic—an artifact of the 1980s that is no longer useful? I do not take the space to review such issues here. My own view is that the term has by no means exhausted its progressive possibilities, though its limitations, addressed in a number of the essays here, are important to ponder. For just three of the many illuminating discussions about the usefulness and desirable domains of the concept of postcolonialism, see Goldberg and Quayson 2002; Loomba et al. 2005; and early issues of the journal *Postcolonial Studies*.

4. For fuller discussions of these histories, see Seth 2009b; Anderson 2009; Anderson and Adams 2007.

5. Anderson 2002; McNeil 2005; Schiebinger 1989.

6. Discussions with Gail Kligman helped me see the importance of the Cold War in masking for Westerners the work of Third World science and technology intellectuals.

7. This may suggest some reasons in addition to sexism for the hysterical demonization the feminist theorists frequently encountered. However, one could argue that the manliness of the militaries certainly was at issue in the Cold War, as it is in every war. Readers "of a certain age" will remember how the newspapers' front pages regularly featured charts depicting two piles of missiles. Representing the West's arms capabilities would be a large pile of big, white missiles; representing the Soviet capabilities would be a small pile of little black ones. The apparent innocence of bygone eras can be startling.

8. The unity-of-science thesis persists today in spite of such philosophers' criticisms, e.g., Dupre 1993; Galison and Stump 1996.

9. Ashcroft, Griffiths, and Tiffin 1989, 1995; Williams and Chrisman 1994.

10. See, e.g., Adas 1989; Blaut 1993; Brockway 1979; Headrick 1981; McClellan 1992.

11. Anderson 2009. An exception to this judgment is the work of Helen Watson-Verran and David Turnbull (1995).

12. As noted earlier, historians of science and medical anthropologists have long examined the knowledge systems of other cultures, though much of this work does not deserve to be called postcolonial. The postcolonial work in these fields has tended to be written for, and to remain primarily the concern of, specialists.

13. Excerpted in Harding 1993.

14. I could identify only a dozen presenters from North America.

15. See, e.g., Eisenstadt 2000; Maffie 2009; Galison and Stump 1996.

16. Several contributors to this collection discuss "polycentric" or "polyvo-cal" epistemologies. For just a few of the many new explorations of distinctive non-Western modernities, see also Mignolo 2000; Ong and Nonini 1997; Shih, forthcoming; Lionnet and Shih 2005; Rofel 1999; Chakrabarty 2000; Eisenstadt 2000; Harding 2008; Prakash 1999.

17. See, e.g., many of the chapters in parts II and IV of this collection.

18. There is no uncontroversial term to refer to this kind of research. *Feminist* is too radical a term for many Westerners who, assisted by persistent media demonization of women's movements, associate the term only with the most ambitious and theatrical parts of the Western women's movements of the 1970s. On the other hand, *feminist* is a conservative term for many people in the United States and around the world who associate it with bourgeois women's rights movements that have had little concern for the lot of poor women, African American women, and other women of color. However, so-called gender stud-ies can seem to lack any awareness of inequalities between men and women, or even between men or between women. I shall alternate between these two inadequate terms, hoping to offend only half of the readers at any given time.

19. This is not to deny either that gender relations vary immensely from one culture to another, as do colonial relations, or that in some colonial contexts gender may not always be the most important variable on which to focus.

20. An early form of the rest of this section appears in Harding 2009.

21. Two comments. First, the insistence on separating the social gender dif-ferences so firmly from the biological relations of sex differences may well be on shaky ground. This is not because biological reductionism was right (it wasn't) but because the binary of gender versus sex is a form of the culture-versus-nature binary that has come under severe criticism in several branches of bi-ology (Fausto-Sterling 1994; Keller 1984). Moreover, the culture-versus-nature binary is no longer legitimate in science studies where cultures and their con-cepts of nature are seen as coconstituting each other. Nor do other cultures' knowledge systems tend to find it appropriate. The solution is to be found in inventing a new kind of biology that does not depend on such a severe separa-tion between the social and the natural, but this we do not yet have. Second, it is also worth thinking about the effects of transgender and transsexuality theories and practices on ways of analyzing gender differences (Valentine 2007).

22. These studies can appear to focus only on women, but they have always been concerned with how economic, political, and social gender inequality unfairly limited women's interactions with sciences and technologies while overadvantaging men's interactions, and with how women resisted such dis-crimination. It is not so much that men are perceived to be the problem for women. Rather, the social institutions that exclusively men have designed and managed do not serve women's interests and desires well.

23. Examples of this relatively early work include Boston Women's Health Collective 1970; Fausto-Sterling 1994; Haraway 1989; Harding 1986, 1991; Harding and Hintikka [1983] 2003; Hubbard, Henifin, and Fried 1982; Keller 1984; Longino 1990; Merchant 1980; Rossiter 1982–95; Schiebinger 1989, 1993; Tobach and Rosoff 1978–84; Wajcman 1991. The medical establishment's disparaging and often erroneous opinions about women's bodies were the object of much early gender and science work. For recent reviews of these issues, see Harding 2008, chap. 4; and Subramaniam 2009.

24. See, e.g., the otherwise excellent, widely used reader for gender and science studies courses, Wyer et al. 2001.

25. My argument is not that the feminist theorists themselves are unaware of these matters but that the theoretical and methodological frameworks that they deploy obscure or marginalize such issues.

26. See, e.g., Denzin, Lincoln, and Smith 2008; Goonatilake 1984; Hess 1995; Hoppers 2002; Nader 1996; Third World Network 1993; Turnbull 2000; Watson-Verran and Turnbull 1995.

27. I am being a bit disingenuous here. Men in every society have resisted giving up their gender privileges. They usually mask their interests in male supremacy with arguments about the trivial or dangerous nature of any challenges to their control over women's lives. This is so even when they have good reasons to resist continued Western intrusions in their societies. Arriving at the best strategies for overcoming inequalities at least requires a lot of public dialogue between all the relevant stakeholders.

28. Many more appear in the larger field of postcolonial studies. Moreover, many additional significant postcolonial feminist science and technology scholars are working in the West and around the world. See those included in this collection, and the many citations throughout this book.

29. Poor people and grassroots activists from around the globe tend to have what they understandably see as more urgent projects than to write essays or books to be published in the West. But there are many other ways for their voices to be heard in the West, as many of the essays here demonstrate.

30. But see D. Smith 2005 for a critical institutional ethnography.

31. See, e.g., Anzaldúa 1981; and Walter Mignolo's articulation of the "colonial difference" (2000).

32. For a highly visible polemic about a standpoint claim, see the discussion of the U.S. Supreme Court (at that time) candidate Sonia Sotomayor's statement in a speech some years earlier that she hoped a wise Latina would make better decisions in some cases than a wise white man (*Los Angeles Times* 2009a, 2009b, 2009c).

## References

Adas, Michael. 1989. *Machines as the Measure of Man*. Ithaca: Cornell University Press.

Amin, Samir. 1989. *Eurocentrism*. New York: Monthly Review Press.

Anderson, Warwick. 2002. "Postcolonial Technoscience." Special issue, *Social Studies of Science* 32.

———. 2009. "From Subjugated Knowledge to Conjugated Subjects: Science and Globalisation, or Postcolonial Studies of Science?" In "Colonialism, Postcolonialism, and Science," ed. Suman Seth, special issue, *Postcolonial Studies* 12 (4).

Anderson, Warwick, and Vincanne Adams. 2007. "Pramoedya's Chickens: Postcolonial Studies of Technoscience." In *The Handbook of Science and Technology Studies*, 3rd ed., ed. Edward J. Hackett et al. Cambridge: MIT Press.

Anzaldúa, Gloria. 1981. *Borderlands/La Frontera*. San Francisco: Spinsters/Aunt Lute.

Ashcroft, Bill, Gareth Griffiths, and Helen Tiffin, eds. 1989. *The Empire Writes Back*. London: Routledge.

———, eds. 1995. *The Post-colonial Studies Reader*. New York: Routledge.

Beck, Ulrich. 1992. *Risk Society: Towards a New Modernity*. London: Sage.

Biagioli, Mario, ed. 1999. *The Science Studies Reader*. New York: Routledge.

Blaut, J. M. 1993. *The Colonizer's Model of the World: Geographical Diffusionism and Eurocentric History*. New York: Guilford Press.

Boston Women's Health Collective. 1970. *Our Bodies, Ourselves*. Boston: New England Free Press.

Brockway, Lucille H. 1979. *Science and Colonial Expansion: The Role of the British Royal Botanical Gardens*. New York: Academic Press.

Canizares-Esguerra, Jorge. 2005. "Iberian Colonial Science." *Isis* 96:64–70.

Chakrabarty, Dipesh. 2000. *Provincializing Europe: Postcolonial Thought and Historical Difference*. Princeton: Princeton University Press.

Crenshaw, Kimberlé, et al. 1995. *Critical Race Theory*. Philadelphia: Temple University Press.

Curtin, Philip. 1961. "'The White Man's Grave': Image and Reality, 1780–1850." *Journal of British Studies* 1:94–110.

Denzin, Norman K., Yvonna S. Lincoln, and Linda Tuhiwai Smith, eds. 2008. *Handbook of Critical and Indigenous Methodologies*. Los Angeles: Sage.

Dupre, John. 1993. *The Disorder of Things*. Cambridge: Harvard University Press.

Eisenstadt, S. N., ed. 2000. "Multiple Modernities." Special issue, *Daedalus* 129 (1).

Fanon, Frantz. 2002. "Medicine and Colonialism." In *A Dying Colonialism*, 121–45. New York: Grove Press.

Fausto-Sterling, Anne. 1994. *Myths of Gender: Biological Theories about Women and Men*. New York: Basic Books.

———. 2005. "The Bare Bones of Sex: Sex and Gender." *Signs: Journal of Women in Culture and Society* 30 (2): 1491–528.

Galison, Peter, and David J. Stump, eds. 1996. *The Disunity of Science*. Stanford: Stanford University Press.

Gibbons, Michael, et al. 1994. *The New Production of Knowledge: The Dynamics of Science and Research in Contemporary Societies*. Thousand Oaks, Calif.: Sage.

Goldberg, David Theo, and Ato Quayson, eds. 2002. *Relocating Postcolonialism*. New York: Blackwell.

Goonatilake, Susantha. 1984. *Aborted Discovery: Science and Creativity in the Third World*. London: Zed.

Gupta, Akhil. 1998. *Postcolonial Developments: Agriculture in the Making of Modern India*. Durham: Duke University Press.

Hackett, Edward J., et al. 2007. *The Handbook of Science and Technology Studies*. 3rd ed. Cambridge: MIT Press.

Haraway, Donna. 1989. *Primate Visions: Gender, Race, and Nature in the World of Modern Science*. New York: Routledge.

———. 1991. "Situated Knowledges: The Science Question in Feminism and the Privilege of Partial Perspectives." In *Simians, Cyborgs, and Women*. New York: Routledge.

Harding, Sandra. 1986. *The Science Question in Feminism*. Ithaca: Cornell University Press.

———. 1991. *Whose Science? Whose Knowledge?* Ithaca: Cornell University Press.

———, ed. 1993. *The "Racial" Economy of Science: Toward a Democratic Future*. Bloomington: Indiana University Press.

———. 1998. *Is Science Multicultural? Postcolonialisms, Feminisms, and Epistemologies*. Bloomington: Indiana University Press.

———. 2004a. "Rethinking Standpoint Epistemology: What Is 'Strong Objectivity'?" In *The Feminist Standpoint Theory Reader*, ed. Sandra Harding. New York: Routledge.

———, ed. 2004b. *The Feminist Standpoint Theory Reader*. New York: Routledge.

———. 2008. *Sciences from Below: Feminisms, Postcolonialities, and Modernities*. Durham: Duke University Press.

———. 2009. "Postcolonial and Feminist Philosophies of Science and Technology." In "Colonialism, Postcolonialism, and Science," ed. Suman Seth, special issue, *Postcolonial Studies* 12 (4): 401–21.

Harding, Sandra, and Merrill Hintikka, eds. [1983] 2003. *Discovering Reality: Feminist Perspectives on Epistemology, Metaphysics, Methodology, and Philosophy of Science*. 2nd ed. Dordrecht: Reidel/Kluwer.

Harding, Sandra, and Elizabeth McGregor. 1996. "The Gender Dimension of Science and Technology." In UNESCO *World Science Report*, ed. Howard J. Moore. Paris: UNESCO.

Harris, Steven J. 2005. "Jesuit Scientific Activity in the Overseas Missions, 1540–1773." *Isis* 96:71–79.

Harrison, Mark. 2005. "Science and the British Empire." *Isis* 96:56–63.

Hartsock, Nancy. [1983] 2003. "The Feminist Standpoint: Developing the Ground for a Specifically Feminist Historical Materialism." In *Discovering Reality: Feminist Perspectives on Epistemology, Metaphysics, Methodology,* and Philosophy of Science, 2nd ed., ed. Sandra Harding and Merrill Hintikka. Dordrecht: Reidel/Kluwer.

Headrick, Daniel R., ed. 1981. *The Tools of Empire: Technology and European Imperialism in the Nineteenth Century*. New York: Oxford University Press.

Heilbrun, Johan, Lars Magnusson, and Bjorn Wittrock, eds. 1998. *The Rise of the Social Sciences and the Formation of Modernity*. Dordrecht: Kluwer.

Hess, David. 1995. *Science and Technology in a Multicultural World: The Cultural Politics of Facts and Artifacts*. New York: Columbia University Press.

———. 2007. *Alternative Pathways in Science and Industry: Activism, Innovation, and the Environment in an Era of Globalization*. Cambridge: MIT Press.

Hollinger, David. 1996. *Science, Jews, and Secular Culture*. Princeton: Princeton University Press.

Hoppers, Catherine A. Odora, ed. 2002. *Indigenous Knowledge and the Integration of Knowledge Systems*. Claremont, South Africa: New Africa Books.

Hubbard, Ruth, M. S. Henifin, and Barbara Fried, eds. 1982. *Biological Woman: The Convenient Myth*. Cambridge: Schenkman.

Jameson, Fredric. 2004. "*History and Class Consciousness* as an 'Unfinished Project.'" Excerpted and revised in *The Feminist Standpoint Theory Reader*, ed. Sandra Harding. New York: Routledge.

Jasanoff, Sheila, ed. 2004. *States of Knowledge: The Co-production of Science and Social Order*. New York: Routledge.

Keller, Evelyn Fox. 1984. *Reflections on Gender and Science*. New Haven: Yale University Press.

———. 2010. *The Gap between Nature and Culture*. Durham: Duke University Press.

Kellert, Stephen H., Helen E. Longino, and C. Kenneth Waters, eds. 2006. *Scientific Pluralism*. Minneapolis: University of Minnesota Press.

Kelly-Gadol, Joan. 1976. "The Social Relations of the Sexes: Methodological Implications of a Women's History." *Signs: Journal of Women in Culture and Society* 1 (4): 810–23.

Knorr-Cetina, Karin. 1981. *The Manufacture of Knowledge*. New York: Pergamon.

Kuhn, Thomas S. 1970. *The Structure of Scientific Revolutions*. 2nd ed. Chicago: University of Chicago Press.

Latour, Bruno. 1987. *Science in Action*. Cambridge: Harvard University Press.

Latour, Bruno, and Steve Woolgar. 1979. *Laboratory Life: The Social Construction of Scientific Facts*. Beverly Hills: Sage.

Lionnet, Françoise, and Shu-mei Shih, eds. 2005. *Minor Transnationalism*. Durham: Duke University Press.

Longino, Helen. 1990. *Science as Social Knowledge*. Princeton: Princeton University Press.

Loomba, Ania, et al., eds. 2005. *Postcolonial Studies and Beyond*. Durham: Duke University Press.

*Los Angeles Times*. 2009a. "Judging from a Personal History." June 4, B1.

———. 2009b. "Sotomayor Speeches Reveal More on Race." June 5, A14.

———. 2009c. "Words of Wisdom." July 17, A29.

Maffie, James, ed. 2001. "Truth from the Perspective of Comparative World Philosophy." *Social Epistemology* 15:4.

———. 2009. "'In the end, we have the Gatling Gun, and they have not': Future Prospects of Indigenous Knowledges." *Futures* 41:53–65.

McClellan, James E. 1992. *Colonialism and Science: St. Domingue in the Old Regime*. Baltimore: Johns Hopkins University Press.

McClintock, Anne. 1995. *Imperial Leather*. New York: Routledge.

McNeil, Maureen. 2005. "Postcolonial Technoscience." Special issue, *Science as Culture* 14 (2).

Merchant, Carolyn. 1980. *The Death of Nature: Women, Ecology, and the Scientific Revolution*. New York: Harper and Row.

Mignolo, Walter D. 2000. *Local Histories/Global Designs*. Princeton: Princeton University Press.

Nader, Laura. 1996. *Naked Science: Anthropological Inquiry into Boundaries, Power, and Knowledge*. New York: Routledge.

Nandy, Ashis. 1983. *The Intimate Enemy: Loss and Recovery of Self under Colonialism*. Delhi: Oxford University Press.

———, ed. 1990. *Science, Hegemony, and Violence: A Requiem for Modernity*. Delhi: Oxford University Press.

Narayan, Uma. 1989. "The Project of a Feminist Epistemology: Perspectives from a Non-Western Feminist." In *Gender/Body/Knowledge*, ed. Susan Bordo and Alison Jaggar. New Brunswick, N.J.: Rutgers University Press.

Needham, Joseph. 1956–2004. *Science and Civilisation in China*. 7 vols. Cambridge: Cambridge University Press.

Ong, Aihwa, and Donald Nonini, eds. 1997. *Ungrounded Empires: The Cultural Politics of Modern Chinese Transnationalism*. New York: Routledge.

Pateman, Carole. 1988. *The Sexual Contract*. Palo Alto: Stanford University Press.

Petitjean, Patrick, et al. 1992. *Science and Empires: Historical Studies about Scientific Development and European Expansion*. Dordrecht: Kluwer.

Potter, Elizabeth. 2001. *Gender and Boyle's Law of Gases*. Bloomington: Indiana University Press.

Prakash, Gyan. 1999. *Another Reason: Science and the Imagination of Modern India*. Princeton: Princeton University Press.

Rhodes, Elizabeth. 2005. "Join the Jesuits, See the World: Early Modern Women in Spain and the Society of Jesus." In *The Jesuits*, vol. 2, *Cultures, Sciences, and the Arts, 1540–1773*, ed. John W. O'Malley et al. Toronto: University of Toronto Press.

Rofel, Lisa. 1999. *Other Modernities: Gendered Yearnings in China after Socialism*. Berkeley: University of California Press.

Rossiter, Margaret. 1982–95. *Women Scientists in America*. Vols. 1–2. Baltimore: Johns Hopkins University Press.

Sachs, Wolfgang, ed. 1992. *The Development Dictionary: A Guide to Knowledge as Power*. Atlantic Highlands, N.J.: Zed.

Said, Edward. 1978. *Orientalism*. New York: Pantheon.

Sardar, Ziauddin, ed. 1988. *The Revenge of Athena: Science, Exploitation, and the Third World*. London: Mansell.

Schiebinger, Londa. 1989. *The Mind Has No Sex: Women in the Origins of Modern Science*. Cambridge: Harvard University Press.

———. 1993. *Nature's Body: Gender in the Making of Modern Science*. Boston: Beacon Press.

———, ed. 2005. "Colonial Science." Special issue, *Isis* 96:52–87.

Selin, Helaine, ed. 2008. *Encyclopaedia of the History of Science, Technology, and Medicine in Non-Western Cultures*. 2nd ed. 2 vols. Dordrecht: Springer.

Seth, Suman, ed. 2009a. "Colonialism, Postcolonialism, and Science." Special issue, *Postcolonial Studies* 12 (4).

———. 2009b. "Putting Knowledge in Its Place: Science, Colonialism, and the Postcolonial." *Postcolonial Studies* 12 (4): 373–88.

Shih, Shu-mei. Forthcoming. *Sinophone Studies: A Critical Reader*. New York: Columbia University Press.

Shiva, Vandana. 1989. *Staying Alive: Women, Ecology, and Development*. London: Zed.

———. 1993. *Monocultures of the Mind: Perspectives on Biodiversity and Biotechnology*. New York: Zed.

Smith, Dorothy E. 1990. *The Conceptual Practices of Power: A Feminist Sociology of Knowledge*. Boston: Northeastern University Press.

———. 2005. *Institutional Ethnography: A Sociology for People*. Lanham, Md.: Rowman and Littlefield.

Smith, Linda Tuhiwai. 1999. *Decolonizing Methodology: Research and Indigenous Peoples*. New York: St. Martin's Press.

Spivak, Gayatri. 1987. "French Feminism in an International Frame." In *In Other Worlds*. New York: Routledge.

———. 1988. "Can the Subaltern Speak?" In *Marxism and the Interpretation of Culture*, ed. Cary Nelson and Lawrence Grossberg. Urbana: University of Illinois Press.

Stepan, Nancy. 1986. "Race and Gender." *Isis* 77.

Subramaniam, Banu. 2009. "Moored Metamorphoses: A Retrospective Essay on Feminist Science Studies." *Signs: Journal of Women in Culture and Society* 34 (4).

Third World Network. 1993. "Modern Science in Crisis: A Third World Response." In *The Racial Economy of Science*, ed. Sandra Harding. Bloomington: Indiana University Press. (Orig. pub. Penang, Malaysia: Third World Network and Consumers' Association of Penang, 1988.)

Tobach, Ethel, and Betty Rosoff, eds. 1978–84. *Genes and Gender*. 4 vols. New York: Gordian Press.

Traweek, Sharon. 1988. *Beamtimes and Lifetimes: The World of High Energy Physicists*. Cambridge: MIT Press.

Turnbull, David. 2000. *Masons, Tricksters, and Cartographers: Comparative Studies in the Sociology of Science and Indigenous Knowledge*. New York: Harwood Academic Publishers.

Valentine, David. 2007. *Imagining Transgender: An Ethnography of a Category.* Durham: Duke University Press.

Waast, Roland, ed. 1996. *Sciences in the South: Current Issues.* Paris: ORSTOM Éditions.

Wajcman, Judy. 1991. *Feminism Confronts Technology.* University Park: Penn State University Press.

Watson-Verran, Helen, and David Turnbull. 1995. "Science and Other Indigenous Knowledge Systems." In *Handbook of Science and Technology Studies,* ed. Sheila Jasanoff, G. Markle, T. Pinch, and J. Petersen, 115–39. Thousand Oaks, Calif.: Sage.

Williams, Patrick, and Laura Chrisman. 1994. *Colonial Discourse and Postcolonial Theory.* New York: Columbia University Press.

Wyer, Mary, et al. 2001. *Women, Science, and Technology: A Reader in Feminist Science Studies.* New York: Routledge.

Yanagisako, Sylvia, and Carol Delaney. 1995. *Naturalizing Power: Essays in Feminist Cultural Analysis.* New York: Routledge.

Young, Robert J. C. 2001. *Postcolonialism: An Historical Introduction.* Oxford: Blackwell.

# I. Counterhistories

The story [of science and colonialism] is a dual one. One of its aspects concerns how science and the scientific enterprise formed part of and facilitated colonial development. The other deals with how the colonial experience affected science and the contemporary scientific enterprise. . . . Science and organized knowledge did not come to Saint Domingue as something separate from the rest of the colonizing process but, rather, formed an inherent part of French colonialism from the beginning. In other words, the French did not colonize Saint Domingue and then import science and medicine as cultural afterthoughts. French science and learning came part and parcel with French colonialism, virtually as a "productive force."—JAMES E. MCCLELLAN III, *Colonialism and Science*

Consider . . . the formation of Western modernity, for which the benefit of empire was crucial; its identity and authority were forged on the stage of colonial and imperial domination. . . . The disciplines did not simply depend upon Europe's prior self-generated cultural and political resources, rather their development in the course of trade, exploration, conquest, and domination instantiated Western modernity.—GYAN PRAKASH, *Another Reason: Science and the Imagination of Modern India*

Modernity and coloniality are two sides of the same coin. . . . The discourse of modernity has embedded in it, as Las Casas and Locke illustrate, the logic of coloniality. . . . Between the triumphant rhetoric of modernity and that which remains beyond its frontiers (tradition, underdevelopment, barbarism, and the like) there is a difference. That difference . . . is colonial, that is, a difference that is constantly built and rebuilt assuming the necessary values implied in modernity and the values to be displaced or replaced when they are different.—WALTER MIGNOLO, *The Darker Side of the Renaissance*

Among the many important events of the nineteenth century, two were of momentous consequence for the entire world. One was the progress and power of industrial technology; the other was the domination and exploitation of Africa and much of Asia by Europeans. Historians have carefully described and analyzed these two phenomena, but separately, as though they had little bearing on each other. It is the aim of this book to trace the connections between these great events.—DANIEL R. HEADRICK, *The Tools of Empire: Technology and European Imperialism in the Nineteenth Century*

PEOPLES OF EUROPEAN DESCENT have been schooled to think of ourselves as uniquely responsible for most of humanity's finest intellectual, political, ethical, and economic achievements. It is our civilization and our unique characters, abilities, and skills from which emerged modern scientific rationality and technical innovation. These also produced the democratic political processes and capitalist relations of economic production, both dependent on scientific rationality and technical expertise, that together replaced feudalism's oppressive forms of governance and economic provision. Yet recent historical accounts provide compelling evidence that this is not an objective-enough account of how our scientific and technological achievements came about.

Modern European sciences began to emerge in the sixteenth century. Yet some of the older traditions of China, India, and other cultures in Asia and the Middle East were far more sophisticated than European ones until Europe's industrial revolution in the nineteenth century (Blaut 1993; Hobson; Needham 1956–2004). Moreover, Europeans had encountered and learned from these older traditions at least since Marco Polo's day. The Islamic world, with its rich scientific and technological traditions, had a visible presence in southern Europe at least until the Muslims were expelled from Spain in 1492. How much did European sciences and technologies borrow from these older traditions? John Hobson proposes that there has been such an explosion of historical scholarship documenting the extensive importation of elements of Eastern traditions into Europe that we should consider thinking in terms of our currently ongoing "discovery" of the "Oriental West."

Critics argue that it takes a lot of work to create and nourish such Eurocentric ignorance (Schiebinger and Proctor 2008). The geographer J. M. Blaut (1993) pointed out that scientists, historians, and philosophers have been beguiled by the image of a mythical "tunnel of time" that enables peoples of European descent to trace their lineage back to ancient Greece while conveniently ignoring the scientific and technological achievements of Asian cultures, many of which European sciences borrowed. Moreover, concepts such as the Dark Ages, the European scientific revolution, and the so-called Renaissance miracle also contribute to Western ignorance. It is Europeans who have been "dark" (blind) to the many achievements of the Middle Ages made by people today counted as non-European. As for the supposed European scientific revolution, it was not a revolution but a slow and steady process over many centuries. Furthermore, science then did not look much like today's science, since it was overtly infused with Christian and Islamic elements (Jacob 1988; Needham 1969; Hobson). Additionally, it is misleading to refer to either science or the scientific revolution as European, since they owed great debts to the Asian cultures that nourished them, especially Arabic and Islamic cultures. Additionally, Europe was not a functioning economic or political unit until later than the early modern period. However, this myth of a Western science that owes no intellectual, methodological, or technological debts to any other society (apart from the safely ancient Greeks) is not itself ancient. Rather, it was constructed largely in the late nineteenth century, precisely when European powers were expanding, solidifying, and justifying their empires and colonial projects around the world (Bernal 1987; Blaut 1993; Hobson).

Another such conceptual device appears in the explicit or implicit denial that European expansion (the so-called voyages of discovery) and the development of modern sciences in Europe had any significant causal relations to each other—except for the now-discredited view that the voyages enabled already existing European achievements to be diffused around the globe. It turns out that the voyages and the flourishing of European modern sciences each needed the achievements of the other for its own successes (Adas 1997). Moreover, Steve Harris argues that historians have been preoccupied with the wrong sciences when considering how Europe was able to advance so rapidly. It is these "long-distance sciences"—that is, the explorations and inquiries that were part of the voyages themselves—that created important kinds of changes in knowledge-seeking practices to which we remain indebted today, but which have been neglected by mainstream histories of science (see also Latour 1987). The three great corporate sponsors of the voyages—the powerful European trading companies, the Jesuits, and the would-be European empires themselves—did not want to lose ships, valuable cargo, sailors, travelers, or settlers to the natural dangers of such trips (not to mention to pirates and unfriendly indigenes encountered by voyagers). The voyagers and their sponsors needed astronomies of the southern hemisphere, better oceanography, climatology, marine engineering, cartography, and botanical, medical, and ethnographic knowledge, as such fields would be referred to today, to travel successfully to and in foreign lands as well as out of sight of Europe's shores.

Did gender play any role in the scientific and technological work that the European voyages needed and themselves produced? Some might think this was not the case, since few women participated in this work or the voyages. However, all-male groups also have gender relations. In the seventeenth and eighteenth centuries, as Mary Terrall points out, risking life and health on such perilous voyages came to be regarded as a criterion for the high quality of the scientific achievements that resulted. The heroic masculine figure of the risk-taking explorer who survives perilous trips into the unknown to discover nature's secrets ensured that at home scientists would be regarded with the nationalist esteem and acclaim accorded to great explorers and military leaders. And science would be perceived to be an epitome of a manly job and therefore unsuitable for a woman. These scientific travel narratives also feminized the audience at home who would listen with bated breath to the scientists' tales of derring-do (Terrall).

Yet one woman who engaged in perilous travels was a widely recognized scientist of her day. The engraver-scientist Maria Sibylla Merian (1647–1714) traveled with her daughter—that is, without the usual male protection of an accompanying merchant or administrator husband or father—to spend two years investigating the life cycles of caterpillars in the Dutch tropical colony in Suriname in South America. Merian produced carefully observed "true-to-life" engravings of the life-stages of caterpillars and butterflies she encountered and also cultivated there, each depicted on the particular plant that provided its nourishment. She thereby decisively overturned the still-prevailing view that caterpillars and butterflies belonged to different species, as well as the not yet thoroughly discredited view that insects generated from rotting organic materials through "spontaneous generation." She said that she learned much from indigenous women about the insects and plants she depicted, and she brought one such indigene back to Germany to assist her in preparing the influential engravings. She became a significant figure in the scientific circles of Amsterdam, Germany, Russia, and elsewhere in Europe and was cited by Linnaeus (Reitsma).

In Merian's era, economic botany, as it was called, was the "big science" of the day (Schiebinger). Kew Gardens outside London was the center of a global network of more than a hundred botanical collection and research stations (Brockway). It developed suitable crops for British plantations in India, Africa, and the Caribbean from plant materials originating in the Americas. Moreover, the indigenous farmers, botanists, and pharmacologists made important though unrecognized contributions to European sciences in these areas, as did African slaves (Schiebinger). The slaves brought to the Americas rice seeds (and probably other botanical resources) and knowledge about how to grow and use them. Indeed, Africans who had farmed rice in West Africa were especially prized by slavers for the high prices they would bring for their rice-growing knowledge and skills. In South Carolina especially, these slaves adapted the African rice to the local growing conditions, producing food for their own use as well as for their owners' plantation production. Rice growing was women's specialty in Africa, on the slave ships, and in America (Carney).

So the development of modern sciences and technologies depended extensively on the knowledge of native informants from other cultures around the globe. David Hess (see part IV) refers to this era of modern science as its period of "epistemic primitive accumulation," echoing the Marxian account of early capitalism. Furthermore, research done offshore, in the European empires and colonies, was central to the development of European sciences and technologies; it was fully part of European sciences. Europeans turned the world into a laboratory for European sciences (Kochhar 1992; McClellan 1992). The steady development and management of that laboratory depended on European control of the seas, shores, lands, and peoples through which Europeans observed nature and gathered information. Moreover, that command of global travel routes enabled the combination and transformation of information about nature gathered from different parts of the world. Darwin's great theory could not have been developed without the access to nature provided by European empire (Latour 1987). The development of the European empire blocked the scientific development of other cultures (Rodney 1982).

These arguments are controversial, as they should be. Yet they are grounded in solid historical research. They raise further intriguing questions about what it means to provide a postcolonial account. Do these still recuperate Europe's others in a universal history—a new kind of history, to be sure, but still a unitary, totalizing account? Should the scientific revolution be regarded as progress for our species or for civilization when its costs and benefits have been so inequitably distributed? Moreover, are civilizations the appropriate historical units to be comparing in light of their internal heterogeneity and continual encounters and exchanges with other cultures? (Hart 1999). Is "European" the appropriate way to think about modern science in light of the "global economy in the Asian Age"? (Frank 1998). What would an epistemology and philosophy of science look like if they were grounded also in these kinds of economic, political, and cultural processes rather than in supposedly only abstract "logics of scientific inquiry"?

## References

Adas, Michael. 1997. "Colonialism and Science." In *Encyclopedia of the History of Non-Western Science, Technology, and Medicine*, ed. Helaine Selin, 214–20. Dordrecht: Kluwer.

Bernal, Martin. 1987. *Black Athena: The Afroasiatic Roots of Classical Civilization*. Vol. 1. New Brunswick, N.J.: Rutgers University Press.

Blaut, J. M. 1993. *The Colonizer's Model of the World: Geographical Diffusionism and Eurocentric History*. New York: Guilford Press.

Chakrabarty, Dipesh. 2000. *Provincializing Europe*. Princeton: Princeton University Press.

Frank, Andre Gunder. 1998. *ReOrient: Global Economy in the Asian Age*. Berkeley: University of California Press.

Hart, Roger. 1999. "On the Problem of Chinese Science." In *The Science Studies Reader*, ed. Mario Biagioli. New York: Routledge.

Headrick, Daniel R., ed. 1981. *The Tools of Empire: Technology and European Imperialism in the Nineteenth Century*. New York: Oxford University Press.

Jacob, Margaret. 1988. *The Cultural Meanings of the Scientific Revolution*. New York: Knopf.

Kochhar, R. K. 1992. "Science in British India: Colonial Tool." *Current Science* (India) 63 (11): 689–94.

Latour, Bruno. 1987. *Science in Action*. Cambridge: Harvard University Press.

McClellan, James E. 1992. *Colonialism and Science: Saint Domingue in the Old Regime*. Baltimore: Johns Hopkins University Press.

Mignolo, Walter. 1995. *The Darker Side of the Renaissance: Literacy, Territoriality and Colonization*. Ann Arbor: University of Michigan Press.

Needham, Joseph. 1956–2004. *Science and Civilisation in China*. 7 vols. Cambridge: Cambridge University Press.

———. 1969. "The Laws of Man and the Laws of Nature." In *The Grand Titration: Science and Society in East and West*. Toronto: University of Toronto Press.

Prakash, Gyan. 1999. *Another Reason: Science and the Imagination of Modern India*. Princeton: Princeton University Press.

Rodney, Walter. 1982. *How Europe Underdeveloped Africa*. Washington: Howard University Press.

Schiebinger, Londa, and Robert Proctor. 2008. *Agnatology: The Making and Unmaking of Ignorance*. Stanford: Stanford University Press.

JOHN M. HOBSON

# 1. Discovering the Oriental West

History cannot be written as if it belonged to one group [of people] alone. Civilization has been gradually built up, now out of the contributions of one [group], now of another. When all civilization is ascribed to the [Europeans], the claim is the same one which any anthropologist can hear any day from primitive tribes—only they tell the story of themselves. They too believe that all that is important in the world begins and ends with them . . . We smile when such claims are made [by primitive tribes], but ridicule might just as well be turned against ourselves . . . Provincialism may rewrite history and play up only the achievements of the historian's own group, but it remains provincialism.
—RUTH BENEDICT

We have been taught, inside the classroom and outside of it, that there exists an entity called the West, and that one can think of this West as a society and civilization independent of and in opposition to other societies and civilizations [i.e., the East]. Many of us even grew up believing that this West has [an autonomous] genealogy, according to which ancient Greece begat Rome, Rome begat Christian Europe, Christian Europe begat the Renaissance, the Renaissance the Enlightenment, the Enlightenment political democracy and the industrial revolution. Industry, crossed with democracy, in turn yielded the United States, embodying the rights to life, liberty and the pursuit of happiness . . . [This is] misleading, first, because it turns history into a moral success story, a race in time in which each [Western] runner of the race passes on the torch of liberty to the next relay. History is thus converted into a tale about the furtherance of virtue, about how the virtuous [i.e. the West] win out over the bad guys [the East].—ERIC WOLF

MOST OF US NATURALLY ASSUME that the East and West are, and always have been, separate and different entities. We also generally believe that it is the "autonomous" or "pristine" West that has alone pioneered the creation of the modern world; at least that is what many of us are taught at school, if not at university. We typically assume that the pristine West had emerged at the top of the world by about 1492 (think of Christopher Columbus), owing to its uniquely ingenious scientific rationality, rational restlessness, and democratic/progressive properties. From then, the traditional view has it, the Europeans spread outward conquering the East and Far West while simultaneously laying down the tracks of capitalism along which the whole world could be delivered from the jaws of deprivation and misery into the bright light of modernity. Accordingly, it seems entirely natural or self-evident to most of us to conflate the progressive story of world history with the Rise and Triumph of the West. This traditional view can be called "Eurocentric." For at its heart is the notion that the West properly deserves to occupy the centre stage of progressive world history, both past and present. But does it?

The basic claim of *The Eastern Origins of Western Civilisation* is that this familiar but deceptively seductive Eurocentric view is false for various reasons, not the least of which is that the West and East have been fundamentally and consistently interlinked through globalisation ever since 500 CE. More importantly, and by way of analogy, Martin Bernal argues that Ancient Greek civilisation was in fact significantly derived from Ancient Egypt.[1] Likewise, the book argues that the East (which was more advanced than the West between 500 and 1800) provided a crucial role in enabling the rise of modern Western civilisation. It is for this reason that I seek to replace the notion of the autonomous or pristine West with that of the oriental West. The East enabled the rise of the West through two main processes: diffusionism/assimilationism and appropriationism. First, the Easterners created a global economy and global communications network after 500 along which the more advanced Eastern "resource portfolios" (e.g., Eastern ideas, institutions and technologies) diffused across to the West, where they were subsequently assimilated, through what I call oriental globalisation. And second, Western imperialism after 1492 led the Europeans to appropriate all manner of Eastern economic resources to enable the rise of the West. In short, the West did not autonomously pioneer its own development in the absence of Eastern help, for its rise would have been inconceivable without the contributions of the East. The task of the book, then, is to trace the manifold Eastern contributions that led to the rise of what I call the oriental West.

The book feeds into the debate between Eurocentrism and anti-Eurocentrism. In recent years a small band of scholars have claimed that the standard theories of the rise of the West —Marxism/world-systems theory, liberalism, and Weberianism—are all Eurocentric.[2] They all assume that the "pristine" West "made it" of its own accord as a result of its innate and superior virtues or properties. This view presumes that Europe autonomously developed through an iron logic of immanence. Accordingly, such theories assume that the rise of the modern world can be told as the story of the rise and triumph of the West. Importantly, the Eurocentric account has enjoyed a new lease of life or fresh reinvigoration, particularly with the 1998 publication of David Landes's *The Wealth and Poverty of Nations*,[3] a book that implicitly harks back to John Roberts's *The Triumph of the West*.[4] Landes's book in particular launches a passionate and pejorative attack against some of the recent anti-Eurocentric

analyses (though for all this it is done with verve and wit and is an especially enjoyable read). Perhaps Landes's most significant service is that he
has helped transform the old theoretical debate conducted between Marx-
ism/world-systems theory, liberalism, and Weberianism into a new one of
"Eurocentrism versus anti-Eurocentrism." This, it seems to me, is where
the real intellectual action lies. For arguably the old debate is something of
a non-debate given that all these approaches now appear as but minor or
subtle variations on the same Eurocentric theme (see the next section below). Accordingly, *The Eastern Origins of Western Civilisation* enters this
new debate and contests each of the major claims made by mainstream
Eurocentrism, while simultaneously proposing an alternative account.

It could, however, be replied that the "Eurocentric versus anti-
Eurocentric" framework that the book operationalises is an oversimplification and is itself a "non-debate." Presuming a kind of Manichean
struggle between two coherent ideologies is problematic mainly because,
it could be claimed, there is no coherent paradigm called "Eurocentrism."
Indeed, I believe it would be wrong to assume that most scholars are
fighting to defend an explicitly Eurocentric "triumphalist" vision of the
West. And while there are some who explicitly associate themselves with
Eurocentrism (such as Landes and Roberts), most do not. Nevertheless,
I firmly believe that Eurocentrism infuses *all* the mainstream accounts
of the rise of the West, even if this mostly occurs behind the back of the
particular scholar (see the next section below). Accordingly, I believe it
to be legitimate to develop my own account by critically evaluating the
many claims made by Eurocentrism.

The main argument of *The Eastern Origins of Western Civilisation* counters one of Eurocentrism's most basic assumptions—that the East has been
a passive bystander in the story of world historical development as well as
a victim or bearer of Western power, and that accordingly it can be legitimately marginalised from the progressive story of world history. Although
differing in various ways from Felipe Fernández-Armesto's phenomenal
book, *Millennium*, nevertheless I share with him his empathic belief that:

> For purposes of world history, the margins sometimes demand more atten-
> tion than the metropolis. Part of the mission of this book is to rehabilitate
> the overlooked, including places often ignored as peripheral, peoples mar-
> ginalized as inferior and individuals relegated to bit-parts and footnotes.[5]

Or in a narrower context, as W. E. B. Du Bois explained in the foreword
to his important book, *Africa in World History*:

> there has been a consistent effort to rationalize Negro slavery by omitting
> Africa from world history, so that today it is almost universally assumed

that history can be truly written without reference to Negroid peoples . . . Therefore I am seeking in this book to remind readers . . . of how critical a part Africa has played in human history, past and present.[6]

Likewise, my major claim in *The Eastern Origins of Western Civilisation* is that the Eurocentric denial of Eastern agency and its omission of the East in the progressive story of world history is entirely inadequate. For not only do we receive a highly distorted view of the rise of the West, but we simultaneously learn little about the East except as a passive object, or provincial backwater, of mainstream Western world history.

This marginalisation of the East constitutes a highly significant silence because it conceals three major points. First, the East actively pioneered its own substantial economic development after about 500. Second, the East actively created and maintained the global economy after 500. Third, and above all, the East has significantly and actively contributed to the rise of the West by pioneering and delivering many advanced "resource portfolios" (e.g., technologies, institutions and ideas) to Europe. Accordingly, we need to resuscitate both the history of economic dynamism in the East and the vital role of the East in the rise of the West. Nevertheless, as we shall also see, this does not mean that the West has been a passive recipient of Eastern resources. For the Europeans played an active role in shaping their own fate (especially through the construction of a changing collective identity, which in turn partially informed the direction of Europe's economic and political development). In sum, these two interrelated claims—Eastern agency and the assimilation of advanced Eastern "resource portfolios" via oriental globalisation on the one hand, entwined with European agency/identity and the appropriation of Eastern resources on the other—constitute the discovery of the lost story of the rise of the oriental West.

In this context it is especially noteworthy that our common perception of the irrelevance of the East and the superiority of Europe is reinforced or "confirmed" by the Mercator world map. This map is found everywhere—from world atlases to school walls to airline booking agencies and boardrooms. Crucially, the actual landmass of the southern hemisphere is exactly twice that of the northern hemisphere. And yet on the Mercator, the landmass of the North occupies two-thirds of the map while the landmass of the South represents only a third. Thus while Scandinavia is about a third the size of India, they are accorded the same amount of space on the map. Moreover on the Mercator, Greenland appears almost twice the size of China, even though the latter is almost four times the size of the former. To correct for what he saw as the racist privileging of Europe, in 1974 Arno Peters produced the Peters projection (or the Peters–Gall projection), which sought to represent the countries of the world according to their actual surface area. Here the South properly looms much larger, while Europe is considerably downgraded. Although no perfect map of the world exists, his representation is certainly free of the implicit Eurocentric distortion found in the Mercator. Not surprisingly, when the Peters projection first appeared there was a political storm, for as Marshall Hodgson points out, "Westerners understandably cling to a projection [the Mercator] which so markedly flatters them."[7]

*The Eastern Origins of Western Civilisation* in effect attempts to correct our perception of world history in the same way that the Peters projection seeks to correct our perception of world geography, by discovering the relative importance of the East vis-à-vis the West. More specifically, I have presented a variant of this projection (the "Hobo-Dyer") at the beginning of the book but have reconfigured it so as to place China at the centre, given its pivotal role in the rise of the West. No less importantly, the USA and Europe now properly occupy the diminished peripheral margins of the Far North-east and Far North-west respectively. And while Africa also occupies the Far West, its upgraded size corrects for its downgraded marginalisation in the Eurocentric model.

The book proceeds in two sections. The first begins by very briefly tracing the construction of the Eurocentric discourse as it emerged during the eighteenth and nineteenth centuries. It then proceeds to show how the major explanations of the rise of the West, found specifically in the work of Karl Marx and Max Weber, became grounded within this discourse. The second section then briefly fleshes out my own two-prong argument as a remedy to the prevailing Eurocentrism of mainstream accounts.

## Constructing the Eurocentric/Orientalist Foundations of the Mainstream Theories of the Rise of the West

### European Identity Formation and the Invention of Eurocentrism/Orientalism

In 1978 Edward Said famously coined the phrase "Orientalism," though in fairness a number of other scholars, including Victor Kiernan, Marshall Hodgson and Bryan Turner, were already thinking along such lines.[8] Orientalism or Eurocentrism (I use them interchangeably) is a worldview that asserts the inherent superiority of the West over the East. Specifically Orientalism constructs a permanent image of the superior West (the "Self") which is defined negatively against the no less imaginary "Other"—the backward and inferior East. It was mainly during the eighteenth and nineteenth centuries that this polarised and essentialist construct became fully apparent within the European imagination. What then were the specific categories by which the West came to imagine its Self as superior to the Eastern Other?

Between 1700 and 1850 European imagination divided, or more accurately forced, the world into two radically opposed camps: West and East (or the "West and the Rest"). In this new conception, the West was imagined as superior to the East. The imagined values of the inferior East were set up as the antithesis of rational Western values. Specifically, the West was imagined as being inherently blessed with unique virtues: it was rational, hard-working, productive, sacrificial and parsimonious, liberal-democratic, honest, paternal and mature, advanced, ingenious, proactive, independent, progressive, and dynamic. The East was then cast as the West's opposite Other: as irrational and arbitrary, lazy, unproductive, indulgent, exotic as well as alluring and promiscuous, despotic, corrupt, childlike and immature, backward, derivative, passive, dependent, stagnant and unchanging. Another way of expressing this is to say that the West was defined by a series of progressive presences, the East by a series of absences.

Particularly important is that this reimagining process stipulated that the West had always been superior (in that this construct was extrapolated back in time to Ancient Greece). For the West has allegedly enjoyed dynamically progressive, liberal and democratic values, and rational institutions from the outset, which in turn gave birth to the rational individual, whose flourishing life enabled economic progress and the inevitable breakthrough to the blinding light and warmth of capitalist modernity. By contrast, the East was branded as permanently inferior. It has allegedly endured despotic values and irrational institutions, which meant that in the very heart of darkness, a cruel collectivism strangled the rational individual at birth, thereby making economic stagnation and slavery its eternal fate. This argument formed the basis of the theory of oriental despotism and the Peter Pan theory of the East, which conveyed an eternal image of a "dynamic West" versus an "unchanging East" (see table 1).

It can hardly escape notice that these binary opposites are precisely the same categories that constitute the patriarchally constructed identity of masculinity and femininity. That is, the modern West is akin to the constructed male, the East the imagined female. This is no coincidence, because during the post-1700 period Western identity was constructed as patriarchal and powerful, while the East was simultaneously imagined as feminine—as weak and helpless. This led to the Orientalist representation of an Asia "lying passively in wait for Bonaparte," for only he could

TABLE 1 The Orientalist and patriarchal construction of
the "West versus the East"

| The dynamic West | The unchanging East |
| --- | --- |
| Inventive, ingenious, proactive | Imitative, ignorant, passive |
| Rational | Irrational |
| Scientific | Superstitious, ritualistic |
| Disciplined, ordered, self-controlled, sane, sensible | Lazy, chaotic/erratic, spontaneous, insane, emotional |
| Mind-oriented | Body-oriented, exotic and alluring |
| Paternal, independent, functional | Childlike, dependent, dysfunctional |
| Free, democratic, tolerant, honest | Enslaved, despotic, intolerant, corrupt |
| Civilised | Savage/barbaric |
| Morally and economically progressive | Morally regressive and economically stagnant |

liberate her from her enslaved existence (an act of liberation, which was subsequently dubbed "the white man's burden"). And this theory was vitally important because branding the East as exotic, enticing, alluring, and above all passive (i.e., as having no initiative to develop of her own accord), thereby produced an immanent and ingenious legitimating rationale for the West's imperial penetration and control of the East.

But this was not just a legitimating idea for imperialism and the subjugation of the East. For by depicting or imagining the East as the West's passive opposite it was but a short step to make the argument that *only* the West was capable of independently pioneering progressive development. Indeed, the outcome of the European intellectual revolution was the construction of the "proactive" European subject, and the "passive" Eastern object, of world history. Moreover, European history was inscribed with a progressive temporal linearity, while the East was imagined to be governed by regressive cycles of stagnation. In particular, within the Eurocentric discourse this divide implied a kind of "intellectual apartheid regime" because the superior West was permanently and retrospectively quarantined off from the inferior East. Or, in Rudyard Kipling's felicitous phrase, "Oh, East is East, and West is West, and never the twain shall meet." This was crucial precisely because it immunised the West from recognising the positive influence imparted by the East over many centuries, thereby implying that the West had pioneered its own development in the complete absence of Eastern help ever since the time of Ancient Greece. And from there it was but a short step to proclaiming that the history of the world can only be told as the story of the pioneering and triumphant West from the outset. Thus the myth of the pristine West was born: that the Europeans had, through their own superior ingenuity, rationality, and social-democratic properties, pioneered their own development in the complete absence of Eastern help, so that their triumphant breakthrough to modern capitalism was inevitable.

It is no coincidence that the social sciences emerged most fully in the nineteenth century at the time when this process of reimagining Western identity reached its apogee. For by then the Europeans had intellectually divided the whole world into the two antithetical compartments. But rather than critique this Orientalist and essentialist West/East divide, orthodox Western social scientists from the nineteenth century down to the present not only accepted this polarised separation as self-evidently true, but inscribed it into their theories of the rise of the West and the origins of capitalist modernity. How did this occur?

Most generally, as the quote from Eric Wolf (posted at the beginning of this chapter) points out,[9] within the mainstream theories we can detect a latent—though occasionally explicit—triumphalist teleology in which all of human history has ineluctably been leading up to the Western endpoint of capitalist modernity. Thus conventional accounts of world history assume that this all began with Ancient Greece, progressing on to the European agricultural revolution in the low middle ages, then on to the rise of Italian-led commerce at the turn of the millennium. The story continues on into the high middle ages when Europe rediscovered pure Greek ideas in the Renaissance which, when coupled with the scientific revolution, the Enlightenment and the rise of democracy, propelled Europe into industrialisation and capitalist modernity.

Pick up any conventional book on the rise of the modern world. The West is usually represented as the *mainstream civilisation* and is enshrined with a *Promethean* quality (to paraphrase the titles of two prominent books).[10] While Eastern societies are sometimes discussed, they clearly lie outside the mainstream story. And it is often the case that if the East is discussed at all, it is discussed in separate sections. Accordingly, one could focus only on the Western sections and get the main story. Thus Eastern societies basically appear as an aside or as an irrelevant footnote. But this aside is important not because it says little about the East but because it describes only the inherent, regressive properties that blocked its progress. Once more, this provides a very powerful confirmation of Western superiority and why the "triumph of the West" was but a fait accompli.

Two main points are of note here. First, this story is one that imagines Western superiority from the outset. And second, the story of the rise and triumph of the West is one that can be told without any discussion of the East or the "non-West." Europe is seen as autonomous or self-constituting on the one hand, and rational/democratic on the other, making the breakthrough all by itself. This is what I refer to as the Eurocentric iron logic of immanence. Both these views underpin the triumphalist Eurocentric notion of the "European miracle" conceived as a "virgin birth." Accordingly, the story of the origins of capitalism (and globalisation) is conflated with the rise of the West; the account of the rise of modern capitalism and civilisation *is* the Western story. It is precisely this notion that Ruth Benedict had in mind when she described "our" conception of world history as "provincial."[11] Or as Du Bois put it:

> It has long been the belief of modern men that the history of Europe covers the essential history of civilization, with unimportant exceptions; that the progress of the white [Europeans] has been along the one natural, normal path to the highest possible human culture.[12]

Nevertheless, it remains to be ascertained just how the categories of Orientalism became endogenised within the mainstream accounts of the rise of the West. Because other anti-Eurocentric writers have deconstructed a range of modern prominent scholars,[13] I shall concentrate here on revealing the Orientalist foundations of the classical theories of Marx and Weber. This focus is legitimate because most subsequent theories have been derived from Marx and especially Weber in one way or another.

## The Orientalist Foundations of Marxism

It might be thought that Marxism would *not* fit the Orientalist mould, given that Karl Marx was one of Western capitalism's most strident critics. But the fact is that Marx privileged the West as the active subject of progressive world history and denigrated the East as but its passive object. And in the process Marx's theory demonstrated all the hallmarks of Eurocentric world history. How so?

Karl Marx's theory assumed that the West was unique and enjoyed a developmental history that had been absent in the East. Indeed, he was explicit that the East had had *no* (progressive) history. This was reiterated in numerous pamphlets and newspaper articles. For example, China was a "rotting semicivilization . . . vegetating in the teeth of time."[14] Consequently, China's only hope for progressive emancipation or redemption lay with the Opium Wars and the incursion of British capitalists who would "open up backward" China to the energising impulse of capitalist world trade.[15] India too was painted with the same brush.[16] This Willis formula was most famously advanced in *The Communist Manifesto* where we are told that the Western bourgeoisie, draws all, even the most barbarian, nations into civilization . . . It compels all nations, on pain of extinction, to adopt the [Western] bourgeois mode of production; it compels them to introduce what it calls civilization into their midst, i.e., to become [Western] themselves. In one word, it [the Western bourgeoisie] creates a world after its own image.[17]

Marx's dismissal of the East was not confined to his numerous newspaper articles (no fewer than seventy-four between 1848 and 1862) and various pamphlets, but was fundamentally inscribed into the theoretical schema of his historical materialist approach. Crucial here was his concept of the "Asiatic mode of production" in which "private property" and hence "class struggle"—the developmental motor of historical progress—were notably absent. As he explained in *Capital*, in Asia "the direct producers . . . [are] under direct subordination to a state which stands over them as their landlord . . . [Accordingly] no private ownership of land exists."[18] And it was the absorption of, and hence failure to produce, a surplus for reinvestment in the economy that, "supplie[d] the key to the secret of the *unchangeableness* of Asiatic societies."[19] In short, private property and class struggle in part failed to emerge because the forces of production were owned by the despotic state. Thus stagnation was inscribed into this publicly owned land system because rents were extracted from the producers, in the form of "taxes wrung from them— frequently by means of torture—by a ruthless despotic state."[20]

This scenario was fundamentally contrasted with the European situation. In Europe the state did not stand above society but was fundamentally embedded within, and cooperated with, the dominant economic class. In turn, being unable to squeeze a surplus through high taxation the state allowed a space to emerge through which capitalists could accumulate a surplus (i.e., profits) to be reinvested in the capitalist economy. Accordingly, economic progress was understood as the unique preserve of the West. Thus what we have in Marx's theoretical understanding of the East and West is the theory of oriental despotism (which subsequently found its most famous voice in Karl Wittfogel's neo-Marxist book).[21] It is true that Marx's notion of the Asiatic mode of production oscillated between the choking powers of the despotic state on the one hand and the stifling role of rural communal production on the other. But whichever factor was crucial does not detract from his abiding belief that the East had no prospects for progressive self-development and could, therefore, only be rescued by the British capitalist imperialists.

No less importantly, Marx's whole theory of history faithfully reproduces the Orientalist or Eurocentric teleological story. In *The German Ideology* Marx traces the origins of capitalist modernity back to Ancient

Greece—the fount of civilisation (and in the *Grundrisse* he explicitly dismissed the importance of Ancient Egypt).[22] He then recounts the familiar Eurocentric story of linear/immanent progress forward to European feudalism and on to European capitalism, then socialism before culminating at the terminus of communism.[23] Thus Western man was originally born free under "primitive communalism" and, having passed through four progressive historical epochs, would eventually emancipate himself as well as the Asian through revolutionary class struggle. For Marx the Western proletariat is humanity's "Chosen People" no less than the Western bourgeoisie is global capitalism's "Chosen People." Marx's inverted Hegelian approach gave rise to a progressive/linear story in which the (Western) species edged closer to freedom through class struggle with each passing historical epoch.

No such progressive "linearity" was possible in the Orient, where growth-repressive "cycles" of despotic political regimes and regressive rural production systems did no more than mark time. Underlying this whole approach is a clear denial of Eastern agency. To paraphrase Marx's discussion of the difference between a proletarian "class-in-itself" (representing inertia and passivity) and a "class-for-itself" (representing a proactive propensity for emancipation), it is as if Marx saw the East as a "being-in-itself" that was inherently incapable of becoming a "being-for-itself." By contrast, the West was from the outset a "being-for-itself." Moreover, it seems no coincidence that the Hegelian influence in Marx's work should have produced this binary "progressive West/regressive East" couplet, precisely because for Hegel the superior Spirit of the West is progressive freedom, whereas the inferior Spirit of the East is regressive, unchanging despotism.[24] In short, for Marx the West has been the triumphant carrier of historical progress, the East but its passive recipient.

All in all it seems fair to dub Karl Marx's approach as "Orientalism painted red."[25] However, none of this is to say that Marxism is moribund, for it undoubtedly remains useful and insightful. But it is to say that as an overall framework it remains embedded firmly within an Orientalist discourse.

## *The Orientalist Foundations of Weberianism*

Nowhere is the Orientalist approach clearer than in the works of the German sociologist, Max Weber. Weber's whole approach was founded on the most poignant Orientalist questions: what was it about the West that made its path to modern capitalism inevitable? And why was the East predestined for economic backwardness? The Orientalist cue in Weber is found both with the initial questions and the subsequent analytical methodology that he deployed in order to answer them. Weber's view was that the essence of modern capitalism lay with its unique and pronounced degree of "rationality" and "predictability," values that were to be found only in the West. From there, as Randall Collins points out,

> the logic of Weber's argument is first to describe these characteristics; then to show the *obstacles to them that were present in virtually all societies of world history until recent centuries in the West*; and finally, by the method of comparative analysis, to show the social conditions responsible for their [unique] emergence [in the West].[26]

This is pristine Orientalist logic, given that Weber selected or imputed a series of progressive features that were allegedly unique to the West. And he simultaneously insisted on their absence in the East, where a series of imaginary blockages ensured its failure to progress. That is, he did not objectively select the key aspects that made the West's rise possible. He in fact imputed them no less than he imputed a series of imaginary blockages that supposedly made the East's failure inevitable (a claim which I demonstrate throughout *The Eastern Origins of Western Civilisation*). The Orientalist character of his analytical template is revealed most clearly in his depiction of the East and West (see table 1).

The crucial comparison here is between tables 1 and 2. This comparison confirms that Weber perfectly transposed the Eurocentric categories into his central social scientific concepts. Thus the West was blessed with a unique set of rational institutions which were both liberal and growth permissive. The growth-permissive factors are striking for their presence in the West and for their absence in the East.[27] Here, the division of East and West according to the presence of irrational and rational institutions respectively very much echoes the Peter Pan theory of the East. In particular, the final two categories located at the bottom of the table deserve emphasis. First, the differences in the two civilisations are summarised in Weber's claim that Western capitalist modernity is characterised by a fundamental separation of the public and private realms. In traditional society (as in the East) there was no such separation. Crucially, only when there is such a separation can formal rationality—the leitmotif of modernity—prevail. This supposedly infuses all spheres—the political, military, economic, social and cultural.

The second general distinguishing feature between the Orient and Occident was the existence of a "social balance of power" in the latter and its absence in the former. Taking their cue from Weber, neo-Weberian analyses commonly differentiate "multi-power actor civilizations" or the European multi-state system from Eastern single-state systems or "empires of domination."[28] And they, like some Marxian world-systems

TABLE 2  Max Weber's Orientalist view of the "East" and "West":
the great "rationality" divide

| Occident (modernity) | Orient (tradition) |
| --- | --- |
| Rational (public) law | *Ad hoc* (private) law |
| Double-entry bookkeeping | Lack of rational accounting |
| Free and independent cities | Political/administrative camps |
| Independent urban bourgeoisie | State-controlled merchants |
| Rational-legal (and democratic) state | Patrimonial (oriental despotic) state |
| Rational science | Mysticism |
| Protestant ethic and the emergence of the rational individual | Repressive religions and the predominance of the collectivity |
| Basic institutional constitution of the West | Basic institutional constitution of the East |
| Fragmented civilisation with a balance of social power between all groups and institutions (i.e., multi-state system or multi-power actor civilisation) | Unified civilisations with no social balance of power between groups and institutions (i.e., single-state systems or empires of domination) |
| Separation of public and private realms (rational institutions) | Fusion of public and private realms (irrational institutions) |

theorists as well as a number of non-Marxists,[29] emphasise the vital role
that warfare between states played in the rise of Europe (which, "by definition," did not exist in the single-state empires in the East). It is here
where the theory of oriental despotism becomes pivotal. Only the Occident enjoyed a precarious balance of social forces and institutions where
none could predominate.[30] European secular rulers could not dominate
on a despotic model. They granted "powers and liberties" to individuals in civil society, initially to the nobles and later on to the bourgeoisie.
By 1500 rulers were anxious to promote capitalism in order to enhance
tax revenues in the face of constant, and increasingly expensive, military
competition between states. By contrast, in the East the predominance
of "single-state systems" led to empires of domination, in large part because a lack of military competition released the state from the pressure
of having to nurture the development of society. Thus in contrast to the
fief (hereditary land tenure) that Western rulers had granted the nobility
before about 1500, Eastern nobles were stifled by the despotic or patrimonial state which imposed prebendal rights (rights which prevented the
consolidation of this class's power). Moreover, the Eastern bourgeoisie
was thoroughly repressed by the despotic or patrimonial state and was

confined to "administrative camps" as opposed to the "free cities" that were allegedly found only in the West. In addition, European rulers were also balanced against the power of the Holy Roman empire as well as the papacy, which contrasted with Eastern caesaropapism (where religious and political institutions were fused). Finally, while Western man became imbued with a "rational restlessness" and a transformative "ethic of world mastery," in part because of the energising impulse of Protestantism, Eastern man was choked by regressive religions and was thereby marked by a long-term fatalism and passive conformity to the world. Accordingly, the rise of capitalism was as much an inevitability in the West as it was an impossibility in the East.

In sum, although the Weberian argument has a different content from Marx's, both worked within an Orientalist framework. And the obvious link here lies in the centrality that both accord to the absence of oriental despotism in the West on the one hand, and the imputed European logic of immanence on the other. Accordingly, as noted earlier, when seen through an anti-Eurocentric lens these so-called radically opposed perspectives appear as but subtle variations on the exact same Orientalist theme.

Probably the most significant consequence of Max Weber's construction of the Eurocentric theoretical template is that it has permeated almost all Eurocentric accounts of the rise of the West even if, as James Blaut also notes, many of the relevant authors would recognise themselves as neither Weberian nor Orientalist.[31] This should hardly be surprising, given that all mainstream scholars begin their analysis by asking the standard Weberian question: why did *only* the West break through to modern capitalism, while, conversely, the East was doomed to remain in poverty? When expressed in this way, an Orientalist story was made inevitable because the question led the enquirer (often unintentionally) to impute an inevitability to both the rise of the West and the stagnation of the East. How so? Applying the Orientalist conception of the binary "West–East divide" furnished Western scholars with the inevitable answer: that only the West had the ingenuity and progressive properties to make the breakthrough—values that were deemed to be entirely absent in the East from the outset. Posed in this way, the question begged the answer: how did the ingenious and progressive liberal West advance to capitalist modernity as opposed to the regressive, despotic East, whose eternal fate lay with stagnation and slavery? Thus the essential causal categories had already been assigned in advance of historical enquiry.

But it might be replied that it is reasonable to begin by noting the present situation of an advanced West and a backward East and then exploring the past to "reveal" the factors that made this so. The problem is that in extrapolating retrospectively the notion of a backward East a subtle but erroneous slippage is made: in "revealing" the various blockages that held the East back, Eurocentrism ends up by imputing to the East a permanent "iron law of *non*-development." And above all, because Eurocentrism appraises the East only through the lens of the West's final breakthrough to modern capitalism, any technological or economic developments that were made in the East are immediately dismissed as inconsequential. In contrast, by taking present-day Western superiority as a fact and then extrapolating this conception back through historical time, the enquirer necessarily ends up by imputing to the West a permanent "iron law of immanent development." This is rendered problematic by the central argument of this book: that there was nothing inevitable about the West's rise, precisely because the West was nowhere near as ingenious or morally progressive as Eurocentrism assumes. For without the helping hand of the more advanced East in the period from 500 to 1800, the West would in all likelihood never have crossed the line into modernity.

Thus much of our Western thinking is not scientific and objective but is orientated through a one-eyed perspective which reflects the prejudiced values of the West, and which necessarily prevents the enquirer from seeing the full picture. This is equivalent to what Blaut calls "Eurocentric tunnel history."[32] What happens, then, when we view the world through a more inclusive two-eyed perspective?

## The Illusion of Eurocentrism:
## Discovering the Oriental West

It is important to note that the Eurocentric and implicit "triumphalist" bias of our mainstream theories does not necessarily make them incorrect. Indeed, as the self-proclaimed Eurocentric scholar, David Landes, has recently argued, there is actually very good reason for Eurocentrism because it *is* the West and *not* the East that has triumphed because, he claims, only the Europeans managed to pioneer the breakthrough to capitalist modernity. Accordingly, Landes dismisses the anti-Eurocentric account as "politically correct goodthink" or "Europhobic" or simply "bad history."[33] But my central argument is that the Eurocentric story is problematic not because it is politically incorrect but because it does not square with what really happened. David Landes, in his self-proclaimed Eurocentric book, forcefully disagrees. As he puts it:

> A third school [in which the present chapter would be included] would argue that the West–Rest [West–East] dichotomy is simply false. In the large

stream of world history, Europe is a latecomer and free rider on the earlier achievements of others. That is patently incorrect. As the historical record shows, for the last thousand years, Europe (the West) has been the prime mover of development and modernity. That still leaves the moral issue. Some would say that Eurocentrism is bad for us, indeed bad for the world, hence to be avoided. Those people should avoid it. As for me, I prefer truth to goodthink. I feel surer of my ground.[34]

But the historical empirical record that I consult reveals that for most of the last thousand years the East has been the prime mover of world development. Conventional scholars assign the leading edge of global power in the last thousand years, without exception, to Western states. But the immediate problem is that Western powers only appear to have been dominant because a Eurocentric view determined from the outset that no Eastern power could be selected in. As *The Eastern Origins of Western Civilisation* shows, all the so-called "leading Western powers" were inferior, economically and politically, to the leading Asian powers. It was only near the very end of the period (c. 1840) that a Western power finally eclipsed China.

Nevertheless, Landes would still claim that even if all this were true, the fact remains that only the Europeans managed to single-handedly break through to capitalist modernity. Or as Lynn White put it: "One thing is so certain that it seems stupid to verbalize it: both modern technology and modern science are distinctively *Occidental*."[35] But as I stated earlier, the West only got over the line into modernity because it was helped by the diffusion and appropriation of the more advanced Eastern resource portfolios and resources. Because the success of my account must lie with the empirical evidence that it marshals rather than because it is simply "goodthink" what then are some of the empirical facts that support my alternative anti-Eurocentric account? Let us take the diffusion and as-similation of Eastern resource portfolios through oriental globalisation first, before turning to the appropriation of Eastern resources through European imperialism.

One revealing example lies with what I call the "myth of Vasco da Gama." We in the West generally pride ourselves on the fact that it was the Portuguese discoverer, Vasco da Gama, who was the first man to have made it round the Cape of Good Hope and sail on to the East Indies where he made first contact with a hitherto isolated and primitive In-dian race. But sometime between two and five decades earlier the Islamic navigator, Ahmad ibn-Mājid, had already rounded the Cape and, hav-ing sailed up the West African coast, had entered the Mediterranean via the Strait of Gibraltar. Moreover, the Sassanid Persians had been sailing across to India and China from the early centuries of the first millennium CE, as did the Black Ethiopians and, later on, the Muslims (after about 650). And the Javanese, Indians, and Chinese had all made it across to the Cape many decades, if not centuries, before Da Gama. It has no less been forgotten that Da Gama only managed to navigate across to India because he was guided by an unnamed Gujarati Muslim pilot. No less irksome is the point that virtually all of the nautical and navigational technologies and techniques that made Da Gama's journey possible were invented (and certainly refined further) in either China or the Islamic Middle East. These were then assimilated by the Europeans, having diffused across the global economy via the Islamic Bridge of the World. And when we add the point that cannon and gunpowder were discovered in China and also diffused across, there is almost nothing left to indicate that the Portuguese had anything to genuinely claim for their own. Finally, the Indians were not primitive barbarians. In fact, they were considerably more advanced than their Portuguese "discoverers"—itself a misnomer precisely because India had long been in direct trading contact with much of Asia, East Africa, and indirectly with Europe, many centuries before Da Gama disingenuously claimed to have discovered it.

More generally it is important to note that Eastern resource portfolios had a significant influence in each of the major European turning points. Most of the major technologies that enabled the European medieval agricultural revolution after 600 CE seem to have come across from the East. After 1000, the major technologies, ideas, and institutions that stimulated the various Western commercial, production, financial, military, and navigational revolutions, as well as the Renaissance and the scientific revolution, were first developed in the East but later assimilated by the Europeans. After 1700, the major technologies and technological ideas that spurred on the British agricultural and industrial revolutions all diffused across from China. Moreover, Chinese ideas also helped stimulate the European Enlightenment. And it is precisely because the East and West have been linked together in a single global cobweb ever since 500 that we need to dispense with the Eurocentric assumption that these two entities can be represented as entirely separate and antithetical.

It is no less important to note that to each of my points a series of counter-measures are deployed which enable (usually unwittingly) the retention of the Eurocentric vision. Thus when Eurocentric writers concede that a certain idea or technology originated in the East, they often resort to what might be called a specific "Orientalist clause." Such clauses dismiss the significance of any particular Eastern achievement, thereby returning us to the Orientalist status quo. This process is rarely undertaken in a conscious way, given that most scholars are not fighting to defend an explicitly Eurocentric vision of the world. More often they deploy Orientalist clauses in order to retain their own theoretical perspective (e.g., Marxist, liberal, Weberian, etc.) rather than Eurocentrism per se. But whether intended or not, the outcome is still the maintenance of the Eurocentric vision if only because these approaches are inherently Orientalist.

Two examples of how such clauses are employed will suffice to illustrate my point. To my claim that China achieved an industrial miracle during the Sung (eleventh century), Eurocentric historians often reply by invoking one of the "China clauses" (or what Blaut calls the "China formula").[36] This clause dismisses its significance by insisting that it was but an "abortive revolution," with the Chinese economy subsequently reverting back to its normal state of relative stagnation. In this way, such theorists are able to preserve their claim that the British industrial revolution was truly the first (the "British clause"). Second, to answer the claim that the Middle East transmitted original scientific thoughts and texts to Europe that enabled the Western Renaissance and scientific revolution, the "Islamic clause" is immediately invoked. This dismisses the Eastern input on the grounds that these texts were in fact pure Greek works and that the Muslims had added nothing of intellectual value—all they did was return the original Greek works to the Europeans. This then overlaps with the "Greek clause," which stipulates that the Ancient Greeks were the original fount of modern (i.e., Western) civilisation. From these two examples alone it should be clear that there are many Orientalist clauses which all overlap to provide a logically coherent "Orientalist text." Thus, to make my case as plausible as possible, it is incumbent upon me—or anyone else who seeks to challenge Eurocentrism—to confront and dismantle every one of these interlinked Orientalist clauses or formulae. It is this task that informs the main narrative of *The Eastern Origins of Western Civilisation*. So much for the diffusion process.

The second major way in which the East enabled the rise of the West was through the European imperial appropriation of Eastern resources (land, labour and markets). Here I emphasise the role of European agency or identity. All the major anti-Eurocentric scholars seek to entirely discount the agency of the West. To include it, they reason, would be to fall back into the Eurocentric trap of emphasising European exceptionalism or uniqueness. But by erasing the notion of European agency we risk several dangers. First, we run the risk of representing the European achievement as truly miraculous.[37] Second, given that my main argument comprises the positive contribution of the East to the Western breakthrough, I risk falling into the trap of Occidentalism, in which the East is privileged and the West is denigrated. In the end, this would be no more appropriate than an Orientalist approach. And third, by denying European agency we run the risk of falling into a kind of structural-functionalist trap, in which human agency becomes replaced by the notion of the individual as a "passive bearer" of material structures. This in effect conceives of humans as receptors of the gift or burden of change rather than as creative directors of change.

My conception of European agency also diverges from the pure materialist approaches of the extant anti-Eurocentric (as well as Eurocentric) literature because it is grounded in the notion of identity, which in turn is a socially constructed phenomenon. And herein lies a link with the first prong of my argument, given that European identity has always been forged in a global context. Thus I pay attention to the various phases in which European identity was constructed and reconstructed in an ever-changing global context, while at all times relating this to the economic progress of the West. Nevertheless, this is by no means to say that material factors are unimportant; indeed, they form a major part of my overall argument. Here I merely note that identity is an important aspect of agency. My notion of agency begins from the premise that the way we think of, or imagine, ourselves and our place in the world to a very important extent informs the way that we act in it. How then did the Europeans construct an imperial identity, and how did this in turn enable the later phase of the rise of the West?

During the early medieval period the Europeans came to define themselves negatively against Islam. This was vital to the construction of Christendom, which in turn enabled the consolidation of the feudal economic and political system as it emerged around the end of the first millennium CE. It was also this identity that led on to the Crusades. Subsequently, European Christian identity prompted the so-called "voyages of discovery"—or what I call the "second round" of medieval Crusades—led by Vasco da Gama and Christopher Columbus. Having arrived in the Americas, various Christian ideas led the Europeans to believe in the inferiority of the American Natives as well as the Negro Africans. This in turn legitimised in their eyes the super-exploitation and repression of the Native Americans and Africans as well as the appropriation of American gold and silver, which in turn assisted European economic development in manifold ways. Then, during the eighteenth century, European identity reconstruction led to the creation of what I refer to as "implicit racism" which led on to the idea of the moral necessity of the imperial "civilising mission." Imagining the East to be backward, passive, and childlike in contrast to the West as advanced, proactive, and paternal was vital in prompting the Europeans to engage in imperialism. For the European elites sincerely believed that they were civilising the East through imperialism (even if many of their actions belied this noble conception). And in turn, the appropriation of many non-European resources through imperialism underwrote the pivotal British industrial revolution.

All in all, this enables me to reintroduce European agency as part of my anti-Eurocentric account of the rise of the West. Scholars such as Blaut might denounce this aspect of my argument principally because it seems to fall back into a Eurocentric argument that emphasises European exceptionalism. But this would be the case *only* if this formed the linchpin of my explanation. Thus it is vital to appreciate my overall explanatory framework: that European identity constitutes a necessary though not sufficient explanatory variable. For without the diffusion of Eastern material and ideational resources through oriental globalisation, no amount of cupidity and appropriationism exhibited by the Europeans could have got them "over the line." This also necessarily means that materialist causes must be factored in alongside the role of identity if we are to craft a satisfactory explanation for the rise of the West.

In sum, when we reveal the larger picture that Eurocentrism obscures, then its pristine picture of Western civilization—as autonomous, ingenious and morally progressive—appears more like Oscar Wilde's picture of Dorian Gray, whose real image has been hidden away from the viewer. My task, therefore, is to reveal this hidden picture and simultaneously resuscitate the Eastern story. In this way, I seek to undermine the Eurocentric notion of the triumphant West that lies, either latently or explicitly, at the heart of the mainstream accounts of the rise of the West. In the process we necessarily discover the origins of the oriental West. Thus, to use the language of Western positivist social science adopted by Landes and others, it is for these empirical reasons (discussed above) that we should avoid Eurocentrism. For only then can we provide a satisfactory account of the rise of the West.

One final point is noteworthy. I have clearly set myself a very ambitious task, which requires a revisionist history of virtually the whole world in the last fifteen hundred years! Clearly it is not possible to provide all the details in one book. Though desirable, my task must be more circumspect. My central objective is to paint the outlines of an alternative picture and to thereby provide just enough evidence to undermine the major tenets of the Eurocentric approach. Put differently, the "intellectual success," I feel, should be appraised not by whether the reader is wholly convinced by the particularities of my own account, but rather by whether (s)he is persuaded by my claim that the Eurocentric explanation and vision of the rise and triumph of the West is a myth that needs to be countered.

# Notes

1. Martin Bernal, *Black Athena*, I (London: Vintage, 1991).

2. Ibid.; Samir Amin, *Eurocentrism* (London: Zed Books, 1989); Janet L. Abu-Lughod, *Before European Hegemony* (Oxford: Oxford University Press, 1989); James M. Blaut, *The Colonizer's Model of the World* (London: Guilford Press, 1993); Bryan S. Turner, *Orientalism, Postmodernism and Globalism* (London: Routledge, 1993); Jack Goody, *The East in the West* (Cambridge: Cambridge University Press, 1996); Andre Gunder Frank, *ReOrient* (Berkeley: University of California Press, 1998); Kenneth Pomeranz, *The Great Divergence* (Princeton: Princeton University Press, 2000); Clive Ponting, *World History* (London: Chatto & Windus, 2000). See also the earlier works of Marshall G. S. Hodgson, *The Venture of Islam*, 3 vols. (Chicago: Chicago University Press, 1974); Eric R. Wolf, *Europe and the People Without History* (Berkeley: University of California Press, 1982).

3. David S. Landes, *The Wealth and Poverty of Nations* (London: Little, Brown, 1998).

4. John M. Roberts, *The Triumph of the West* (London: BBC Books, 1985).

5. Felipe Fernández-Armesto, *Millennium* (London: Black Swan, 1996), 8.

6. W. E. B. Du Bois, *Africa and the World* (New York: International Publishers, 1975 [1946]), vii.

7. Marshall G. S. Hodgson, *Rethinking World History* (Cambridge: Cambridge University Press, 1993), 33.

8. Edward W. Said, *Orientalism* (London: Penguin, 1991 [1978]); Victor G. Kiernan, *The Lords of Mankind* (New York: Columbia University Press, 1986 [1969]); Hodgson, *Venture*, I; Bryan S. Turner, *Marx and the End of Orientalism* (London: Allen & Unwin, 1978).

9. Wolf, *Europe*, 5.

10. E.g., Joseph R. Strayer and Hans W. Gatzke, *The Mainstream of Civilization* (New York: Harcourt Brace Jovanovich, 1979); David S. Landes, *The Unbound Prometheus* (Cambridge: Cambridge University Press, 1969).

11. Ruth Benedict, *Race: Science and Politics* (New York: Modern Age Books, 1940), 25–6.

12. Du Bois, *Africa*, 148.

13. See especially James M. Blaut, *Eight Eurocentric Historians* (London: Guilford Press, 2000).

14. Karl Marx in Shlomo Avineri, *Karl Marx on Colonialism and Modernization* (New York: Anchor, 1969), 184, 343; see also Brendan O'Leary, *The Asiatic Mode of Production* (Oxford: Blackwell, 1989), 69.

15. Karl Marx, "Chinese Affairs" (1862), in Avineri, *Marx*, 442–4.

16. E.g., Karl Marx, "The Future Results of British Rule" (1853), in Avineri, *Marx*, 132–3; Karl Marx, *Surveys from Exile* (London: Pelican, 1973), 320.

17. Karl Marx and Friedrich Engels, *The Communist Manifesto* (Harmondsworth: Penguin, 1985), 84.

18. Karl Marx, *Capital*, III (London: Lawrence and Wishart, 1959), 791, 333–4; Marx, *Capital*, I (London: Lawrence and Wishart, 1954), 140, 316, 337–9.

19. Marx, *Capital*, I, 338, my emphasis.

20. Karl Marx, *Capital*, III, 726.

21. Karl Wittfogel, *Oriental Despotism* (New Haven: Yale University Press, 1963).

22. Karl Marx, *Grundrisse* (New York: Vintage, 1973), 110.

23. Karl Marx, *The German Ideology* (London: Lawrence and Wishart, 1965).

24. Georg W. F. Hegel, *The Philosophy of History* (New York: Dover Publications, 1956).

25. Teshale Tibebu, "On the Question of Feudalism, Absolutism, and the Bourgeois Revolution," *Review* 13 (1) (1990), 83–5.

26. Randall Collins, *Weberian Sociological Theory* (Cambridge: Cambridge University Press, 1986), 23, my emphasis.

27. See especially Weber's *The Religion of China* (New York: The Free Press, 1951); *The Religion of India* (New York: Don Martindale, 1958); *General Economic History* (London: Transaction Books, 1981); *The Protestant Ethic and the Spirit of Capitalism* (New York: Charles Scribner's Sons, 1958).

28. E.g., Anthony Giddens, *The Nation-State and Violence* (Cambridge: Polity, 1985).

29. E.g., Immanuel Wallerstein, *The Modern World System*, I (London: Academic Press, 1974); Giovanni Arrighi, "The World according to Andre Gunder Frank," *Review* 22 (3) (1999), 348–53; Jared Diamond, *Guns, Germs and Steel* (London: Vintage, 1998).

30. Max Weber, *Economy and Society*, II (Berkeley: University of California Press, 1978), 1192–3.

31. Blaut, *Colonizer's Model*, ch. 2.

32. Ibid., 5.

33. Landes, *Wealth*, ch. 29.

34. Ibid., xxi.

35. Lynn White cited in Blaut, *Eight Eurocentric Historians*, 39 (emphasis in the original).

36. Blaut, *Colonizer's Model*, 115–19.

37. Immanuel Wallerstein, "Frank Proves the European Miracle," *Review* 22 (3) (1999), 356–7.

STEVEN J. HARRIS

# 2. Long-Distance Corporations, Big Sciences, and the Geography of Knowledge

> Nor must it go for nothing that by the distant voyages and travels which have become frequent in our times many things in nature have been laid open and discovered which may let in new light upon philosophy. And surely it would be disgraceful if, while the regions of the material globe—that is, of the earth, of the sea, and of the stars—have been in our times laid widely open and revealed, the intellectual globe should remain shut up within the narrow limits of old discoveries.—FRANCIS BACON[1]

BACON'S APHORISM WAS OF COURSE part of his general plan to reform natural knowledge through his novel method of knowledge acquisition. The science Bacon advocated was, he thought, a new science laid upon a new foundation; it was to be a body of knowledge about nature, and especially about the control of nature. It was not to be gained from books but from nature itself (or, as Bacon would have it, herself), not through argument but through art, not via contemplation but in action. Nor was the study of nature, according to his proposal, to be confined to small and familiar places. Rather, it was to be pursued as an "active philosophy" ranging across the entire width and breadth of the terraqueous and heavenly globes in search of "Heteroclite" and "Traveling Instances" of nature.[2] And if philosophy itself was to be renewed, the world of the mind must be thrown open and made to keep abreast of the ever-expanding boundaries of the geographical world.

While Bacon's admonition was originally directed toward those of his contemporaries who would let their minds rest within the established limits of received knowledge when all about them lay new worlds, I believe we may extract from his rhetoric something of use regarding our current situation. . . . It is his linking of long-distance travel to the growth of science and his explicit identification of geographical and intellectual discovery that may help us expand "the narrow limit of old discoveries" with regard to the Scientific Revolution. In the former association Bacon has given us a clue to the importance of travel in the making of scientific or natural knowledge, and in the latter he has pointed to contemporary awareness of what might be called the geography of knowledge. . . .

## The Geography of Knowledge

While geography and cartography have certainly received considerable scholarly attention outside the historiography of the Scientific Revolution, my concern here is not simply to append those histories to our canon. Rather than rehearsing the otherwise well-known story of contemporary knowledge of geography, I wish to direct attention to the more general problem of the geography of knowledge. While the latter in some ways depends upon the former, a study of the geography of knowledge is really a very different undertaking since it entails the spatial and temporal distribution of people, graphics,[3] and objects required in the making of all sorts of natural knowledge, not just geographical knowledge. And since it is not just the distribution but also the movement of things that is of interest, we might distinguish three related approaches to the geography of knowledge. In the first instance, it means a static geography of place: where did people "do science"? where were they when they aimed telescopes and recorded observations, logged positions and sketched in charts, performed dissections or prepared medicaments, executed experiments and calculations, or wrote and published the accounts of their activities? In the second, it means a kinematic geography of movement: whence came the constituents of scientific practice and knowledge, the measuring or cutting instruments, the authoritative texts or latest correspondence, the exotic natural curiosity or well-wrought experimental apparatus, the returning botanist or navigator? In the third sense, geography of knowledge also means the dynamics of travel: why and by what means did all these movements take place? what was the *anima motrix* responsible for the multiple peregrinations of the elements of knowledge? If we are to understand the role of travel in the making of scientific knowledge, we must somehow bring into historiographic order these geographies of place, movement, and social organization. While the burden of this essay will be to explicate the dynamics of travel in the context of long-distance corporations, I first need to illustrate briefly the static and kinematic pictures.

To this end let me propose the following simple-minded but nonetheless useful exercise. Imagine having attached a long piece of thread to each object in all the curiosity cabinets that existed in Europe between the sixteenth and eighteenth centuries; or to each plant in all the pleasure, medical, and herbal gardens and in all herbaria across the continent; or to the contents of the many vials of earths, salts, powders, and simples found in alchemical laboratories and apothecary shops; or to the exotic contents of physicians' materia medica, among dyers' supplies, or in metallurgists' storerooms. Then we ask what paths these threads traced through space as each object moved from its place of origin in the natural world to its artificial resting place within the human world. Ginseng and rhubarb from the Far East, cloves and mace from the Moluccas, snakestone from the Malabar Coast, nautilus shells from the Maldives, ambergris from Madagascar, tobacco from the Caribbean, obsidian from Mexico, bezoar and cinchona from Peru, ipecac from Bahfa, guaiac and brazilwood from Brazil, potato plants from Chile, armadillo shells from Argentina, walrus tusks from Newfoundland.

But since knowledge of the natural world depends on more than natural objects themselves, we need also to imagine the motion of the artificial objects—the instruments, texts, and graphical representations—that enabled the movement of those natural objects in the first place. Thus we must also have thread attached to sea charts, captains' instruction books and navigators' manuals, ephemerides and lunar tables, and to quadrants, astrolabes, cross-staffs, chronometers, dials, and compasses—both magnetic and geometric—as well as to the ships and sails that carried them all across the seas and back again. And then of course there are the records of observation themselves, and so there ought to be thread for the paper and parchment upon which were recorded the reports, logs, descriptions, correspondence, coastal charts, and inventories of travel.

Plants, instruments, and reports do not of course travel by themselves; they are carried about by people moving about. And so we also need to attach thread to the trained navigators, remote observers, peripatetic collectors, and wandering recorders of nature who traveled to distant reaches for whatever purpose and brought back to Europe the things they carried. Finally, if either the itinerant or the sedentary practitioner should decide to weave together the observational, natural, textual, and instrumental threads feeding into his (occasionally her) own local world in order to produce a new description or explanation of some part of the natural world (typically in the form of a book, but also in the form of a map, globe, table, illustration, or collection), then we would need even more thread: for the published manuscript would not only have thousands of miles of thread trailing behind it, it now would begin its own multilineal trace through space as its many printed copies traveled from printer to bookseller, from bookseller to book fair, from book fair to reader, and from reader to reader.

Surely the actual mapping of provenance for all the constituents of a scientific text would be an exceedingly tedious task. As an exercise in imagination, however, such "thread maps" alert us to the importance of travel in the geography of knowledge. But of course that geography, when drawn with an eye toward kinematic patterns and not just static connections, would take on different shapes depending on the nature of the knowledge-project we have in mind. We can easily imagine that some maps—say, for Newton's *Principia*—would show traces covering only a very small region of the globe, and others, like Mercator's atlas, would necessarily have lines running literally around the world. Jesuit shipments of cinchona from Lima to Rome would produce a map with a dense cluster of threads running along well-defined channels, while the map for the traffic in natural curiosities might resemble a filigree of greatest delicacy. The mid-seventeenth-century musaeum Kircherianum in Rome, because it depended almost exclusively on the far-flung but strongly centralized network of Jesuit missionaries, would show lines converging to a central node rather like a spider's web. On the other hand, the circulation of eclipse observations among the several major astronomical observatories attached to scientific academies, universities, courts, and municipalities in the eighteenth century would likely produce a diffuse and multimodal— or perhaps even an anodal—map more like a cobweb.[4] *In each case the pattern of traces would represent a record of movement coincident with scientific practice and antecedent to the making of a scientific text, collection, or object.* Thus we can use the geography of scientific knowledge, as we use the geography of the earth's surface, to ask about the spatial distribution and movement of the natural and human referents whose interrelationships we seek to elucidate; in this case, the places of origin and paths of communication of the people, graphics, and objects necessary for the production of scientific knowledge.

Insofar as our thread maps help us visualize the kinematics of scientific practice, they are worth the modest investment of imagination it takes to construct them. But of course the threads themselves have no independent existence apart from the motions they record, nor do they maintain themselves in the absence of human travel. Like a ship's silent wake or the silk route across the steppes, if these paths were not regularly remade through repeated cycles of travel, they would be quickly reabsorbed by nature and thus vanish entirely from the world of human intercourse. Our threads, in other words, depend ultimately upon the shuttling to and fro of people going about their business. And if we ask about their business, how they shuttled about and why they traveled where they did, then we are essentially asking about the dynamics of travel and, by extension, the dynamics of scientific practice.

The early modern period is of course distinguished by an unprecedented explosion in mobility, in regard to both increased range and frequency of travel and also the sheer number of trips and travelers, and there were certainly innumerable reasons motivating travel. Yet when we consider the ways in which mobility was organized in the sixteenth through eighteenth centuries, much of it fell under the jurisdiction of legally constituted corporations engaged in overseas activities—that is, East and West Indies trading companies, colonial administrative bureaus, religious orders with overseas missions, and (eventually) the larger scientific academies capable of launching foreign expeditions. This is not to say that all travel was corporate travel, or that all corporate travel was conducive to work we would deem scientific. Rather, what I am arguing is that by the very nature of their enterprise, long-distance corporations were obliged to organize the movements of their members (both human and nonhuman) in certain ways if they were to succeed in the execution of institutional programs. Moreover, I am claiming that such "corporate" or "organized travel" played a crucial role in the making of knowledge in a number of historically important scientific fields. As John Law and Bruno Latour have argued, the operation of long-distance networks depended fundamentally on their ability to sustain the circulation of people and things from center to periphery and back again: in reciprocating patterns of decision and action, intelligence and directive, the knowledge gained through repeated cycles of travel to the periphery strengthened the network and allowed it to extend its reach, project its power, and (in Latour's phrase) "act at a distance."[5]. . .

## Long-Distance Corporations

Let me try to make this all a bit more concrete by briefly considering four disparate corporations: the Spanish colonial administrative bureaus in Seville known as the Casa de la Contratación de las Indias (also called the House of Trade or Indian House) and, with a somewhat different jurisdiction, the Consejo Real y Supremo de las Indias (or Council of the Indies); the Dutch joint-stock trading corporation based in Amsterdam known variously as the Dutch East India Company, the Verenigde Oostindische Compagnie (voc), or, more colloquially, Jan Compagnie; and the Society of Jesus, the religious order of clerks regular founded by St. Ignatius of Loyola in 1540, with its headquarters in Rome. As manifestly different as these four corporations are with regard to their mandates, fields of action, centers of operation, and forms of corporate life, what they shared were certain features of internal organization and broadly comparable methods of gathering and distilling the information necessary for their operation as long-distance corporations. More importantly, while none of these corporations was established for the purpose of pursuing "disinterested scientific knowledge" (that would come only with the foundation of state-supported scientific academies in the latter half of the seventeenth century), each assembled a corpus of natural knowledge that was, in the first instance, of direct use to its internal operations; and in the second, of intense interest to learned audiences outside the corporate fold.

## Casa de la Contratación and Consejo de las Indias

The Casa de la Contratación was founded by royal decree in 1503 and charged with the oversight of the newly conquered territories of the Caribbean and coastal regions of Central and South America. Located in the old Alcázar guarding the port of Seville, the Casa soon became the administrative center for the regulation of all trade and commerce with the "Indies" and at the same time a center for "cosmography." This latter term embraced the training and examination of pilots (in principle, only licensed pilots could undertake voyages sanctioned by the Spanish Crown); the drafting and enforcement of laws pertaining to navigation; and the collation and standardization of all charts, geographical records, and logbook accounts of newly discovered or recently revisited lands.[6]

The geographical and cartographic activities in the Alcázar grew so rapidly that in 1508 a second royal decree established a separate Cosmographical Department (sometimes also called the Hydrographic Office) and provided the newly created office with explicit instructions concerning its duties and obligations. These instructions specified that pilots should receive training in the use of navigational instruments and the making of navigational charts. The decree also established a new office, the *piloto mayor* or chief pilot, whose main tasks, in addition to the examination of pilots, were supervision of the manufacture of standardized navigational instruments, review of the qualifications of professors of cosmography, and oversight of the compilation of the *Padrón Real*, or pattern map.[7]

The idea of a master map arose in response to the problem of collating the various coastal charts and logbooks brought back to Seville by Spanish pilots returning from the New World. In the absence of standard training procedures and clear instructions governing the observation of position and time, or set ways of describing winds, currents, coastlines, and headlands, maps compiled before 1508 contained troublesome discrepancies regarding the exact location and description of landfalls. Just as in the case of pilot training and instrument construction, the authority of the piloto mayor over the *Padrón Real* was supposed to guarantee control through standard procedure. At regular intervals the piloto mayor was to meet with the chief cartographers of the Casa and the most capable of the ships' pilots to discuss additions to the master map. To ensure a regular procedure for cartographic representation, every modification of old information and insertion of new information had to be "properly attested and sworn to" by informants (i.e., returning pilots and navigators) and ratified by the leadership of the Casa (i.e., the president, the judges of the Casa, the piloto mayor, and royal cosmographers). Finally, the decree insisted that pilots were allowed to make copies for their voyages only from the *Padrón* itself, and that any new findings or notable differences between the map and the territory discovered in the course of their voyages should be recorded directly on their copies of the *Padrón*.[8]

Although few maps survive from the earliest years of the Cosmographical Department, and none (to my knowledge) positively identified as a *Padrón Real*, there is reason to believe that several of the earliest surviving world maps were directly associated with the making of the *Padrón*. Indeed, by 1513 the proliferation of maps produced outside of the regimen of the Casa caused the new piloto mayor, Juan Díaz de Solis, to file a complaint against unlicensed chart makers who "were employing different measures for a degree" (though in some cases the extradepartmental maps were of quite good quality).[9] In an attempt to regain control of cartographic standards, the Casa set about in 1514 to produce a new *Padrón*, and permission was now given for copies to be made not only for Spanish pilots but also for general sale. In 1524 the Casa convened a Commission of Pilots, the so-called Badajos Junta, in the face of ongoing territorial tensions with Portugal and in response to Sebastian del Cano's return to Seville after his circumnavigation of the world. At least three surviving world maps are connected to this commission: one by Juan Vespucci (1526) and two by Diogo Ribeiro (1527 and 1530), who also was a member of the Badajos Junta.

A related mapping and intelligence-gathering project attempted some years later, though not strictly under the jurisdiction of the Casa, casts additional light on the power and constraints that Spanish bureaus faced in the administration of the Crown's remote territories. In 1524 a royal decree established a new overseas corporation, the Consejo Real y Supremo de las Indias, as a clearinghouse for information pertaining to Spanish territories. Whereas the purpose of the Casa was primarily focused on questions of transport (trade, navigation, and cartography), the institutional task of the Consejo concerned the day-to-day administration of Spanish overseas territories.[10] Although legal, political, and related commercial matters predominated, there was a collateral need for information that in an academic context would be classified as astronomy, geography, natural history, and botany. The best-known example of this was the attempt to establish systematically the longitudinal positions of towns and territories in the New World. In the early 1570s, Juan López de Velasco, recently appointed the new *cronista-cosmógrafo* of the Consejo, hit upon the idea of reducing the procedure of making observations of a lunar eclipse to a series of simple steps executable (in principle) by those untrained in astronomy. He incorporated those instructions in a printed questionnaire sent to Crown officials scattered throughout Central America and Peru.[11] Although the questionnaire included directions for making simple angle-measuring instruments and clear instructions for would-be colonial astronomers on how to conduct observations of two lunar eclipses (one in 1577 and one in 1578), a host of problems rendered the project a failure; chief among these were delayed or undelivered forms, incomprehension of instructions, observational and clerical errors, substitution of irrelevant data, and—by far the biggest problem—outright noncompliance.

Despite the less-than-satisfactory outcome of the eclipse survey, López de Velasco did not abandon the strategy of information-gathering by questionnaire but instead lowered his expectations regarding the technical competence of Crown officials and made the questionnaire into his chief instrument of remote reconnaissance. A series of printed questionnaires, most of which contained between 100 and 200 questions, were sent from Seville in 1569, 1571, 1573, and 1577 to the governors of the seventy or so *alcaldías mayores* or colonial districts of Peru and New Spain and to the more than two hundred *corregidores* (local county magistrates) serving under them. In general terms, the questions solicited information regarding chorography (place-names, descriptions of local climate, soils and crops, distances between towns, town plans), natural history (descriptions of plants and animals, medicinal and aromatic herbs, minerals or precious metals, mountains, rivers, and valleys), and "moral" or civil history (local history, indigenous languages, religious foundations, commercial activities, etc.). In addition to prose responses to explicit questions, the *Instrucción* of 1577 also requested *pinturas*, or "picture-maps," to be made of every town and region and returned along with the completed questionnaire. Collectively, the responses to this cycle of questionnaires are known as the *Relaciones Geográficas*, and they constitute the most important body of information about the New World in the sixteenth century.

In the case of the Casa de la Contratación, the regulation of travel through the licensing of pilots and the laws of navigation were the first steps in the Spanish Crown's attempt to manage systematically the information gathered from the New World. The authority invested in the newly created office of piloto mayor led to the enforcement of the standardization of pilot training, navigational instruments, and methods for recording observations at sea. The creation of the *Padrón Real*, or rather of the protocols instituted for its construction, provided a means for the systematic collation and cartographic representation of the disciplined information now flowing back to Seville. Copies made from the improved *Padrón* enhanced the effectiveness of those very same captains and navigators as they set out on subsequent rounds of travel to the Caribbean, Central and South America, and beyond. The chartrooms of the Casa thus served not only as a center for the concentration of cartographic knowledge but also as a center for the dissemination (in this case, largely intra-corporate dissemination) of corporate-certified knowledge. While documentary sources lack the density of coverage necessary to recover the knowledge-negotiating process involved in the actual construction of the *Padrón Real*, it is clear that the process took place within the corporate space of the Casa and that the knowledge produced there was of direct relevance to the ongoing success of the overseas colonial enterprise.

In the case of the Consejo de las Indias, the emissaries to remote regions were not trained pilots but printed questionnaires. While printing was sufficient to guarantee the standardization of instructions, questions, and response forms, the *Instrucción* itself was unable to instill adequate levels of expertise in the instruments—both human and nonhuman—necessary for its execution. And what was surely even more troubling to the leadership of the Consejo, the authority of the *Real Cédula* that accompanied the *Instrucción* was wholly unable to extract anything but the most minimal compliance from the Crown's remotely located agents. In the absence of formal training procedures and mechanisms of enforcement, the execution of the eclipse observations for 1577 was a haphazard matter that failed to produce usable information. By the same token, the heterogeneous conventions used in the construction of local picture-maps rendered those works unintegrable, either with standard maps in Seville or even among themselves. In the absence of standardization of instruments, procedures, and recording conventions, observations could not be combined, collated, or concentrated at the center. And in the absence of concentration there was no cumulative cartographic knowledge issuing from the questionnaires, as there had been in the case of the Casa and the *Padrón Real*. What *could* be integrated, however, were the textual descriptions. Thus, repeating the cycle of questionnaires with much-lowered expectations regarding technical competence improved compliance and procured for the *cronistas* of the Consejo a great quantity of useful intelligence. And distillations of this textual information fed directly into several volumes of "histories of the Indies" written by authors directly associated with the Consejo.

## The Jan Compagnie and the Company of Jesus

One could hardly imagine two corporations more different than the Dutch East India Company and the Society of Jesus. The former was a joint-stock company with a monopoly on all seaborne trade between Cape Town and Tierra del Fuego, headquarters in the great citadel of Calvinism, and a commercial empire unrivaled in the seventeenth century. The latter was a religious order of clerks regular with a near monopoly on higher education in Catholic Europe, headquarters in the great citadel of the Counter-Reformation, and a missionary enterprise that stretched from the Malabar Coast to China in the east and from Canada to Patagonia in the west. Yet despite the self-evident differences in mandate, confession, territory, and governance, both companies faced similar challenges with regard to personnel: how to keep their members healthy, motivated, and loyal when stationed in remote and often hostile environments. While a systematic comparison of methods of discipline and morale in the two organizations would lead us away from the question at hand, one facet of the problem—maintaining the health of members—may be used to illustrate the ways in which even markedly diverse corporate programs required the pursuit of similar kinds of natural knowledge.

High morbidity and mortality rates among the crew-members of Dutch East Indiamen represented a constant threat to profits, especially on ships noted for their efficiency. Precisely because fewer crew-members were needed to sail a VOC ship than any other ship engaged in the eastern trade, the loss of even a small number of men through death or incapacitating illness could jeopardize the entire voyage. The constant exposure to extremes of weather, the spoilage of food and drink caused by tropical heat and humidity, the filth of the ships, shipboard accidents, and injury from armed conflict made the health of the crews a constant problem. Mortality rates on the long voyages, which often meant six to eight months at sea, could approach 50 percent, and the morbidity rates 75 percent. And with the establishment over the course of the seventeenth century of trading factories in coastal towns from the Cape of Good Hope to the Malabar Coast, Goa, Indonesia, and Japan, the health of resident governors and their staff also became a pressing issue as company personnel were exposed to alien diets and novel forms of infectious and parasitic disease.[12]

While there is little evidence to suggest that the health of the average crew-member improved over the course of time, the VOC did attempt to institute measures in an effort to ameliorate the health of shipboard and overseas personnel. Simon van der Stel, who was Compagnie governor of the Cape of Good Hope from 1679 to 1699, succinctly stated one aspect of the problem in a letter to the *Heeren XVII* in Amsterdam: crews stopping off in Cape Town on their way to and from India "lose heart from want of nourishment . . . and all germs of strength failing them, they die."[13] Both the leadership in Amsterdam and the Cape governor saw the importance of expanding the agricultural production of the hinterland so as to make Cape Town into a "victualling station" for East Indiamen. Van der Stel himself vigorously encouraged agricultural activity (in part because he was one of the biggest landowners in the area and saw personal advantage in supplying VOC ships), and Cape Town eventually became a thriving way station. With the establishment of similar stations on the Cape Verde Islands, Sierra Leone, St. Helena, Madagascar, Mauritius, Colombo, and Batavia, provisions of fresh fruit and water could be more or less regularly obtained.[14]

A second measure, commencing after about 1625, was the recruiting and dispatching of ships' surgeons, apothecaries, and (when available) medically licensed doctors to serve on ships bound for the Far East. Medical personnel of the VOC in Amsterdam oversaw the selection and training of young surgeons and the printing of handbooks to be consulted during voyages.[15] Ships' stores were now to be augmented with various medicaments in sufficient variety and quantity to meet the needs of crews of whatever size and however long the voyage. And while apothecaries and medical doctors stationed in overseas factories brought with them Western medicaments, they quickly learned to avail themselves of local herbal remedies, and in some instances even published pharmacopoeia of mixed Western and indigenous recipes. Thus VOC physicians gradually began to expand their botanical horizons, and some became as well known for their botanical work as for their medical expertise. Perhaps the most famous of the company's physician-botanists was Georg Eberhard Rumpf (or Rumphius), the "blind Pliny of India," who spent nearly fifty years in the service of, or associated with, the VOC in Ceylon and the Moluccas. His *Herbarium Amboinense* (Amsterdam, 1741–50) was an exhaustive survey of the plants, herbs, fruits, and trees of Amboina and adjacent islands; first printed some forty years after his death in 1702, it is recognized as one of the best early accounts of tropical botany. Less famous but with a similar trajectory was Paul Hermann, who was stationed in both Ceylon and Cape Town as the Company's medical officer for several years in the 1660s and 1670s. During this time, he pursued his botanical interests in the hinterlands and sent seeds and plant specimens back to Amsterdam. He himself was recalled to Amsterdam in 1679 and made director of the Leiden Botanical Garden, a position he held until 1695. The botanical gardens of the Netherlands, especially those in Amsterdam and Leiden, flourished from the late seventeenth century onward and themselves became important sites for botanical research. The Commelin brothers, Caspar and Joannis, were both associated with the Amsterdam gardens and together indexed existing collections of plants and published descriptions of plants and seeds sent by Simon van der Stel, H. A. van Reede tot Drakenstein, and L. Pijl (VOC governors of the Cape of Good Hope, Malabar, and Ceylon, respectively), as well as by Andreas Cleyer, a company doctor stationed in Batavia, and other VOC officials in Java, Bonaire, and Curaçao.

The botanical gardens of Amsterdam and Leiden served as centers of concentration for plant specimens and plant-related knowledge obtained by company officers working in the remote outposts of the VOC. Although these gardens contained botanical specimens of both medicinal and commercial value that played direct roles in the operation of the VOC, they were not wholly under the control of the VOC. Indeed, they seemed to function as mutually beneficial interfaces between the overseas network of the VOC and the medical faculties of regional universities: professors of medicine and botany gained access to exotic plant and seed specimens, while the VOC received in return the benefit of their expertise. The symbiotic relationship between two nominally distinct corporations not only allowed for the easy exchange of plants and personnel, the governors of the VOC could also recruit, directly or indirectly, the medical staff and knowledge it needed to sustain its members.[16] And while much less regulated than the VOC chartrooms—which, like the Casa de las Indias, served as centers for the concentration of geographical knowledge, and with which they might be favorably compared—the Dutch botanical gardens were nevertheless the disciplined product of corporate travel. That Dutch botanical knowledge was not as tightly controlled as Spanish cartographic knowledge matters little since both were manifestly bound to corporate practices. The fact that both VOC-obtained botanical specimens and Casa-produced cartographic representations eventually passed beyond their respective corporate membranes simply points up the ease of knowledge-dissemination as compared with the difficulty of knowledge-collection and -concentration.

Concentration and dissemination also played a role in the botanical interests of the Society of Jesus, though here there is evidence of Jesuit attempts to control the latter. The Jesuit leadership faced medical and health-care problems roughly comparable to those of the VOC. Especially for Jesuits sent to the overseas missions, the rigors of travel, harsh climates, and the strains imposed by an alien diet worked great hardship on the Society's personnel. But since canon law forbade clerics from either studying or practicing medicine, and since the Society's own *Constitutions* strongly discouraged even the teaching of medicine and surgery at its colleges and universities, the Jesuit leadership was much more constrained in its options than were the directors of the voc. And yet the health of Jesuits had been a chief concern for the Society from its earliest days.[17] The few Jesuits who had studied medicine before entering the Society were much valued but of insufficient numbers to alleviate the general problem of health care.

Unable to train or freely recruit physicians, the Society turned to its temporal coadjutors, the "invisible rank" of unordained lay-brothers typically consigned to the nonpriestly and nonbookish offices of the Order. Because they were not bound for the priesthood, temporal coadjutors could both practice medicine and train others of their rank in the healing arts. From the late sixteenth century onward, Jesuit pharmacies were a standard part of Jesuit college and university compounds; and from the mid-seventeenth century onward, Jesuit apothecaries and surgeons were part of the regular contingent of overseas missionaries. The decision ca. 1680 by the Portuguese and Spanish patronates to open their colonial possessions to missionaries from the Holy Roman Empire encouraged the Jesuit leadership to recruit German apothecaries (widely considered to be the best in Catholic Europe) as lay-brothers and send them to the overseas missions. By the beginning of the eighteenth century, a small but talented group of German Jesuit apothecaries found themselves doing brisk business in the Jesuit provinces of Mexico, Peru, Brazil, Paraguay, Goa, and the Philippines. In most of these provinces, they held virtually unchallenged pharmaceutical monopolies throughout the first half of the eighteenth century. Confronted with a number of new illnesses and literally surrounded by a cornucopia of herbal remedies, they flourished in their temporal calling as never before. While Jesuit apothecaries were for the most part neither well published nor of lasting reputation outside their time and place, they not only successfully ministered to the health needs of their confrères in a dozen different provinces, they also collectively provided detailed descriptions of more than 1,500 plant species (several score of which were first-time descriptions, and several hundred of which were accompanied by either illustrations or dried specimens), including such medicinally important species as cinchona, ginseng (both the Chinese and Canadian varieties), curare, kosso, and ipecac.[18]

But for an apothecary the chief task was to provide neither literary descriptions nor graphic representations, but efficacious preparations— and certainly the most successful of all preparations produced by Jesuit apothecaries was cinchona, the wonder-working antifebrile once known as "Jesuits' bark." Emerging fairly early in the development of the Society's missionary-apothecaries, the story of cinchona not only illustrates how the Jesuit network was able to identify, procure, transport, prepare, and distribute a useful herbal remedy, it also shows how that remedy, once securely recruited, itself helped to strengthen and extend that network.

The earliest accounts of European awareness of cinchona are many and often divergent on matters of detail. Most, however, tend to credit an anonymous Jesuit missionary with the observation that Indians of the Loja region of the Quito province used a powdered bark to lessen their shivering when crossing cold mountain streams, and with the inference that the powder might also work against the chills of fever.[19] Found effective in trials in Lima, the bark eventually made its way to Europe. . . .

The purpose of the *Schedula* was to establish standards for the preparation and administration of a drug of great potential value. Indeed, for the Society it appeared to have a double and even triple value: it could be used as an effective antifebrile for the Society's own; it could be a handsome source of income if sold to non-Jesuits; and it could be deployed as a means of securing the favor of feverish patrons and benefactors. The Society, however, would be able to exploit the full worth of cinchona only if it could control the source (and shipment) of the bark, discipline its properties, and gain access to sickly elites. For a time it succeeded on all three counts. By mid-century, Jesuit missionaries had established a strong presence in Peru, and their friendly relations with the Indians who knew the tree from which the bark came secured, if not quite a monopoly, then at least a steady, high-quality supply. The *Schedula*, along with the regular training that Jesuit apothecaries received within the Society's pharmacies, helped Jesuits' bark from succumbing to the pharmaceutical equivalent of Gresham's law;—that is, a good drug's being driven out of circulation by counterfeits, admixtures, adulterations, and dilutions. And the Society's practice of ministering to the spiritual needs of elites, already well established at the beginning of the century, gave Jesuits access to virtually every court in Catholic Europe. . . .

Cashing in on the full value of cinchona, however, depended not only on its horizontal or geographical distribution but also on its vertical rise through social ranks. But of course the latter strategy entailed great risk. In the autumn of 1652 Archduke Leopold Wilhelm, brother of Holy Roman Emperor Ferdinand III, fell ill with fever. His physician, Jean-Jacques Chiflet, reluctantly administered what he felt to be an insufficiently tested drug as the fever worsened. The Archduke recovered, but suffered either a relapse or the onslaught of a new disease a month later. The emperor forbade the use of the powder a second time, and Leopold Wilhelm died. Chiflet immediately published a tract in 1653 critical of the new drug, and a long and messy literary war ensued after Fabri's counterattack in 1655. Despite compromised reputations, both of Jesuits and of their bark, in the following decades cinchona gained gradual acceptance and eventually worked its way even into the pharmacopoeias of the English and Dutch despite their enduring suspicions of papist plots. It took, however, two stunning cures in the late 1680s to remove the bitter memory of the Archduke's ill-timed demise. The first was the remarkable recovery of Louis XIV in 1686 from severe ague (the king subsequently favored dissemination of the drug in France), and the even more spectacular recovery of the Chinese Emperor K'ang Hsi in 1687. Joachim Bouvet, one of the Jesuit mathematicians sent to China by Louis XIV in 1686, published (for European consumption) the following dramatic account:

> A large number of sick persons, among whom were many officers of the household and even one of the Emperor's own sons-in-law, were cured by taking medicines that we had brought from Europe. A short while afterward the Emperor himself fell into a dangerous illness. After having tried without success the medicines of his physicians, he had recourse to ours, which rescued him from the danger in which he had been. His physicians wanted to have the honor of effecting a cure but they were not too fortunate in this respect. The emperor could be cured only by means of *kinkina* [cinchona], which Fathers de Fontaney and Visdelou, who fortunately arrived at that time, had brought with them.[20]

Bouvet was eager to enhance the reputation of Jesuits' bark, since it in turn enhanced the reputation of Jesuits. Claiming a cure for the king of France helped French Jesuits to "arrive in time" to minister to the sick Chinese emperor. Effecting a Jesuit cure in the presence of, and in competition with, Chinese doctors was, as it had been in Paris before French physicians, a double demonstration of the usefulness of both drug and clerics at court. And making known that latter cure through a French publication served to complete the cycle: French critics of the Society—who were neither few in number nor ill placed—learned that the greatest rulers of Europe and Asia had been restored to physical health by their spiritual doctors. After appropriate "trials and experiences" in the world, cinchona, like a Jesuit in training, eventually proved itself to be a reliable agent in the execution of the Society's multipronged apostolic program.

If we choose not to abstract these maps, books, gardens, and drugs from the conditions of their production, we find that each not only arose directly from corporate activities but also figured in the ongoing operation of the corporation in question. Whether large and immovable like a garden, or small and swift like cinchona, each of the final products had trailing behind it a history of travel organized under corporate aegis. In each case, the final representation, collection, or object was the product of "local negotiations," yet the practices antecedent to that end product were distributed across the corporate network. And while the means for producing the knowledge contained in or represented by each production cannot be extracted from the complex of social relations of each corporation, the products themselves can be. That is, a copy of the *Padrón Real* can be used, botanical illustrations can be consulted, and preparations of cinchona may be administered by those outside the producing or procuring corporation. In this sense, then, our exemplars of corporate production are at once local and distributed, scientific and social, natural and artificial, objective and constructed.

## Matters of Scale

Before taking up the question of historiography alluded to at the beginning of this essay, let me first return to the notion of geography of knowledge. Recall that in connection with our imaginary thread maps I suggested that different knowledge-making projects would result in differing patterns of traces: some diffuse, some channeled, some centripetal, and so on. I have tried to argue that if we wish to understand the dynamics of scientific practice, if we are to understand the causes behind the motion of things, then the sort of movement required by long-distance corporations might be as good a place as any to look. And as we have seen in the case of the *Padrón Real* and Dutch botanical gardens, the corporate network of data or objects feeding into them could be quite large. Putting aside the question of the scale of corporate practices, let me consider a related question regarding the scale of scientific practice. Just as we imagined a simple spatial metric for the operation of a corporation, so too can we imagine a similar metric for a given scientific practice: that is, what was the geographical extent and temporal duration of the practices required in a particular set of astronomical observations, the construction of an instrument, or the writing of a scientific treatise? Without repeating the exercise of imagining thread maps, it should be self-evident that the development and initial use of the astronomical telescope and the experimental air pump, say, were "small" enterprises, since their construction was the work of just a few people working over a short period of time in a restricted geographical setting. Mapping South America, assembling a curiosity cabinet, and constructing a taxonomy of quadrupeds were, on the other hand, "big" enterprises, since constitutive knowledge, observations, and/or objects required the long-term labor of large numbers of people scattered across wide geographical fronts. A systematic examination of scale of practice among the various fields of early modern science would thus yield a spectrum, rather like the one for corporations, ranging from "big sciences" on one end to "small sciences" on the other.

Without getting caught up in the details of the scale of practice, we could make the general claim that the category of "big sciences" would embrace a good part of stellar and planetary astronomy, cartography, mathematical and descriptive geography (including what we would now call anthropology and ethnography), natural history, meteorology (which, like parts of astronomy, would require a long time-line as well as geographically dispersed observers), pharmacy and medical botany, and some parts of mixed mathematics (surveying, navigation, hydrology, etc.). Conversely, the "small sciences" would include virtually all of experimental philosophy, anatomy and surgery, most of "pure" mathematics, and a surprising number of the classics of the Scientific Revolution. "Small" in this context, of course, in no way implies insignificance. Rather, it refers to the number of observations (Copernicus's innovation in planetary theory depended upon almost no new observations), their geographical range (the data for Newton's "moon test" required no overseas—or even cross-channel—expedition), the length of time needed to gather pertinent information (Galileo's initial round of telescopic observations extended over no more than a few months), or the space in which the crucial observations were gathered (Boyle's air pump experiments were "bench-top" science).

What do we gain by this curious division of the sciences? Certainly the notion of scale of practice could lead us back to long-distance corporations, since these were very often the contexts in which large-scale scientific practices had their greatest relevance. But let me put aside corporations for the moment and instead make a historiographic point. If we compare the body of literature organized under the rubric of "the Scientific Revolution" with that of the social constructivist program, what we find is that both focus almost exclusively on the small sciences and neither has done justice to the big sciences. There is more than a little irony in this state of affairs. On the one hand, the Scientific Revolution is supposed to be a grand narrative: it seeks to tell a story of broad geographical scope, long duration, and great import. Yet for all its *grandeur*, the standard plot hinges on the biographies of a handful of great men and the analysis of a shelfful of their most seminal texts. The "first rank" of heroic figures totals no more than a dozen and the second scarcely two score, and yet members of that pantheon have secondary bibliographies running to hundreds, sometimes thousands, of published items. Moreover, just as a few conceptual markers (heliocentrism, mathematical realism, mechanistic explanation, and experimental method) have been used to identify and isolate key players, so too have they been used to segregate "important" from "unimportant" disciplines. The defining stories of the Scientific Revolution have typically followed core developments in celestial and terrestrial mechanics, and even when they are expanded to include, say, the rest of Thomas Kuhn's "classical sciences" or anatomy and physiology, the number of scientific fields treated has remained quite small. What is truly striking, however, is that almost all the scientific disciplines chosen for inclusion in the story of revolution were small-scale practices, while the large-scale practices listed above have been given short shrift or have been neglected altogether.

When we turn to the microhistories informed by social constructivist approaches, we encounter a major distinction in the scale of narration but not much difference in the scale of scientific practices treated. Whether one speaks of "local," "situated," or "embedded" knowledge, the implication is that the narrative is somehow confined to a small "space"—if not in the literal sense of a geographical metric, then at least in the sense of restricted social, cultural, and temporal metrics. The "localist thrust," in other words, has not only predisposed researchers to choose research sites that are spatially and temporally circumscribed, it has also encouraged the selection of scientific practices that were themselves spatially and temporally circumscribed. Consequently, the constructivist program has made a selection of sciences that is, if anything, more restrictive than that found in the historiography of the Scientific Revolution. Indeed, while generally eschewing the notion of a pantheon of revolutionaries, constructivist scholarship has not been averse to selecting its subjects from that pantheon.[21] However justifiable such choices may be, the net effect is that social constructivists have brought few new players into the game and, most ironically, have omitted from their accounts of early modern science roughly the same large tracts of early modern science as have the historiographers of the Scientific Revolution, from whom they otherwise have sought to distinguish themselves. Thus we would seem to have a grand narrative blind to big sciences and microhistories unacquainted with scientific practices that extended beyond the laboratory, court, or academy.

But must this be so? Is it a necessary consequence of the epistemological and narratological assumptions of either camp, the intellectual "revolutionaries" or the social constructivists, that the big sciences should be left on the margins? While one might argue that this mismatch between scales of practice and scales of narration is really nothing more than an accident of the recent history of our discipline (i.e., Cold War Platonism versus postmodern skepticism), matters of epistemology are in fact not easily separated from matters of narrative. As the debates between internalists and externalists, positivists and constructivists, intellectual and social historians have made clear, knowledge of the past and knowledge of nature are joined at the hip. We are continually compelled to see our beliefs regarding the grounds of scientific knowledge exemplified in our historical research, and so we continually attempt to translate our episte-mological stances into the framing conditions of our historical projects. This might be all well and good, except when our epistemological claims lead to egregious acts of omission with regard to well-attested and incon-trovertibly significant segments of the historical record.[22] If we wish to tell *longue durée* stories of early modern sciences, and not just the ultimate origin story of modern science or a series of unconnected microhistories trapped in their respective "black holes of context," then we will need to find both an epistemology and a narrative format capable of moving across scale.

The problem is one not simply of the scale of narration but also of the scale, or rather "scalability," of the epistemology guiding that narrative. To take a single example of an epistemological stance that has helped struc-ture plots and subplots within the historiography of the Scientific Revolu-tion, consider the applicability of Kuhn's model of scientific revolutions. In recalling our list of big sciences (i.e., observational astronomy, geogra-phy, natural history, meteorology, navigation, etc.), we may note that each field enjoyed robust development in the period after 1500; that collectively they engaged hundreds, if not thousands, of people; that constitutive knowledge, graphics, and/or objects were gathered through geographi-cally and temporally extended practices; and that most of these tangible productions came together in specific, localized centers of concentration (observatories, chartrooms, botanical gardens, apothecary shops, *Wun-derkammer*, physical cabinets, libraries, etc.). By the end of the seven-teenth century, most of the big sciences could claim to be cumulative enterprises with ever-increasing stores of treatises, manuals, tables, maps, charts, reports, handbooks, globes (both celestial and terrestrial), instru-ments, and specimens. Yet it is difficult to identify in this rich historical record a single instance of a dramatic theoretical transformation, a "revo-lution" of the Kuhnian sort. While we can identify cases of consensus-based knowledge (recall the periodic meetings called by the piloto mayor in the Casa de las Indias), paradigms (e.g., the *Padrón Real*), and evidence of surprising transformations (e.g., the charting of the Western Hemi-sphere), it is difficult to isolate crisis-producing anomalies, arational paradigm shifts, or debilitating incommensurabilities. Moreover, few of these practices depended directly on Kuhn's "classical sciences," and they only partially—very partially—overlap with his "Baconian sciences." In a word, the big sciences were not Kuhnian in any sense of the term: they did not have to "wait until later" to undergo "delayed" conceptual trans-formations, nor did they experience discontinuous transformation dur-ing their period of greatest change. "Revolution" simply does not capture the practice-driven, cumulative, big-scale sciences associated with long-distance networks. And yet it would seem that if the history of a scientific practice fails to conform to the topoi of revolution and overthrow, it is marginalized and left to develop narrative coherence on its own. This at least would seem to be the historiographic fate of most of the big sciences given above.

Kuhn's model of Scientific Revolution was derived from the gestalt switch of the individual researcher who suddenly saw the world differently, and it was "scaled up" to fit a small handful of crucially placed consensus-makers who suddenly described the world differently. Whether or not that model works with any degree of precision for any specific episode in the history of early modern science, it seems it cannot be "scaled up" to the level of the big sciences, let alone to the level of grand narrative.[23] Conversely, the Scientific Revolution as origin story, while it has frequently been "scaled down" to serve as the justification and framework for a small-scale story of discovery or invention, has failed to capture the diffuse discoveries and communal labor characteristic of the big sciences. And finally, the localist hermeneutic of the constructivist program would seem to require that we forgo the grand narrative altogether, just as its internal logic has implicitly discouraged the investigation of spatially and temporally extended practices. Such is our current historiographic dilemma; but, in my view, it is an unstable dilemma that cannot endure.

The problem of scale is essentially the problem of travel, and, as I have suggested above, attention to the geography of knowledge—its acquisition, transport, and concentration—is the key. More precisely, it is in Law's and especially Latour's model of long-distance networks that we find an epistemology of mobility, or rather of mobilization, that may provide a partial solution to our dilemma. Since networks can attach either to the charismatic person or, as I have argued, to the enduring "legal person" of a corporation, and since networks must grow to live, the diachronic narratives of networks must perforce be scalable narratives. If the story of Galileo as an operator of networks of instruments and patrons spans two decades and half of Italy, the story of the Society of Jesus as a corporate operator of networks of missions and colleges spans more than two centuries and the entire globe.

Insofar as we may identify a "proper unit of analysis" for scalable research sites,[24] the corporation is perhaps the most promising from both a historical and a sociological perspective. Early modern Europe was paved with a mosaic of overlapping and interpenetrating corporations (it is difficult to find a historical figure of relevance to the history of science who did not participate in corporate life in one form or another), and corporations large and small provided a deep social matrix through their machinery of rule making and rule enforcing, and a well-defined social grouping by virtue of their explicit membership and codes of conduct. Situated knowledge may thus be studied in a rich social and cultural context without sacrificing either the heroic figure (i.e., the "actor" as operator of a charismatic network) or extended practices (i.e., the members as agents of a corporate network). Situating knowledge and its means of acquisition in the context of corporations allows knowledge production to be viewed both as "local" and as "distributed" without privileging the former over the latter or, more generally, the micro over the macro. The move from the micro to the macro can, in principle, be made without sacrificing the embedded character of scientific knowledge and while retaining—and indeed foregrounding—contemporary claims of utility and efficacy in pursuit of agendas of control and projection of power.

## Notes

1. Francis Bacon, *The New Organon and Related Writings,* ed. Fulton H. Anderson (Indianapolis: Bobbs-Merrill, 1960), Aphorism 84, 81.

2. Bacon, *The New Organon*, Book 2, Aphorisms 28 and 41, 177–78, 215–16.

3. The word "graphics" here includes both written and pictorial representations (e.g., texts, tables, lists, *schedula*, charts, drawings, plans, graphs), as connoted in the original Greek *graphikos*. See Svetlana Alpers, "The Mapping Impulse in Dutch Art," in *Art and Cartography*, ed. David Woodward (Chicago: University of Chicago Press, 1987), 67–9.

4. A typology of thread maps might include, in addition to the geographical extent of the network, the number of gathering sites required; the length of time it takes to gather such information; and the frequency of exchange of information between periphery and center. That is, in analogy with Karl Marx's notion of "density of communication," one might think of the "density of practice" of early modern science as well as the intensity and extensiveness of practice.

5. John Law, "On the Method of Long-Distance Control: Vessels, Navigation, and the Portuguese Route to India," *Sociological Review Monographs* 32 (1986): 234–63; Bruno Latour, *Science in Action: How to Follow Scientists and Engineers through Society* (Cambridge, Mass.: Harvard University Press, 1987), 219–37.

6. Much of the following description of the activities of the Cosmographical Department is taken from Edward L Stevenson, "The Geographical Activities of the Casa de la Contratación," *Annals of the Association of American Geographers* 17 (1927): 39–59.

7. The first holder of this office was Amerigo Vespucci, who himself had sailed much of the eastern coast of South America under the authority of the Casa (but also under the Portuguese flag) some years before. The second piloto mayor, Juan Díaz de Solis, assumed the office upon Vespucci's death in 1512.

8. Stevenson, "Geographical Activities," 42. See also Gonzalo Menéndez-Pidal, *Imagen del mundo hacia 1570: Segun noticias del Consejo de Indias y de los tratadistas españoles* (Madrid: Gráficas Ultra, 1944), 4–5.

9. Stevenson, "Geographical Activities," 44.

10. Menéndez-Pidal, *Imagen del mundo*, 4–7.

11. Clinton R. Edwards, "Mapping by Questionnaire: An Early Spanish Attempt to Determine New World Geographical Positions," *Imago Mundi* 23 (1969): 17–28. See also Barbara E. Mundy, *The Mapping of New Spain: Indigenous Cartography and the Maps of the Relaciones Geográficas* (Chicago: University of Chicago Press, 1996), 18–19.

12. C. R. Boxer, *The Dutch Seaborne Empire, 1600–1800* (London: Penguin Books, 1965), 82–8.

13. Ibid, 87.

14. Ibid., 83–4, 279–80, 285.

15. See, for example, Cornells Herls, *Examen der Chyrugie* (Amsterdam, 1625), with later editions in 1645, 1660, and 1672, as well as the much-augmented edition by Johannes Verbrugge in 1680: Peter van der Krogt, ed., *voc: A Bibliography of Publications Relating to the Dutch East India Company, 1602–1800* (Utrecht: HES 1991), 473–75.

16. Here we might note in passing that the Society of Jesus also enjoyed the benefit of having full corporate control over both an overseas network (in the form of its missions) and an academic network (in the form of its many colleges and universities), an aspect of its organization that accounts in large part for its remarkable record of knowledge production.

17. A. Lynn Martin, *Plague? Jesuit Accounts of Epidemic Disease in the Sixteenth Century* (Kirksville, Mo.: Sixteenth-Century Journal Publishers, 1996), 59–61.

18. Emil Bretschneider, "Early European Researches into the Flora of China," *Journal of the North-China Branch of the Royal Asiatic Society,* November 1880: 1–195.

19. Saul Jarcho, *Quinine's Predecessor: Francesco Torti and the Early History of Cinchona* (Baltimore: Johns Hopkins University Press, 1993), 4–5.

20. From Bouvet's *Portrait historique de l'Empereur de la Chine* (Paris, 1697), 160–61, quoted in Jarcho, *Quinine's Predecessor,* 102. On the cure of Louis, see ibid., 65–6.

21. See, for example, Mario Biagioli, *Galileo Courtier: The Practice of Science in the Culture of Absolutism* (Chicago: University of Chicago Press, 1993); Rivka Feldhay, *Galileo and the Church: Political Inquisition or Critical Dialogue?* (Cambridge: Cambridge University Press, 1995); Simon Schaffer, "Glass Works: Newton's Prisms and the Uses of Experiment," in *The Uses of Experiment,* ed. David Gooding, Trevor Pinch, and Simon Schaffer (Cambridge: Cambridge University Press, 1989), 67–104; Shapin and Schaffer, *Leviathan and the Air Pump: Hobbes, Boyle, and the Experimental Life* (Princeton: Princeton University Press, 1985), 22, 77–80.

22. To take but one example, there is no mention whatsoever of either geography or cartography, which may be viewed as "paradigmatic" of the big sciences, either in the "classics" that have defined the field (like those by Marie Boas, Herbert Butterfield, Allen Debus, E. J. Dijksterhuis, A. Rupert Hall, and Richard S. Westfall), or in the recent "challenges" and "reappraisals" that have appeared by Peter Barker and Roger Ariew, David C. Lindberg and Robert S. Westman, Roy Porter and Mikulás Teich, or Steven Shapin. The story of geography has of course been well told; it simply has had no place in the telling of the Scientific Revolution

23. Ian Hacking, "Was There a Probabilistic Revolution 1800–1930?" in *The Probabilistic Revolution*, ed. Lorenz Krüger, Lorraine Daston, and Michael Heidelberger (Bielefeld: Universität Bielefeld and B. Kleine Verlag, 1983), 45–55.

24. Stephen Fuchs, "Positivism Is the Organizational Myth of Science," *Science in Perspective* 1 (1993): 11n16.

MARY TERRALL

# 3. Heroic Narratives of Quest
# and Discovery

IN THE ENLIGHTENMENT, what came to be known as the Scientific
Revolution was usually touted as the work of great men illuminating
the darkness of superstition and authority with the light of reason and
experience.[1] As d'Alembert told it, the work of his intellectual forebears
prefigured the full glare of enlightenment: "[These] great men, lacking
the dangerous ambition to strip the blindfolds from the eyes of their
contemporaries, prepared from afar, working in silence and darkness, the
light which would illuminate the world little by little, by imperceptible
degrees.[2] The metaphor of light brought with it the language of vision,
and discovery was represented as a process of bringing the dark corners
of the world out into the light of reason. Great men had the ability to show
others how to see, by illuminating and displaying. The new science of the
moderns, assimilated and appropriated by Enlightenment thinkers, had
replaced words with things, and authority with reason and calculation,
based on the direct experience of nature. The men cast as founding fathers
of this science—Galileo, Descartes, Bacon, Locke, Newton, sometimes
Leibniz—were not always admired uncritically, but they were applauded
for dispelling the fog of previous error. They were, in fact, portrayed
as forerunners of the Enlightenment. Even in hotly contested debates
over concepts like gravity or *vis viva*, the founding fathers escaped the
vicious attacks reserved for contemporaries, since they had not had the
advantage of living in an enlightened age. So when Cartesian physics
was maligned, Descartes himself escaped censure, as when d'Alembert
reminded his readers: "The genius that [Descartes] displayed in seeking
out a new, albeit mistaken, route in the darkest night was his alone. Those
who first dared to follow him into these shadows at least showed courage;
but there is no longer any glory in losing one's way following [Descartes's]
path once it has become light."[3]

Descartes's accomplishment was to mark a trail through a dark laby-
rinth. The metaphor of seeing is here compounded with the metaphor of
moving through uncharted territory. Descartes's confrontation with philosophical authority took place in the safety of his study, but d'Alembert
represents it here as a physical exploration of the unknown. Descartes's
greatness and glory derive from his courageous challenge to darkness;
d'Alembert's image also shows how the first tentative path was superseded

by a different, and now clearly visible, road to natural knowledge. Where, then, were philosophers to find glory in the century after Descartes, once the initial illumination of darkness had been achieved? Physical exertion and exploration came to be associated with discovery and understanding, and played an important part in establishing scientific reputations. A few men of science, looking for glory as well as truth, cultivated literally the character of intrepid explorer that d'Alembert used figuratively to describe Descartes. In their own discovery narratives, these Enlightenment men of science portrayed themselves as heroic followers of the great men of previous generations, and especially of Newton. They did so through accounts of a kind of focused exploration of nature that combined physical effort and daring with mathematical and instrumental prowess. Their firsthand stories were then incorporated into a heroic account of the scientific adventure and discovery that fed into grand narratives of the progress that follows from courageous battles against ignorance. These narratives are all implicitly gendered, and the expedition accounts I will be analyzing in this paper raise the question of how such representations of discovery and exploration applied gendered categories to the practice of science and to the construction of histories of science.

Travel provided a supplement to the cerebral means of pursuing truth, a strategy adopted by men who journeyed, with their instruments, far from the cosmopolitan centers of learning. This is a kind of subplot of the canonical large-scale narrative of cumulative enlightenment told by d'Alembert, Condorcet, and their intellectual descendants down to our own century. The characters in this particular subplot gain knowledge and authority through heroic physical effort. The quest for truth, in these stories of scientific accomplishment, required certain manly attributes, including risk-taking and physical toughness, to accompany the intellectual brilliance required of the successful man of science. These men sought glory through the emulation of soldiers and classical heroes. European readers learned about scientific expeditions from texts that chronicled the collection of measurements and calculations from distant locations around the globe. The heroes of these tales went out into the world and brought back data, abstractions that could ground theoretical claims, but that could only be acquired at some physical cost. Once completed successfully, these quests brought considerable rewards from patrons and readers, precisely because of the risks the travelers had faced and survived. The firsthand narratives defined their authors as heroic men of science, and provided the raw material for secondary historical accounts of discovery and the progress of knowledge that grounded later versions of the grand narrative of scientific progress.

Eighteenth-century histories of science fed into a larger narrative of progress in all human endeavors. The history of civilization was told as a story of increasing enlightenment, increasing knowledge, and increasingly useful knowledge. This dynamic of progress described civilization in general, with science as an integral part of the development of humankind. Scientific progress paralleled and enabled social and cultural movement toward an ever-nearer goal of perfection. This vision became entrenched as a cliché of Enlightenment rhetoric, but the connection to an ideology of progress was critical to the practice of eighteenth-century science, a practice that aimed at pushing the boundaries of illumination and discovery out into ever-widening circles. Narratives of discovery, and in particular heroic discovery, promoted the value of science to the state and to enlightened civilization. At the same time, they elevated the heroes of science above the rest of humanity, often in gendered terms. The Enlightenment grand narrative of progress, then, entailed a paradoxical conjunction of universal human reason with the special attributes reserved for a heroic masculine elite, whether social theorists, politicians, or men of science.[4]

The "new science" of the seventeenth century has long been linked to the voyages of discovery that expanded the conceptual and physical horizons of the European world.[5] For Francis Bacon, the very existence of hitherto-unknown places and phenomena inspired a challenge to established ways of seeing the world:

> by the distant voyages and travels which have become frequent in our times, many things have been laid open and discovered which may let in new light upon philosophy. And surely it would be disgraceful if, while the regions of the material globe . . . have been in our time laid widely open and revealed, the intellectual globe should remain shut up within the narrow limits of old discoveries.[6]

In Bacon's view, travel effectively opened up unexpected hidden treasures to the European gaze. The exploration of unfamiliar geographical territory led to discoveries more profound than the particular new phenomena witnessed on the voyage. Bacon's image of the traveler "laying open" distant regions gives the discoverer the power to reveal knowledge to his contemporaries. By the eighteenth century, scientific voyagers dispatched to obtain crucial data were stepping into a well-established tradition of voyages of exploration, even though they were not exploring simply for the sake of finding novelties. Their very willingness to travel beyond the limits of the familiar implied an openness to new ways of seeing and thinking. Their narratives of discovery drew on several generations of voyage literature for some of their literary conventions, while integrating technical accounts of scientific results into their stories.[7]

I look here at several astronomical expeditions mounted by the Paris Academy of Sciences to bring measurements back to France from distant parts of the world. Long-distance journeys to exotic destinations provided scientific travelers in the eighteenth century with a model for narratives of their heroic quest for truth. Set in exotic locales, the accounts invoked famous discoveries of new parts of the globe and linked the discovery of arcane scientific knowledge to the popular genre of travel literature. Heroic accounts of scientific expeditions contributed to the representation of science as the accomplishment of individuals with exceptional physical and moral qualities. In France, this had partly to do with the gradual evaporation of an ideal of corporate collective pursuit of knowledge and the related intensification of individualism. Scientific voyages presented the opportunity to serve the Academy and the state, while simultaneously enhancing individual honor and reputation through tales of heroic feats. These tales instantiated the relation of science to state power, since that power made the expeditions possible. Expedition narratives also underlined the gendered nature of scientific work, as the exploits of heroes were construed in explicitly masculine terms.

## The Shape of the Earth, as Viewed from Paris

Two sets of academicians were sent by the French crown on geodetic expeditions to Lapland and Peru in the 1730s. These expeditions were conducted in an atmosphere of lively controversy inside and outside the Academy, and generated a range of texts, from books to academic addresses to technical papers to pamphlets. I have written elsewhere about the tensions within the Academy in this period between Paris Observatory astronomers, who were continuing the surveying work begun decades earlier for a new map of France, and a group of self-styled mathematicians exploring the applications of Leibnizian mathematics to Newtonian physics.[8] The earth's shape received more attention from academicians than any other single problem in the 1730s. The *Mémoires* of the Academy were filled with reports of observational techniques, instrument design, error calculation, algebraic analysis of curves and force laws, tables of measurements, diagrams of triangulations, and reports from the field. Whatever their theoretical predilections, everything the "Newtonians" said about methods, theory, and calculation was said in the context of the ongoing cartography program under the leadership of the observational astronomer Jacques Cassini.[9] Since the numbers implying an elongated earth had come out of meridian measurements made in France as part of this cartographic project, the call for precise measures taken at distant latitudes was a barely veiled challenge to the accuracy of the Cassini team (and their instruments). This was especially true as the original impetus for the new expeditions came not from the crown, but from an informal group in the Academy, who defined their project in utilitarian terms in order to get the necessary financial support.

The physical effect in question, calculated as a flattening at the poles by Newton and Huygens, and as an elongation by Cassini, was small enough to require measurements with precision instruments for its detection.[10] Traveling to distant latitudes would reveal any distortion of the earth's sphericity more definitively than could local measures, but there was still plenty of room for debate about the accuracy of instruments in unfamiliar climates, about phenomena such as the aberration of starlight, about the estimation of error, and the like. For the instigators—Louis Godin, Charles-Marie de La Condamine, Pierre Bouguer, and Pierre-Louis Moreau de Maupertuis—the controversy presented an opportunity to put themselves, and the Academy, into a spotlight that focused public attention on their enterprise.

## Measurement and Observation in Exotic (and Dangerous) Locales

The Lapland expedition was planned as a mirror image of the Peru expedition, which left first, in 1735. In the event, the experiences of the two groups were not at all symmetrical. For one thing, Maupertuis quite effectively put himself forward as the leader and the spokesman of the Lapland group. Thus, although the team included the accomplished mathematician Alexis Clairaut and the Swedish astronomer Anders Celsius, as well as several younger French astronomers, the whole effort came to be identified with Maupertuis. He campaigned vigorously for its importance beforehand, convinced the Minister of the Navy, the Comte de Maurepas, to use his influence to make the necessary diplomatic and financial arrangements, made strategic decisions while in the field, and defended the results afterward. He also managed to control public accounts of the team's work and adventures. The whole enterprise worked remarkably efficiently, helped along by the cooperation of the Swedish government, Swedish colonists in Lapland, and native providers of food and transport. The academicians were back in Paris after an absence of only sixteen months.[11]

The story was different for the team that traveled south. They were unprepared for the difficulties of climate, terrain, and politics that met them in South America, all of which interfered with their operations, but they also wrangled among themselves about what measurements to take and how to make corrections and calculations. Eventually they split into two groups. Not everyone made it back; those who did traveled to Europe separately. Pierre Bouguer, the first to return, had been gone nine years when he got back in 1744. By this time, the question of the earth's shape had been settled in the Academy. Bouguer's rival La Condamine reached Paris shortly afterward, musing bitterly: "Having departed in April 1735, one year before the expedition to the north, we returned seven years too late to bring the first definitive news to Europe of the earth's flattening.[12] However, he was able to bring back tales of adventures in the Andes and the Amazon rain forest, along with measurements more extensive than those taken in Lapland.

Let us look, then, at the stories of the two expeditions for the representation of the quest for scientific truth and the men who pursued that quest. In addition to some elements familiar from other discovery stories, the metanarrative of courageous heroism distinguishes these accounts from reports of earlier astronomical expeditions. It made manifest a particular version of masculinity that expanded the list of desirable attributes for practitioners of science to include physical courage and fortitude as well as intellectual acumen. In 1735, the Academy's secretary Bernard de Fontenelle described the Peru expedition as "incomparably more arduous" than the domestic cartographical expeditions, especially because the territory to be measured was "still wild [*sauvage*]"; the process of measuring and mapping required an encounter between civilization— the travelers and their instruments— and wilderness: "How many hardships, and fearful hardships at that, accompany such an enterprise? How many unforeseen perils? And what glory must not redound to the new Argonauts?"[13] It took courage to venture into "an unknown country where towns are rare, into vast swamps, forests without roads, and inaccessible mountains."[14]

The mythical figure of the Argonaut resonated nicely with the quest for elusive precision measurements under such conditions. Voltaire flattered Maupertuis by comparing him favorably to the leader of the Argonauts:

> The Greeks made the Argonauts . . . into demigods; but are they worth . . . [even] as much as the assistants who accompanied you? [The Argonauts] became divinities, and you? What is your recompense? . . . Rest assured that the approbation of the thinking beings [*êtres pensants*] of the eighteenth century is far beyond the apotheoses of Greece.[15]

Venturing out into the unknown to find the golden fleece that would resolve a pressing scientific question, the travelers became heroes in search of trophies. They had to overcome obstacles, as Jason and his companions had, and they had to bring their trophies home. This image of masculine prowess depends on cleverness, bravery, and stamina, though Jason's female helper Medea is crucially absent from the enlightened version of the myth.[16] Voltaire indicates that the admiration of contemporary "*êtres pensants*," the literary public, is worth more than the apotheosis of the Greek heroes. Once again, the moderns outdo the ancients, even as they evoke the classical aura of heroism. Voltaire himself wrote for the same public, whose approval was vital to the reputation (often referred to as "glory") of the enlightened man of letters. This audience included an important female contingent; the female presence in the myth of the Argonauts was thus displaced in the Enlightenment quest myth onto the audience for the scientific travelers' exploits. The feminized audience was not thereby rendered inconsequential, however, since its appreciation of scientific accomplishments was crucial to the process of making knowledge.

The numbers calculated from the tabulations of painstaking observations comprised the ostensible goal of the enlightened Argonauts' quest. Another and hardly less important trophy was the narrative of their accomplishments, disseminated to a wide audience. Reports from Lapland centered on the difficult working conditions and the "curiosities" of the local scene.[17] The extremes of climate and latitude required a leap of imagination on the part of Parisian readers, who found it difficult to put themselves into the cold or into the company of the people living in such conditions. Once back in Paris, Maupertuis recounted to a public session of the Academy the challenges of doing science in a strange environment: finding appropriate observing sites, transporting and verifying sensitive instruments, building observatories on remote mountains, living in the most rudimentary of shelters. His triumphal speech to the Academy became the opening section of a full account of the expedition's results, published in 1738.[18] He described the process of deciphering the earth's shape as *physically* difficult, requiring strength and endurance as well as the ability to handle instruments and interpret observations. Long and "painful" calculations, dark winter days spent measuring distances in subfreezing temperatures, forays up snowy mountains pulled by temperamental reindeer, or journeys down white water in strange native craft, all represented trials hardly imaginable from the vantage point of Paris. Even tasks that were conceptually simple—measuring a length on the frozen surface of a river—became monumental feats in the bone-chilling cold of the Arctic. Maupertuis asked his readers to put themselves in the shoes of the men on the icy river:

I will say nothing of the hardships nor of the perils of this operation; you may imagine what it is to walk in two feet of snow, carrying heavy measuring sticks, which must be continually set down in the snow and retrieved. All this in a cold so great that when we tried to drink eau-de-vie, the only drink that could be kept liquid, the tongue and lips froze instantly against the cup and could only be torn away bleeding; a cold so great that it froze the fingers of some of us, and continually threatened us with yet greater accidents.[19]

Far from saying nothing of the dangers, the narrative returns to the risky conditions again and again. The physical strength and perseverance necessary to conquer such obstacles made of the returning men of science not just selfless seekers after truth, but tough adventurers.

The reader of Maupertuis's book cannot help being acutely aware of the bodies of the academicians, as they freeze or struggle up mountains or suffer the torments of flying insects. These physical hardships make the accomplishment real and substantial to a genteel urban audience with little direct experience of anything beyond the boundaries of the French provinces. In his summary of the expedition for the Academy's *Histoire*, Fontenelle dwelt on the physical challenges of the Arctic setting, repeating the story about the perils of drinking eau-de-vie and noting that, "even with the help [of a troop of soldiers], the academicians needed courage as determined as that of the soldiers; sometimes it was up to [the academicians] to set an example as well as to give orders."[20] The soldier facing dangers in the field became the prototype for the man of science far from home. Though they relied on the strength and perseverance of "their" soldiers, the academicians had to outdo the implicitly ignorant soldiers at displaying toughness in the face of adversity. In their relations to the soldiers, the Parisians took on the identity of aristocratic officers ordering the strategic maneuvers of troops under their command—though technically, of course, they were not officers at all.

In keeping with its theme, Maupertuis's book was read as a romance as well as a scientific report. Once again, Voltaire's overblown rhetoric points to a crucial feature of the text:

I have just read a story and a piece of physics more interesting than any novel [*roman*]. . . . Your preface . . . elicits an extreme impatience to follow you to Lapland. As soon as the reader is there with you, he thinks himself in an enchanted fairyland where philosophers are the fairies. . . . In ecstasy and in fear, I follow you across your cataracts and up your mountains of ice. Certainly you know how to paint; only you could be our greatest poet as well as our greatest mathematician. If your operations are worthy of Archimedes and your courage of Christopher Columbus, your description of the snows of Tornea is worthy of Michelangelo.[21]

Voltaire thus welcomes Maupertuis into the club of right-thinking philosophes, where scientific truth merges with enlightenment, amusement, and aesthetic response. The characterization of the book as novel, fairy tale, or poetry, capable of transporting the reader to the frozen north, reminds us of the relatively wide readership that Maupertuis was addressing. Writing the book became as crucial to the success of the venture as the measurements themselves, and writing in the heroic vein was in part a play for the sympathy of women readers, as we have seen. To represent the validation of Newton's theory of gravity as a romance was to claim a significance and an audience for science that reached beyond the confines of the Academy, and this is surely what appealed to Voltaire about the expedition.[22]

The South American expedition provided further evidence of the hardiness of French academicians. La Condamine marked his return to Paris with a public discourse on his harrowing sixteen-month journey down the uncharted waters of the Amazon. He had anticipated the difficulties of taking the long way home from Quito, "but after soberly reflecting on it, I judged that none of the obstacles that people raised to discourage me from my project were invincible, with a bit of resolution."[23] Throughout his narrative, and the others he wrote about his time in South America, La Condamine presented himself as the personification of sober reflection combined with intrepid resolve to overcome whatever obstacles he might encounter.

In accounting for his own long absence, Bouguer pointed out that the equatorial expedition, in measuring an arc three times as long as the one in Lapland, had undertaken a much longer and more arduous task:

> The physical difficulties . . . were so great that it would be impossible to describe them. . . . The great height of the mountains, which in Europe ordinarily has facilitated this sort of operation, was on the contrary completely injurious for us. . . . Storms blew away our signals, and often reduced us to the annoying necessity of thinking of nothing but our own preservation. . . . We had to penetrate almost continually into deserted places where we found nothing but paths made by wild animals.[24]

In addition, they had to contend with extremes of hot and cold weather, poisonous insects, snakes, and uncooperative indigenous people. Again, after disingenuously asserting the impossibility of describing the desperate wilderness conditions, published accounts stressed adventures and the survival of hardships alongside reports of observational data, experimental design, and methods of calculation.

All of these scientific travelers were acutely aware of their stay-at-home audience, including not only colleagues in the Academy but also the wider public who followed their adventures in the press and later read their books. The masculinity of science was defined by contrast to such a feminized readership, whose admiration made the risks worthwhile. Stories of wilderness inhabited by ignorant but fascinating people who lived without the benefit of civilization, and who could therefore not understand what the foreigners were doing with their enormous and apparently precious instruments, reaffirmed the values of a civilization where scientific knowledge could be appreciated.

## Making Heroes

For the public and for the Academy, and presumably for its patrons as well, settling the question of the earth's shape, in itself a delicate technical matter, was inseparable from the conditions under which the work was done and the trials it entailed. This meant recognizing that some kinds of science required exceptional physical effort. Later retellings of these stories retained the heroic tone. In 1774, Condorcet eulogized La Condamine, making him a hero in service to the French nation as well as the Academy. "The Academy recognized how much his zeal and courage could serve the success of the enterprise," even though La Condamine was no astronomer, and not much of a mathematician: he had established a reputation as an adventurer on a voyage to the Orient with the squadron of the notorious privateer DuGuay-Trouin, and his desire to join the voyage to South America "turned him into an astronomer."[25] As Condorcet told it, La Condamine's selfless courage and adventurous spirit made him not a genius, but a hero of enlightened science, "a hero of a new kind, who deployed, to enlighten and perfect humankind, an intrepidity that has almost never been used except to enslave or exterminate."[26] That is, he turned the valor commonly associated with conquering military heroes to the enlightened goal of producing beneficial knowledge. In a bizarre melding of cosmopolitanism and national chauvinism, Condorcet glossed the conflict over Newtonian physics in the Paris Academy as ending in the triumph of French scientific heroism:

> In carrying out the measure of a degree of the meridian, the Frenchmen made it possible for their fatherland [*patrie*] to glory in an honor more justifiable than England's pride in Newton's discoveries. A discovery is the

work of a man born by chance in one place rather than another; but an enterprise like measuring the degree, which requires the protection of the government and the approbation of the public, should be regarded as the work of a whole nation. Thus, while the pamphlet writers seriously accused the Newtonians of being bad citizens, these Newtonians occupied themselves with the glory of France.[27]

In this way, Condorcet transmuted the individual heroism of the scientific travelers into a communal quest for national glory; the men of science are now canonized as heroes of the nation.

A few years later, in his history of astronomy, Jean-Sylvain Bailly reminded his readers that long-distance expeditions demanded physical strength: "to confront the change in climate and dangerous operations, the force of the body must respond to the courage of the spirit."[28] Only some men could qualify for this kind of challenge, due to the extreme conditions and the necessity of strenuous physical effort. "This kind of work demands the cooperation [*concours*] of talents and resources. We must leave our cities, the center of our forces and our power, to be transported to the deserts where nature is wild and poor, the soil arid and underpopulated, and where, left to the inclemency of a rude and foreign climate, man is alone, reduced to himself and to his industry."[29] In Bailly's account, the men who put themselves forward for these ventures were risking life and limb for the glory, not of conquest, but of finding the traces of nature's laws in the wilderness, and of bringing back the evidence to the center of civilization. Alone in the wilderness, they never forgot their connection to the metropolis, as it was that connection that gave their enterprise meaning. When Bailly retold their exploits in his disciplinary history, he made their heroic suffering—figured as solitary, even though the travelers were never truly alone—the cornerstone of their accomplishments.

J. F. Montucla's *Histoire des mathématiques*, which traversed the centuries from ancient times through to the end of the eighteenth century in four volumes, included a heroic account of the geodetic expeditions as well. Montucla identifies the travelers' motivation as "the love of glory and the sciences," a phrase that nicely captures the appetite of these men for personal fame combined with service to the crown and the acquisition of new knowledge; he then reinscribes the heroism of the academicians into his narrative of the progress of mathematics:

One can only admire the zeal and the kind of devotion with which these mathematicians surmounted all obstacles opposed to their plans. We see them [in the published narrative] cross thick forests on foot, where men have perhaps never before penetrated; scale the steepest mountains, and stay there for weeks at a time in spite of the greatest inconveniences.[30]

For Montucla, as for Voltaire before him, the literary style of Maupertuis and La Condamine inspires as much admiration as their actual measurements. Thus the first-person narratives lived on in the history of the discipline of mathematics, where they made the accumulation of mathematical knowledge appear as an unfolding adventure. "Imagine a chain of mountains," Montucla urges his readers, "of which the summits, though in the tropics, are covered in perpetual ice. . . . It was there that they had to establish their stations, sometimes to stay for months on end, waiting for a moment of good weather so they could see the signals placed on other mountains."[31] Without a degree of perseverance that is figured as almost superhuman, the measurements would never have been possible.

## Astronomical Expeditions in the Seventeenth Century

Accounts of earlier French astronomical expeditions did not incorporate the heroic motif. The obvious precursor to the geodetic expeditions was that of Jean Richer, sent to the island of Cayenne, near the equator, in 1672.[32] Richer was an *élève* in the Paris Academy at the time, not in a position to seek glory for himself. He was instructed by Jean-Dominique Cassini to make observations coordinated with analogous measurements being made in Paris at the same time; his observations simply extended the astronomical program of the new Observatory to a distant latitude. His measurements of the length of a seconds pendulum became famous when Newton used them to calculate the reduction in the force of gravity at the equator. Richer made systematic observations over the course of thirteen months, but his account is entirely silent about the details of the journey or the physical hardships of working in the equatorial region. While the mounting of an expedition to this remote colony marked the expansion of French colonialism and of French astronomy, and the links between the two, Richer neither billed himself as a hero nor rewrote the values associated with doing astronomy.[33] He barely mentions the fact that it took him more than a year to reach his destination, let alone set up an observing station. The illness that forced him to return to France before his work was completed, and the death of his assistant, one M. Meurisse, do not appear in his account. Cassini's analysis of the Cayenne observations referred in passing to Richer's illness and Meurisse's death, but only as an explanation for why the series of observations was cut short.[34] The voyagers themselves never made it into the picture, nor did the terrestrial circumstances of the expedition. It is only with Bailly's history of astronomy, in 1779, that Richer's accomplishment is put into the language of risk and peril familiar from the mid-century geodetic accounts: "These dangers [illness and death] . . . prove that human curiosity has its courage as well as its honor. The glory of the sciences is thus rightly due both to the successes of genius and often also to what they cost."[35] Bailly put Richer into the heroic mold, in line with the late-Enlightenment view of the link between heroic effort and the progress of knowledge.

## Manly Heroes and Feminine Admirers

Narratives of heroic accomplishment registered the bodies of scientific travelers as part of the practice of astronomy and geodesy, since results depended on the strength and endurance of men who ventured out of the protected spaces of the study and the salon. But the physical feats did not stand alone; by incorporating the records of observations and calculations, the accounts included the process of calculation and manipulating delicate instruments in the Herculean task of making scientific knowledge. Linking contentious claims about methods and theories to stories of the physical difficulty of obtaining the evidence made science worthy of admiration. And women figured prominently among the admirers of the men, their books, and their science. First-person reports, whether oral or written, were instrumental in creating the public persona of the adventuring man of science. These scientific travelers were still exceptions among their peers, but the stories of their experiences brought science into the public eye and showed the search for knowledge to be a physical effort as well as a cerebral one.

These stories also evoke a picture of masculine camaraderie, sometimes veering off into rivalry. The wilderness engendered solidarity, and when this broke down, as it did for La Condamine and Bouguer, each went off to prove to the other that he was strong enough to do without the collaboration. They ended up by retailing charges and countercharges against each other to their Parisian audience, publishing a series of rival versions purporting to tell what had really happened, and who was the real hero. From Lapland, stories of liaisons between the Frenchmen and local women made their way back to Paris, and fueled gossip in the salons and in the press. Maupertuis remarked to one of his female correspondents, "In the frozen regions, there are very good-looking and very amiable persons; they dance and sing there, and indeed they do everything that is done in Paris."[36] Parisian women were incredulous, but fascinated, to think of their peers making love to "Lapp" women. The accounts of musical evenings and festive dinners in the land of the midnight sun testified to the power of French civilization to captivate a foreign audience. Though not part of the printed accounts, the gallantry of the Frenchmen became part of the heroic persona as the tales were told and retold across Europe.

The effect of all this was to insinuate science into the culture of the French aristocracy. Traveling to remote and potentially dangerous destinations, the academicians took as many of the trappings of civilization with them as possible.[37] The Lapland expedition included a priest-cartographer who could conduct the Catholic mass in Lutheran territory as well as helping with the astronomical observations.[38] The men who risked their lives in pursuit of knowledge were doing so in the name of France and, most importantly, in the service of the king. Though all of them were looking to make names for themselves, they also were quite conscious of contributing to the honor of the nation and the glory of the king. Putting their bodies on the line was a way of linking their service figuratively to that of military men or privateers. Abbé Outhier recognized the heroic nature of science in this context; he noted in his journal of the northern expedition, "a passion for the sciences renders men capable of great enterprises, and like the passion for glory, it can have its own heroes."[39]

## The Scientific Revolution and Heroic Tales

From the perspective of the first generation after Newton, the most problematic aspect of the notion of a revolution in science is the implication of closure following upheaval. Though no one disputed the stature of Newton's accomplishments, many of the same problems that had concerned Descartes, Newton, and Leibniz were still being actively addressed in the eighteenth century. Even some elements of Newton's "synthesis" were open to challenge, at least on the Continent. The cumulative progress of scientific knowledge became a cliché, ubiquitous in texts on astronomy and physics whose authors situated themselves with respect to the great men of earlier generations. To some extent, this meant refining the insights or correcting the mistakes of their predecessors. But the enlightened scientific travelers also portrayed themselves as going beyond their forefathers as well, when they transferred the language of courage, risk-taking, and exploration from the intellectual realm into the physical. Where the great thinkers of the seventeenth century had shown courage by questioning the authority of the ancients, or of the church, in the eighteenth century one way to claim membership in the scientific lineage was to take physical risks in the name of science. These risks became more noteworthy when undertaken far from home. This is not to say that physical effort supplanted intellectual work; rather, it was a way of showing how dexterity with mathematics or precision instruments ought to be accompanied by physical courage and strength. The intellectual or theological risks were no longer as daunting as they had been to Descartes and Newton, but men of science could distinguish themselves by facing the dangers associated with making more sophisticated observations and measurements that would link them to the earlier heroic confrontations with the evil empire of dogmatism and scholasticism.

By the end of the century, historical accounts had fully canonized Newton as the heroic torchbearer of enlightenment *avant la lettre*. Newton, the prescient innovator, did not need to display physical heroism; his enlightened followers did. In their first-person accounts, Maupertuis, La Condamine, and Bouguer portrayed themselves as the heirs of Newton, ready to take physical risks to continue his legacy to civilization. They inserted themselves into a story of the Scientific Revolution already taken for granted as the story of the progress of human knowledge. This story became more entrenched in the following decades, and the scientific expeditions became part of the Enlightenment histories of science.

Condorcet, for example, presents a triumphal vision of the history of mechanics:

> Thus man at last discovered one of the physical laws of the universe, a law that has hitherto remained unique, like the glory of the man who revealed it. A hundred years of labor have confirmed that law which appears to govern all celestial phenomena to a degree that is, so to say, miraculous. Every time that a phenomenon appears not to come under that law, this uncertainty soon becomes the occasion of a new triumph.[40]

In this typical formulation of progress, Newton stands metonymically for humankind, As a result of his revelation of truth, all men "discovered" the laws of nature. Those laws belong to mankind, especially since they were confirmed by "a hundred years of labor." Condorcet conflates physical and intellectual labor, just as Maupertuis did in his account of his laborious calculations in the frigid nights of the wilderness. Scientific expeditions then participate in the Newtonian triumph, by virtue of the exertions of men willing to take themselves and their instruments into risky situations for the sake of the advancement of knowledge.

Physical prowess was not a necessary requirement for the man of science, of course, nor was it sufficient. Stories of expeditions resonated with traditional notions of glory associated with military exploits, but they also situated their heroes with respect to an intellectual lineage of great men who had made their discoveries without ever leaving home. I particularly want to emphasize this conflation of physical strength and daring with the intellectual strength traditionally associated with mathematical ability. Exploration of the abstract world of geometric and algebraic relations paralleled a new kind of exploration of the physical world, where measurements and calculations were the objects of the quest. My purpose in analyzing these narratives of scientific exploits has been to see how the discourse of honor and truth was linked to an old-regime masculinity, making science a possible route to valor. My reading implies the necessity of making gender analysis sensitive to local context. Pioneering work in the 1980s on gender in the Scientific Revolution presented a grand narrative of sorts,with reference to the shift in the ways that nature was gendered by men building the "new science."[41] The work of Carolyn Merchant, Brian Easlea, and Evelyn Fox Keller drew attention to the conceptual links connecting mechanism, the domination and control of nature, and the oppression of women.[42] But none of these writers paid much attention to the daily practice of science, or its institutional and cultural specificity. The eighteenth-century French travelers did not couch their accomplishments in the rhetoric of the conquest of nature; their success is rather a triumphant demonstration of the natural power of human minds and bodies to understand the globe and the cosmos. Bailly reflected just this sense of the naturalness of the progress of knowledge when he described a key moment in the development of scientific understanding of nature thus: "Newton tied everything together in his thoughts, just as nature does in her works. . . . The immortal work [*ouvrage*] of the philosopher became the basis of human knowledge."[43] By measuring the globe with a precision that supplanted all previous efforts, these men linked their own actions with the metaphorical mapping of knowledge that characterized such Enlightenment projects as the *Encyclopédie* of Diderot and d'Alembert.

We have seen Condorcet use his prerogative as secretary of the Academy of Sciences to tell heroic tales in the service of a grand narrative of scientific progress. Men like La Condamine and Maupertuis provided the substantial details that fleshed out this narrative. In telling their own tales, the travelers involved their readers vicariously in the experience of working in potentially life-threatening situations. "One cannot read without astonishment," Montucla notes, "the details of the difficulties they had to suffer in order to accomplish [their measurements]."[44] Using "the charms of a picturesque style, a simple and animated portrayal of adventures and dangers," the travelers created personas in their texts that still moved readers several generations later.[45] These personas defined scientific exploration as intrepid, arduous, and masculine. The femininity of the readers of these texts reaffirmed the gendering of scientific practice. Progress toward enlightenment and toward a more perfect knowledge of the world resulted from heroic acts by heroic individuals. Not everyone can be a hero bringing useful knowledge back to his audience, reinforcing the hierarchies of insider and outsider, masculine and feminine, enlightened and unenlightened.

The heroic narratives of scientific expeditions in the eighteenth century, written by participants and rewritten in eulogies and disciplinary histories at the end of the century, make up a subplot to the familiar overarching narrative of scientific progress. The Enlightenment vision of rational progress buttressed by the heroic efforts of individuals became the foundation of still later grand narratives of Scientific Revolution, from Whewell to Koyré. The stubborn persistence of the heroes of the Scientific Revolution, their defiant confrontations with the forces of darkness in the name of advancement and of light—these images were nurtured by the heroic stance adopted by scientific travelers in the first generation after Newton.

## Notes

1. E.g., Jean Le Rond d'Alembert, *Preliminary Discourse to the Encyclopedia* (1751); M.-J.-A.-N. Caritat de Condorcet, *Sketch for a Historical Picture of the Progress of the Human Mind* (1793).

2. Jean Le Rond d'Alembert, *Discours préliminaire de l'Encyclopédie* (Paris: Gonthier, 1965), 89. (All translations are mine unless otherwise noted.)

3. Ibid., 105–6.

4. On this paradox, see Dorinda Outram, *The Enlightenment* (Cambridge: Cambridge University Press, 1995).

5. R. Hooykaas, "The Rise of Modern Science: When and Why?" *British Journal for the History of Science* 20 (1987): 453–73.

6. Francis Bacon, *The Novum Organon*, Aphorism 84, in *The Works of Francis Bacon*, ed. James Spedding, Robert Leslie Ellis, and Douglas Denon Heath, vol. 1 (London: Longmans, 1857), 191.

7. On the character of the explorer in the eighteenth century, see Marie-Noëlle Bourguet, "The Explorer," in *Enlightenment Portraits*, ed. M. Vovelle (Chicago: University of Chicago Press, 1997), 257–315.

8. Mary Terrall, "Representing the Earth's Shape: The Polemics Surrounding Maupertuis's Expedition to Lapland," *Isis* 83 (1992): 218–37.

9. For the Cassini map, see Josef Konvitz, *Cartography in France, 1680–1848: Science, Engineering, and Statecraft* (Chicago: University of Chicago Press, 1987).

10. Newton and Huygens, each using his own theory of gravity, based their calculations on observations of the slowing of the seconds pendulum at the equator, made by Jean Richer in 1672. Jacques Cassini derived his elongated sphere from geodetic measurements along the meridian running through Paris. See John Greenberg, *The Problem of the Earth's Shape from Newton to Clairaut* (Cambridge: Cambridge University Press, 1995).

11. P.-L. M. de Maupertuis, "Discours qui a été lu dans l'Assemblée publique . . . le 13 novembre 1737," in idem, *La figure de la terre, déterminée par les observations de Mssrs. de Maupertuis, Clairaut, Camus, Le Monnier . . . . faites par ordre du roy au cercle polaire* (Paris, 1738), 6.

12. Charles-Marie de La Condamine, "Relation abrégée d'un voyage fait dans l'intérieur de l'Amérique méridionale . . . ," *Mémoires de l'Académie Royale des Sciences,* 1745, 392.

13. Bernard de Fontenelle, "Discours prononcé à l'Académie des Sciences, dans l'assemblée publique d'après Pâques 1735, sur le voyage de quelques académiciens au Pérou," in idem, *Oeuvres complètes,* vol. 1 (Paris, 1818), 29–30.

14. Bernard de Fontenelle, "Sur la figure de la terre," *Histoire de l'Académie Royale des Sciences,* 1735, 47–8.

15. Voltaire to Maupertuis, May 22, 1738, in *Oeuvres complètes de Voltaire,* ed. T. Besterman et al., vol. 89 (Geneva, 1969), 131. Voltaire used the Argonaut image again ("Argonautes nouveaux") in his *Discours en vers sur l'homme,* 1738.

16. Jason indeed would not have ended up with the golden fleece if it had not been for Medea.

17. Few of the letters written from Lapland survive. For an interesting example, see "Lettre de Maupertuis à Mme. de Vertillac," [April] 1737, in *Mélanges Publiées par la Société des Bibliophiles Français* 6 (1829): 1–10.

18. Maupertuis, "Discours".

19. Ibid., 51–52.

20. Bernard de Fontenelle, "Sur la figure de la terre," *Histoire de l'Académie Royale des Sciences,* 1737, 91.

21. Voltaire to Maupertuis, May 22, 1738, 131.

22. For the link between novels and female readers, see Joan DeJean, *Tender Geographies: Women and the Origins of the Novel in France* (New York: Columbia University Press, 1991); idem, *Ancients against Moderns: Culture Wars and the Making of a Fin de Siècle* (Chicago: University of Chicago Press, 1997).

23. La Condamine, "Relation abrégée d'un voyage," 393.

24. Pierre Bouguer, "Relation abrégée du voyage fait au Pérou par Mssrs. de l'Académie Royale des Sciences, pour mesurer les degrés du Méridien aux environs de l'Equateur, & en conclure la figure de la terre," in idem, *La figure de la terre* (Paris, 1749), v. Part of this text was read to a public session of the Academy in November 1744.

25. M.-J.-A.-N. Caritat de Condorcet, "Éloge de La Condamine," *Histoire de l'Académie Royale des Sciences,* 1774, 93.

26. Ibid., 104.

27. Ibid., 92–3.

28. Jean-Sylvain Bailly, *Histoire de l'astronomie moderne,* vol. 3 (Paris, 1782), 12.

29. Ibid., 13.

30. J. F. Montucla, *Histoire des mathématiques* (completed and published by Jérome Lalande), vol. 4 (Paris: Henri Agasse, 1802), 150.

31. Ibid., 152.

32. John Olmsted, "The Scientific Expedition of Jean Richer to Cayenne (1672–1673)," *Isis* 34 (1942): 117–128.

33. Jean Richer, "Observations astronomiques et physiques faites en l'isle de Caïenne" (1679), *Mémoires de l'Académie Royale des Sciences, 1666–1699*, vol. 7, pt. 1 (Paris, 1729).

34. J.-D. Cassini, "Les élémens de l'astronomie vérifiés par M. Cassini par le rapport de ses tables aux observations de M. Richer faites en l'isle de Caïenne" (1684), *Mémoires de l'Académie Royale des Sciences, 1666–1699*, vol. 8 (Paris, 1730), 96.

35. J.-S. Bailly, *Histoire de l'astronomie moderne*, vol. 2 (Paris, 1779), 366.

36. Maupertuis to Mme. de Vertillac, 9. The local inhabitants with whom the men of science fraternized were Swedish colonials, and not the indigenous Sami ("Lapp"); these latter appear in the accounts, but not as dinner companions or mistresses. For the gossip in Paris, see Émilie du Châtelet to Maupertuis, December 1, 1736, in *Les lettres de la Marquise du Châtelet*, vol. 1, ed. T. Besterman (Geneva, 1958), 125.

37. These items are listed in the invoice prepared for outfitting the ship sailing from Dunkirk to Stockholm on the initial leg of the expedition. "État de la dépense faite en achat de vivres . . . ," cited in C. Nordmann, "L'expédition de Maupertuis et Celsius en Laponie," *Cahiers d'Histoire Mondiale* 10 (1966): 84.

38. The Abbé Outhier had considerable experience with surveying and mapping techniques, had worked with Jacques Cassini at the Paris Observatory, and was a correspondent of the Academy of Sciences.

39. Réginald Outhier, *Journal d'un voyage au nord en 1736 et 1737* (Paris, 1744), "Préface," n.p.

40. M.-J.-A.-N. Caritat de Condorcet, *Sketch for a Historical Picture of the Progress of the Human Mind* (1793), in idem, *Selected Writings*, ed. Keith Baker (Indianapolis: Bobbs-Merrill, 1976), 239.

41. See Carolyn Merchant, *The Death of Nature: Women, Ecology, and the Scientific Revolution* (New York: Harper and Row, 1980); Brian Easlea, *Witch-Hunting, Magic and the New Philosophy: An Introduction to Debates of the Scientific Revolution, 1470–1750* (Brighton: Harvester Press, 1980); Evelyn Fox Keller, *Reflections on Gender and Science* (New Haven, Conn.: Yale University Press, 1985). Keller's book belongs with the others by virtue of her perceptive essays on Bacon; she does not make a case for a grand narrative so much as hint at the possibility of such an interpretation.

42. For example, Easlea equates mechanism with rationalism and a denial of body and sexuality, so that the control of nature has to be rational rather than physical: Easlea, *Witch-Hunting*, 141–42.

43. Bailly, *Histoire*, 2.

44. Montucla, *Histoire*, 152.

45. Ibid., 157.

ELLA REITSMA

# 4. Maria Sibylla Merian

## A Woman of Art and Science

LONG BEFORE MARIA SIBYLLA MERIAN'S BIRTH, scholars in Western Europe were convinced that it was of the utmost importance to study the wonders of nature at first hand—in other words, "in the field." The ideas of classical authors and medieval theologians about mankind, nature, and the origins of life were no longer unquestioningly assumed to be true. The religious wars in Europe and fierce theological debates among proponents of diverse religious faiths rocked centuries-old convictions. People had to find out for themselves. Empirical research, actively investigating and drawing conclusions from one's own observations, was considered to be of paramount importance. Everything that had been written in the past had to be subjected to a critical re-examination.

The first Western academy of sciences was established in Italy in 1603 under the inspiring leadership of Prince Federico Cesi. He and his friends collected minerals, fossils, fungi—everything that was open to examination—in the hills around Rome and Naples. Cesi became the driving force behind various innovative scientific publications which in turn influenced scholars elsewhere; they then founded their own academies. The discoveries in the early seventeenth century were spectacular.

Among the members of Cesi's scientific society, the Academia dei Lincei, was the internationally renowned Galileo Galilei.[1] The astronomer and mathematician was a highly controversial figure in the theological world because he was a fervent supporter of Copernicus's discoveries, by now already some fifty years old, and used his telescope to prove incontrovertibly that the sun, not the Earth, was the centre of the stars and planets. This was a shocking discovery, with far-reaching theological implications, and a threat to the authority of the Church in Rome.

The newly discovered reality could no longer be ignored. The improvement of telescopes was matched by the further development of the microscope, and scholars gradually became convinced that all of insect and animal life derived from an egg.

Between 1660 and 1679 there was almost no one who could say with certainty that *all* the insects, the "bloodless" creatures, also came from an egg. And yet in the introduction to her first Caterpillar Book Maria Sibylla states without the slightest equivocation that *all* caterpillars come from eggs. She must have looked very closely and thought long and hard before she came to this conclusion. Her painstaking methods are revealed

by a note in her Study Book, which contains hundreds of drawn and written observations, indicating that a particular butterfly had laid 158 eggs
on the first day and 78 on the second. When you know how tiny these
eggs are, you understand how precise and tenacious she must have been.[2]
In 1679, for a woman who had not studied and had little or no knowledge
of Latin, the language of scholars, it was a daring statement indeed.

She must have worked night and day to observe, describe, and draw
all those behaviours, feeding patterns, and metamorphoses. If you are to
draw well—and she did—you have to *understand* what you see. In preparing the illustrations, she writes in the introduction to her first Caterpillar Book, she was assisted by her "dearest" Andreas, and she thanks
him for his "successful help."

## Caterpillars

The study of caterpillars was in its infancy. In the sixteenth and seventeenth centuries physicians and apothecaries were the leading researchers
of nature. They had published the odd thing about butterflies and moths,
but almost nothing about caterpillars. The English physician Thomas
Mouffe—who for the most part made use of other people's research—had
already discovered in about 1600 that different butterflies went through
the stages of egg, caterpillar, pupa, and adult insect, and that they were
not created from mud and the heat of the sun. But his findings did not
appear in print until later in the seventeenth century.

This butterfly-mud-sun belief was stubborn and still persisted in Merian's day. And it was not as preposterous as it might seem. She writes in her
Study Book of how it was only after rigorous study that she discovered
that some species of caterpillar pupate in the earth. (Strictly speaking, we
should call the Study Book her entomological working journal, but since
the facsimile edition of 1976 was published, *Study Book* has become standard in the literature.) They remain there all winter and in the spring
emerge through a narrow opening as a butterfly.[3] To a casual observer,
would it not seem obvious that the butterfly has fluttered up out of the
mud in the warm spring sunshine?

In Italy it was the eminent Bolognese physician Ulisse Aldrovandi who
was the first, in 1602, to publish at length on insects and to use illustrations. Some sixty years later, after hundreds of experiments with flies and
rotting flesh ranging from rat to cow, another Italian doctor, Francesco
Redi, was able to prove that spontaneous generation in insects had to be
consigned to the realm of fable.[4] Johannes Goedaert (1617–1668) of Zee-
land, who was also a talented painter of flower still lifes, took a consider-

able stride forward in entomology. His three-volume work *Metamorphosis Naturalis* appeared between 1660 and 1669. Maria Sibylla went a step further, and the number of observations she made is many times greater.

Goedaert's work paid little attention to the caterpillar, and he did not undertake a thorough study of its behaviour. Maria Sibylla did. Her particular fascination was with the outcasts of the animal kingdom—caterpillars, flies, beetles, cockroaches, frogs, slugs, and snails. And, above all, those "bloodless creatures" that cause damage to plants. She never drew a plant with leaves that did not have holes or nibbled edges.

Her religious faith, inspired at first by Calvinism, then by Pietism and ultimately by the teachings of Jean de Labadie, prompted her to study precisely these diminutive and humble beings. God had manifested himself equally in them. She saw a parallel between caterpillar and man, between butterfly and soul. The earthly caterpillar is greedy and ruthless, then seems dead (the pupa or cocoon) and finally rises into the air as a butterfly. And so it is with man's soul, which ascends to the Creator after the Resurrection. Or at least it does if he is one of the chosen. This was not Merian's own interpretation; it was a metaphor frequently used by religious thinkers of the age.

Maria Sibylla, as she tells us in the introduction to her first Caterpillar Book, started to take a serious interest in insects when she was just thirteen. She was eager to study their origins and development, but if you want to find out anything about the behaviour of caterpillars you have to rear them yourself. And this she did, for more than fifty years, in glass jars, wooden boxes covered with gauze, her attic and cellar. The house must often have smelt far from pleasant in the spring and summer, and now and then a hirsute creature, escaped from its prison, must have suddenly been spotted on the seat of a chair or in the middle of the table. It happened to me more than once during two summers of rearing caterpillars. I doubt her family was thrilled by it; even nowadays, it provokes anguished cries.

Merian's older daughter, Johanna Helena, could already paint when she moved to Wieuwerd, and her sister, Dorothea Maria, who was then thirteen or so, was able to learn the trade from her mother in Amsterdam. She would have had few drawing and painting lessons in the Labadist community, with its focus on discipline and religious observance. She would, though, have been able to assist her mother in collecting and rearing. Catching caterpillars in the orchard, kitchen garden, and medicinal herb garden was absolutely essential to Merian, even in Wieuwerd. The Labadist doctrine for children was strict obedience to their parents; they had to be of unquestioning help and support.[5]

In 1692 Johanna Helena, who, like her mother, had been elected a "sister" in the Labadist community, married the Labadist brother Jakob Hendrik Herolt, who earned his living as a merchant after leaving Wieuwerd. He had trading contacts with Paramaribo in Suriname. The Herolts continued to live in Amsterdam, but Jakob Herolt, accompanied by his wife or otherwise, must have made the voyage to South America and back on a number of occasions.[6] The three women evidently set up a sort of family business in Amsterdam and painted hundreds of miniatures on vellum. We can conclude this from the large sale that Merian staged in Amsterdam on 2 February 1699, shortly before she set out for Suriname, and from the sheets that are now in St. Petersburg and London.[7]

In the department of prints at the British Museum there are dozens of coloured drawings pasted into two fat albums—160 drawings from the early Amsterdam period (1691–1699), and 91 from the Suriname and later Amsterdam periods.[8] Those dating from the early Amsterdam period, in particular, are of exceptionally high quality: tiny works of art on vellum. All these drawings once belonged to the physician Hans Sloane (1660–1753), whose huge collection was the basis for the establishment of the British Museum. They have been crucial to my research, both biographically and stylistically.

It was the publication of Maria Sibylla Merian's book of the flora and fauna of Suriname that brought her widespread fame.With this book and various reissues she reached an extraordinarily large readership by the standards of the day—much larger than the readership for her caterpillar books of Western European insects. The drawback of the latter was that they had been printed in small editions and were written in German, a language many people did not know.

The spectacular feature of the Suriname insect book was that it illustrated, life-size and for the first time, previously unknown metamorphoses of butterflies and beetles from a new region of America. The accompanying descriptions were written in a lively style and embellished with all sorts of fascinating details—sometimes ethnographic. What's more, this folio-format book contained illustrations of lizards and snakes. There were also plenty of plants, many of them never before seen by anyone in Europe.

Maria Sibylla spent two years in what was, for her, the all-too-insalubrious climate of Suriname, until ill health forced her to go back to Europe in 1701, very much against her will. Three and a half years after her return she published *Metamorphosis Insectorum Surinamensium. Ofte Verandering Der Surinaamsche Insecten* (*The Metamorphosis of the Insects of Suriname*). In the spring of 1705 a Dutch and Latin edition with sixty prints and descriptions appeared. She could not possibly have achieved this on her own; her daughters must have helped. For years after this book was produced, they continued to make variations and new images and compositions under their mother's name, in many cases under her supervision. Thanks to her daughters, Maria Sibylla Merian's fame spread far and wide. . . .

Merian's research on Western European insects was at least as innovative and interesting as her work on the insects and plants of Suriname. But the flora and fauna in that new, tropical country simply had a more exciting, exotic appeal. The butterflies there were larger and more colourful, the fruit juicier and sweeter, the ants and spiders more fearsome. In the tropics life was extreme. The stay-at-homes avoided the risk of contracting such deadly diseases as malaria, dysentery, and yellow fever that flourished in these regions.

Thirteen years before she went to Suriname, Maria Sibylla's Labadist brothers and sisters had established a plantation there, far from Paramaribo, upstream on the banks of the River Suriname. Although it bore the promising name of La Providence, they had suffered grievously in that hot and humid land, and many of them had died. But still that did not keep Merian at home in Amsterdam.

She wanted to be an artist-pioneer in Suriname. The Frenchman Plumier had been the first to describe and draw plants from French Guyana in 1693. Hans Sloane had been in Jamaica and the surrounding islands from 1687 to 1689 and had plans—which everyone knew—to publish a well-documented and illustrated book on the flora and fauna of the area. He had brought out a small work without illustrations as a sort of trial run in 1695.

Describing, illustrating and identifying plants and animals from the New World was a challenge for many scholars, but most of them could not draw. They were forced to rely on the services of locals or minor artists who accompanied them on their expeditions.

After her book on Suriname came out, Merian's fame extended to English-, Danish-, German- and French-speaking lands. A new Suriname Insect Book with twelve additional illustrations supposedly from Maria Sibylla's workshop appeared two years after her death. Three more editions came out in the course of the eighteenth century. A French translation was published in 1726, while two later editions, in 1730 and 1771, included the Caterpillar Books and the Flower Books as well. . . .

The new image that Merian created—the life cycle from caterpillar to butterfly against the background of the host plant—was emulated by artists until well into the nineteenth century. But as the eighteenth century progressed, the distinction between art and science became increasingly pronounced, and during the nineteenth century science started to split into all sorts of specialist disciplines. Merian's fame as an entomologist waned. When discoveries are improved upon or superseded they are often forgotten or treated with contempt by posterity, as if they had never even been scientific.

Twenty years or so after Maria Sibylla's death, the Swedish physician and naturalist Carl Linnaeus (1707–1778) developed a new classification principle and a simple naming system on the basis of the work done and the discoveries made by his predecessors. They had floundered hopelessly in the sheer volume of new species that were constantly being discovered and, faced with the strings of adjectives applied to these species, had no idea how to tell them apart. It was a mess.

Linnaeus came up with the binomial system that is still being used today. He gave the genus a Latin name, which was followed by the species name.

In the tenth edition of his *Systema Naturae* of 1758 Linnaeus discusses at least a hundred species—both Western European and Surinamese—and refers to Merian's illustrations.[9] Maria Sibylla drew the butterflies so accurately that naturalists went out into the moors, meadows, woods and dunes with her books in tow. The London apothecary James Petiver did. So did the German physician and naturalist Daniel Gottlieb Messerschmidt, who went on an expedition to Siberia at Tsar Peter the Great's behest with Merian's Caterpillar Books in his coat pocket. The Natural History Museum in London has his copy, annotated in 1720, with butterflies he caught pressed between the relevant pages of the book.

## Notes

1. The friends chose the lynx as a highly meaningful symbol because according to Pliny, godfather of natural history, this animal was known for its exceptionally keen sight, unsurpassed by any other creature. Anyone interested in reading about the history of the science of natural history could not do better than Freedberg's book, a splendidly written eye opener. Freedberg 2002.
2. Study Book 1976, 247/53 no. 149.
3. Study Book 1976, 179/20 no. 50 and introduction to Caterpillar Book I.
4. Cobb 2006, 79–93.
5. Saxby 1987, 244–249.
6. Ibid., 381 note 95.

7. Dudok van Heel 1975, 160.

8. British Musuem, inv. nos Sl. 5275 and 5276.

9. Linnaeus used two Suriname butterflies from her Suriname Insect Book for his descriptions in *Systema Naturae* (1758), so much faith did he have in the quality of her images.

## References

Beer 1976. Wolf-Dietrich Beer (ed.), *Maria Sibylla Merian: Schmetterlinge, Käfer und andere Insekten, Leningrader Studienbuch/Butterflies, Beetles and other Insects, The Leningrad Book of Notes and Studies*, Leipzig 1976.

Cobb 2006. Matthew Cobb, *The Egg & Sperm Race, The Seventeenth-Century Scientists Who Unravelled the Secrets of Sex, Life and Growth*, London 2006.

Dudok van Heel 1975. S.A.C. Dudok van Heel, "Honderdvijftig advertenties van kunstverkopingen uit veertig jaargangen van de Amsterdamsche Courant 1672–1711," *Jaarboek Amstelodamum* 67 (1975), 149–173.

Freedberg 2002. David Freedberg, *The Eye of the Lynx: Galileo, His Friends, and the Beginnings of Modern Natural History*, Chicago/London 2002.

Merian 1705. Maria Sibylla Merian, *Metamorphosis Insectorum Surinamensium. Ofte Verandering Der Surinaamsche Insecten. Waar in de Surinaamsche Rupsen en Wormen met alle des zelfs Veranderingen na het leven afgebeeld en beschreven worden, zynde elk geplaast op die Gewassen, Bloemen en Vruchten, daar sy op gevonden zyn; waar in ook de generatie der Kikvorschen, wonderbaare Padden, Hagedissen, Slangen, Spinnen en Mieren werden vertoond en beschreeven, alles in America na het leven en levensgroote geschildert en beschreven ... , Metamorphosis Insectorum Surinamensium. In qua Erucae ac Vermes Surinamenses ... ,* Amsterdam 1705.

Saxby 1987. Trevor John Saxby, *The Quest for the New Jerusalem, Jean de Labadie and the Labadists, 1610–1744*, Dordrecht/Boston/Lancaster 1987.

Study Book 1976. see: *Beer 1976.*

LONDA SCHIEBINGER

# 5. Prospecting for Drugs

## European Naturalists in the West Indies

It is quite by accident and only from savage nations that we owe our knowledge of specifics [medicines]; we owe not one to the science of the physicians.
—P.-L. MOREAU DE MAUPERTUIS, 1752

PIERRE-LOUIS MOREAU DE MAUPERTUIS (1698–1759), president of the Berlin-Brandenburg Akademie der Wissenschaften, surely overstated his case when he claimed that European drug discoveries depended on either "accidents" or the knowledge of non-European peoples, whom the Europeans often characterized as "savages." How were new drugs identified in the eighteenth century? How did useful new medicines such as ipecacuanha, jalapa, and Peruvian bark arrive at London shops and Parisian hospitals?

Historians of the early modern period have begun detailing how botanical exploration—bioprospecting, plant identification, transport, and acclimatization—worked hand in hand with European colonial expansion. While Spanish conquistadors had entered the Americas looking for gold and silver, by the seventeenth century Europeans increasingly turned their attention to "green gold." The riches stored in plants would supply lasting, seemingly ever renewable profits long after the mineral wealth of the New World ran out. The very definition of *Amerique* in Denis Diderot and Jean Le Rond d'Alembert's *Encyclopédie* emphasized the business of plants—trade in sugar, tobacco, indigo, ginger, cassia, gums, aloes, sassafras, brazil wood, and guaiacum.[1] By the eighteenth century sugar was the largest cash crop imported into Europe from the Americas, but Peruvian bark (quinine) was the most valuable commodity by weight shipped out of America into Europe. Identifying profitable plants in the Caribbean was one way European naturalists contributed to the colonial effort.

European naturalists also contributed to Europe's expansion by providing new medicines to keep European troops and planters alive in the colonies. Colonial botany was central to the project of Europe's control of tropical areas where voyagers from temperate zones became sick and died in alarming numbers. Nicolas-Louis Bourgeois (1710–76), who served as secretary of the Chamber of Agriculture during his twenty-eight-year residence in Saint Domingue (Haiti), noted that finding new drugs was not merely a matter of curiosity but was necessary "to cure our maladies and provide new assistance." Europeans moving into the tropics encountered illnesses completely unknown to them; their standard pharmacopoeia proved largely ineffective against new diseases. Bourgeois complained that "apothecaries and surgeons buy their remedies from ship captains coming from France, but the vegetable remedies last hardly one year, and when they are old, they do more harm than good."[2]

This chapter explores European naturalists' field practices as they collected plants in the West Indies either for profit or for medicinal use. Was it scientific training that allowed for new drug discoveries? To what extent did Europeans in the West Indies depend on indigenous populations and African slaves for their superior knowledge in what later came to be known as tropical medicine, as Maupertuis suggested? How did naturalists from Europe wrest this information—on which their very survival often depended—from populations they had conquered and enslaved?

## Drug Prospecting in the West Indies

Before the onset of rampant racism in the nineteenth century, many Europeans valued the knowledges of indigenous Americans, Africans, and peoples of India and the East Indies. In the eighteenth century, too, Europeans in the West Indies often took as their starting point for empirical investigations the drugs, dyes, and foodstuffs suggested to them by native "informants." Before leaving home, learned naturalists pored over documents from all across Europe seeking information concerning the flora they were likely to encounter in the course of their voyages. Hans Sloane (1660–1753), future president of the Royal Society of London, was typical in this regard. Before setting out for Jamaica, where he was to take up the post of physician to the governor there, Sloane collated all data available in Europe concerning plants of tropical areas (in both the East and West Indies) so that he would recognize new plants when he encountered them. On arriving in the West Indies, Sloane turned to locals for information, collecting "the best information" concerning the natural products of the country from "books and the local inhabitants, either European, Indian or Black."[3]

Who were the local informants to whom Europeans turned for information concerning useful foods and medicines? When Christopher Columbus (1451–1506) arrived in Hispaniola in 1492, the island was inhabited by approximately one million Tainos and Caribs.[4] By the eighteenth century native populations in the Caribbean had been decimated by conquest and disease. The Caribs had run the Arawaks out of the Lesser Antilles; the Spanish had crushed both peoples. A 1660 peace agreement between the English, the French, and the Spanish exiled the Caribs to the islands of Saint Vincent and Dominica, leaving the larger islands, such as Jamaica and Hispaniola, heavily inhabited by Europeans and Africans, with only small populations of Amerindians.[5] European physicians nonetheless gleaned what information they could from the few survivors.

In the eighteenth century, as earlier, most botanists were physicians. Medicine and botany were closely allied, and botany remained a required element in European medical education. Jean-Baptiste-René Pouppé-Desportes (1704–48), royal physician in Cap Français, Saint Domingue, from 1732 until his death in 1748, typifies doctors working in the Caribbean. In efforts to increase the efficacy of his cures, he supplemented his mainstay of remedies sent from the Hôpital de la Charité in Paris with local "Carib simples." Because the first Europeans who came to the Americas, Pouppé-Desportes wrote, were afflicted by illnesses completely unknown to them, it was necessary to employ remedies used by "the naturals of the country whom one calls savages."[6] In the third volume of his *Histoire des maladies de Saint Domingue*, he presented what he called "an American pharmacopoeia," offering an extended list of Carib remedies. Europeans had begun producing pharmacopoeia, official compendiums of drugs for major cities, in the sixteenth century in an effort to secure uniformity in remedies. Pouppé-Desportes's pharmacopoeia is one of the first to record Amerindian remedies. As was typical of these works, he cross-referenced plant names in Latin, French, and the vernacular Carib. Offering synonyms (across languages and cultures) was a common European practice, especially in the pre-Linnaean era, but only rarely were Amerindian names systematically included.[7]

Even more important than adopting Amerindian cures, Pouppé-Desportes urged that Europeans emulate the native Caribbean way of life; Europeans living in Saint Domingue would not need medicines, he wrote, if they lived as "frugally and tranquilly as the savages."[8] But, and his is a common refrain among medical men serving in the colonies, since Europeans (and he means males, who made up the majority of the European population in the West Indies) are "given to excesses in both their eating and liquor, they often require strong remedies."[9] During his years on the island Pouppé-Desportes continued to collect and test local remedies that he then classified according to their medicinal properties.

Native populations, however, continued to decline and by the 1780s Nicolas-Louis Bourgeois complained that "of the prodigious multitude of natives of which [the Spanish chroniclers] speak, not a single of their descendants can be discovered whose origin has been conserved pure and without mixture."[10] By the 1790s a chronicler of the revolution in

Saint Domingue bemoaned the fact that no trace of a "single native" remained.[11]

With the decline of indigenous populations, slave medicines took on an unexpected importance in the West Indies, even though Africans were originally no more native to the area than were Europeans. Unlike Europeans, however, Africans knew tropical diseases, their preventions, and their cures. The Scottish mercenary Lt. John Stedman (1744–97), for example, living in Surinam, worked alongside a number of African slaves. One old "Negro" named Caramaca had given him the threefold secret of survival: (1) never wear boots but instead harden your bare feet (which Stedman did by incessantly pacing the deck of his boat); (2) discard the heavy European military jacket and dress as lightly as possible; and (3) bathe twice a day by plunging into the river—something that distressed the water-wary Europeans.[12]

Bourgeois, a longtime resident of Saint Domingue, also appreciated slave medicines. Considering health a matter of state importance, he eulogized the "marvelous cures" abounding in the islands and remarked that *les nègres* were "almost the only ones who know how to use them"; they had, he wrote, more knowledge of these cures than the whites (*les blancs*).[13]

It is impossible to know with any precision how much African herbal knowledge was transferred into the New World. Displaced Africans must have found familiar medicinal plants growing in the American tropics, and they must have discovered—through commerce with the Taino, Arawak, or Caribs or through their own trial and error—plants with virtues similar to those used back home. Bourgeois confirms that there were many "doctors" [*médecins*] among the Africans who "brought their treatments from their own countries," but he did not discuss this point in detail.[14]

Bourgeois also praised the skill of slave doctors. "I could see immediately," he wrote, "that the negroes were more ingenious than we in the art of procuring health. . . . Our colony possesses an infinity of negroes and negresses [*nègres & meme des négresses*] who practice medicine, and in whom many whites have much confidence. The most dangerous [plant] poisons can be transformed into the most salubrious remedies when prepared by a skilled hand; I have seen cures that very much surprised me.[15] Confidence in African cures was so high among whites that when Sir Henry Morgan, lieutenant governor of Jamaica, became dissatisfied with Sloane's treatment of his disease, he sent for a "black doctor."[16]

Attitudes among Europeans across the Caribbean, of course, were not uniform. The French royal botanist Pierre Barrère (1690–1755), working in Cayenne off the coast of French Guiana from 1722 to 1725, did not think much of Amerindian cures. The good health the Indians (he named twenty-four peoples living there) enjoyed he supposed to be the result of their careful diet, frequent bathing, and moderate indulgence in pleasure. He wrote, "our Indians are completely ignorant of how to compound medicines. The few remedies they know they have learned from the Portuguese and other Europeans."[17] He nonetheless preserved several Amerindian plant names along with their medical uses. David de Isaac Cohen Nassy (1747–1806), a Jewish physician working next door in Surinam in the latter part of the eighteenth century, remarked that the "negroes" played a large role in the health of the colony with their "herbs and claimed cures"; however, he also noted that their cures were "more valued among the Christians than among the Jews."[18] Sloane had a similarly poor opinion of slave medicine. Although he took care to collect what the Africans living in Jamaica told him, he did not find their cures in any way "reasonable, or successful"; what they knew, he wrote, they had learned from the Indians.[19]

Even in this era when many Europeans valued Amerindian and African knowledge useful to them, mythologies of drug discoveries suggested that knowledge traveled up a rather anthropo- and Eurocentric chain of being, from animals (with their instinctive cures), to indigenous peoples, to the Spanish, and, according to the French mathematician Charles-Marie de La Condamine (1701–74), ultimately to the French. La Condamine, who traveled extensively in what is now Ecuador and Peru in an effort to procure Peruvian bark (*Cinchona officinalis*—the source of the therapeutically effective alkaloid quinine), recounted the ancient legend that South American lions suffering from fevers found relief by chewing the bark of the cinchona tree. Observing the curative powers of the bark, the Indians too began treating malarial and other recurring fevers with it. The Spanish then learned of the cure from the Indians, and the French, the self-appointed keepers of universal knowledge in the age of Enlightenment, learned of it from the Spanish.[20]

La Condamine's countryman Pouppé-Desportes offered two further examples in this genre—one from Martinique and another from Saint Domingue. In the first, a potent antidote to snakebite was discovered by the lowly grass snake. "Unhappy enough to live on the serpent-infested island of Martinique," wrote Desportes, the snake learned to employ a certain herb when attacked by a venomous serpent. The effect was so wonderful that the natives called the plant *herbe à serpent*, or serpent herb. In similar fashion the native hunters discovered the excellent qualities of the "sugar tree" by observing wild pigs, which are called "maroons," shredding the bark of this tree with their tusks when hurt and then rubbing their injuries with its sap; for this reason this sap was called "wild pig balm."[21] Edward Long (1734–1813) in Jamaica was very much a racist in how he looked at local cures. Although he agreed that Negro cures often worked "wonderfully" where even European art had failed, he did not admit to any creativity among Africans. Negroes, he charged, apply their herbs randomly and, like monkeys, whom he claimed they resembled, they received their skill only from "their Creator, who has impartially provided all animals with means conducive to their preservation."[22]

The process of colonial expansion opened Europeans' minds to new knowledge systems. From the fifteenth century onward, European naturalists underwent an epistemological shift away from relying solely on the "summa of ancient wisdom" (Dioscorides, Pliny, Galen) toward valuing (or at least appreciating) the knowledges of native peoples encountered through colonial expansions.[23] European physicians no longer defined their task as verifying the effectiveness of ancient medicines or identifying local substitutes for those remedies. Their desire—on the one hand, to identify valuable economic plants for themselves and their patrons, and on the other hand, to find useful cures for their countrymen who all too often succumbed to the ravages of tropical diseases—led physicians to rely on colonial locals for information concerning potentially helpful plants. Botanists in the eighteenth century trod a fine line between their nascent prejudices against the peoples of America and Africa and their need to survive in unknown environments.

## Biocontact Zones

In her study of French natural history, Emma Spary has analyzed how botanists and gardeners at the Jardin Royal des Plantes in Paris, such as André Thouin (1747–1824), instructed and manipulated voyagers to speed new specimens into urban centers in order to enhance both the power and the wealth of nations. Successful acclimatization of plants for agriculture, medicine, or the luxury trade required that plants be sent into France with precise information about their cultivation, virtues, and uses. Following Bruno Latour's model, Spary has postulated that botanical men at the center needed a "rigidly structured, unvarying, universally agreed upon method of describing and inscribing" so that the useful plants could be known and successfully positioned within European frameworks of agri- and horticulture.[24]

What the naturalists at the Jardin du Roi, the Hortus Medicus in Amsterdam, or, later, at the Kew Gardens in London could not control and standardize was the "contact zone"—where Europeans negotiated with informants of all kinds and those with medicinal and natural historical

expertise in particular. Mary Louise Pratt has defined this "space of colonial encounters" as "the space in which peoples geographically and historically separated come into contact with each other and establish ongoing
relations, usually involving conditions of coercion, radical inequality, and
intractable conflict."[25] Here I wish to introduce the notion of "biocontact
zones" in order to set the contact between European medical botanists
and African and Native American healers in a context that highlights the
exchange of plants and their cultural uses.

The notion of a "contact zone," as it is often understood, is not unproblematic. To isolate the contact between Europeans and non-Europeans
as the primary site of analysis improperly constructs an overly rigid notion of non-Europeans as "others." Furthermore, contact is not restricted
to a bounded zone; Europeans had "contact" at many junctures over the
course of a voyage. There were the "encounters" among persons of different classes and professions within Europe. There were certainly the
"encounters" of young men who, driven by hunger to seek refuge in urban
port cities, were kidnapped and shipped out as sailors on trading company ships.[26] Abroad, encounters between Europeans and other Europeans were often as troubled as those with non-Europeans. Maria Sibylla
Merian (1647–1717), for example, who traveled to Surinam to study exotic
insects, had unhappy relations with Dutch planters in Surinam. "They
mock me," she wrote, "because I am interested in something other than
sugar."[27] In her turn, Merian criticized the planters for failing to cultivate plants other than sugarcane and for their harsh treatment of the
Amerindians.

Here I will focus (despite my caveats) on the biocontact zones in the
West Indies and look closely at strategies Europeans deployed in their
efforts to learn about useful plants to entice information from their informants and how they were often, in turn, manipulated by them. Inside
Europe naturalists are known to have carefully choreographed complex
systems of patronage—coaxing favors while also disciplining one another
with varying degrees of success. Europeans in the field often found it difficult to manipulate their informants, whom they wished to exploit and
control. These "actors" in the network often simply refused to cooperate.
My account necessarily relies on European sources; neither Amerindians
nor Caribbean slaves left written records documenting these encounters.
As we shall see, botanical exchange in these transcultural zones was fraught
with difficulties.

The loudest "noise"—or intellectual interference—in the biocontact
zone was the cacophony of languages. Europeans usually only scratched
the surface of local peoples' knowledges of plants and remedies because

they were often unable or unwilling to speak local languages. Edward Bancroft (1744–1821—a Massachusetts privateer and double agent who sent missives in invisible ink during the American Revolutionary War) worked in Guiana as a young man in the 1760s. He bemoaned the fact that he was "but little acquainted with the Indian languages," which he judged necessary for "acquiring that knowledge of their properties, and effects of the several classes of Animals, and Vegetables, which experience, during a long succession of ages, must have suggested to these natives." Though he endeavored to overcome these difficulties through the use of interpreters, he remarks that these efforts were largely "in vain."[28]

La Condamine, Pouppé-Desportes, and Alexander von Humboldt (1769–1859) were all keenly interested in local languages. La Condamine spoke what he called "the Peruvian language" and even owned a 1614 "quichoa" (Quechua) dictionary—which he used to study etymologies (of "quinquina," or quinine, for example). Concerning the language of the ancient Peruvians, La Condamine noted that by his day (the 1730s and 1740s) it was "strongly mixed" with Spanish.[29] The French mathematician judged the languages of South America harshly, writing that "many possess energy and bear traces of elegance, but they are universally barren of terms for the expression of abstract and universal ideas, such as time, space, substance, matter, corporeality. . . . Not only metaphysical terms, but also moral attributes are completely absent."[30] One wonders, however, if La Condamine had sufficient knowledge of the language to make such judgments. Humboldt, who traveled extensively in present-day Venezuela and Colombia, prepared a dictionary of the Chaymas language that consisted of only about 140 words, most of which depicted simple things such as fever, hammock, boy, girl, bridegroom, fire, sun, moon or were used in phrases such as "he likes to kill" or "there is honey in my hut."[31]

The problem of language in the West Indies was sometimes ameliorated by the slave and native populations there who, serving as active linguists, created Creole languages. The Caribs, for example, created "a jargon" through which they dealt with the French in Saint Domingue, which was a mixture of "Spanish, French, and Caraïbe pell-mell [all] at the same time."[32] African slaves in Surinam created a language that served as common currency there called "Negro-English." It was composed of Dutch, French, Spanish, Portuguese, and mostly English.

Problems of communication did not arise simply from the lack of knowledge. The Frenchman Charles de Rochefort (1605–83) as early as 1658 placed the problems of language in the context of war and conquest. "Some of the French have observed," he wrote, "that the Caribbeans have an aversion to the English tongue; nay their loathing is so great that some affirm they cannot endure to hear it spoken because they look upon the English as their enemies." He noted that the Caribs had, in fact, assimilated many Spanish words to their own language but that this was done at a time when there were friendly relations between the two nations. De Rochefort noted further that the Caribs shied away from teaching any European their language "out of a fear that their [own] war secrets might be discovered."[33]

"Noise" in the medical contact zone also resulted from the inflexible theoretical frameworks that made Europeans unable to absorb radically new information. Understanding of Carib, Taino, Arawak, or transplanted African medicinal herbs was limited to what made sense to Europeans within a humoral context of heating and cooling drugs.[34] European naturalists thus tended to collect specimens and specific facts about those specimens rather than worldviews, schema of usage, or alternative ways of ordering and understanding the world. They stockpiled specimens in cabinets, put them behind glass in museums, and accumulated them in botanical gardens and herbaria, but as specimens "stripped of narrative," supporting once again the notion that "travelers never leave home, but merely extend the limits of their world by taking their concerns and apparatus for interpreting the world along with them."[35]

In addition to problems involving language, conceptual frameworks, and physical hardships in biocontact zones, Europeans, Amerindians, and slaves each had their own economic interests and cultural aims—all of which further curtailed transcultural exchange. Humboldt complained that his Amerindian guides were interested in trees only as timber for building canoes and as a result paid little or no attention to their leaves, flowers, or fruit. Exasperated, he exploded: "like the botanists of antiquity they deny what they had not taken the trouble to observe. They are tired with our questions, and have exhausted our patience in turn."[36]

Encounters between Europeans and Amerindians or African slaves were not pure meetings between equals but antagonistic struggles full of exaggerations on both sides. Humboldt derided his "Indian" pilot, who spoke to him in Spanish, for exaggerating the dangers of water serpents and tigers. "Such conversations are matters of course when you travel at night with the natives," he wrote. "By intimidating the European travelers, the Indians believe, that they shall render themselves more necessary, and gain the confidence of the stranger." Humboldt attributed the exaggerations of the Amerindians to "the deceptions which everywhere arise from the relations between persons of unequal fortune and civilization."[37]

Even had conditions been ideal, the enormity of the task of understanding the medical value of New World plants was mind-boggling. La Condamine, already exhausted and in Quito, wrote that to detail the plants of the Amazon basin would "require years of toil from the most indefatigable botanist . . . and draughtsman. . . . I speak here merely of the labor which a minute delineation of all these plants, and the reduction of them into classes, genera, and species, would necessarily require, but if to this were superadded an examination of the virtues ascribed to them by the natives of the country, certainly the most interesting part of a study of this nature, how daunting would be the task."[38]

Conditions in biocontact zones were hardly ideal. The excesses of colonial rule brought a Carib rebellion on Saint Vincent and slave uprisings in Saint Domingue and elsewhere in the 1790s. Fear of poisonings led planters to outlaw slave medicine in English colonies in the 1730s and in the French islands in the 1760s. Even as early as the 1680s, Hans Sloane reported the dangers of herbalizing in islands where "run away Negroes lye in Ambush to kill the Whites who come within their reach."[39]

## Secrets and Monopolies

Europeans were often curious about Amerindian and slave medicines and eager to learn, but the indigenes and slaves were less eager to divulge this knowledge to their new masters. Along with miraculous cures came the silence of secrets. Bourgeois characteristically remarked that though "the negroes treat themselves successfully in a large number of illnesses . . . most of them, especially the most skilled, guard the secret of their remedies."[40] A Dutch physician, Philippe Fermin (1729–1813), confirmed that the "negroes and negresses in Surinam know the virtues of plants and offer cures that put to shame physicians coming from Europe . . . but," he continued, "I could never persuade them to instruct me."[41]

Naturalists in the West Indies devised various methods for wresting secrets from unwilling informants. Bourgeois attempted to win slaves' confidences with friendship; failing this, he offered money, but often without success.[42] Fermin in Surinam, anxious to save the colony the cost of foreign drugs and the malfeasance of ill-intentioned slaves, attempted to learn from the "black slaves" their knowledge of plants, "but these people are so jealous of their knowledge that all that I could do," he wrote, "be it with money or kindness [caresses], was of no use."[43]

Europeans also threatened and coerced reluctant informants. Sloane related the story of how Europeans learned about *contra yerva*, the potent antidote to the natives' poisoned arrows. The story was told to him by an English physician named Smallwood, who had been wounded by an Amerindian's poisoned arrow while fleeing the Spanish in Guatemala. Not having much time, he took one of his own Indians prisoner, tied him to a post, and threatened to wound him with one of the venomous arrows if he did not disclose immediately the antidote. At this, the Indian (of an unnamed people) chewed some *contra yerva* and placed it into the doctor's wound, and it was healed.[44]

The effort to secure secrets against enemies or competitors was not unique to the vanquished in the colonies. Europeans, of course, had many secrets of their own. Before patents, guilds kept the techniques of their trades secret. In the medical domain many physicians and apothecaries protected their remedies by keeping recipes secret until they could sell them for a good price. In a celebrated case the apothecary Sir Robert Talbor of Essex (1639–81) garnered fame and wealth from his "marvellous secret" that cured fevers: his *remède de l'Anglais* secured him a knighthood in England and an annual pension of two thousand livres from Louis XIV.[45]

The eighteenth century saw the rise of public health measures that celebrated the free exchange of life-saving remedies and procedures. Sir Hans Sloane, president of the London College of Physicians, walked a fine line between respecting the monetary security bound up in secrets and advancing the emerging ethic of making effective cures available to the public for the "welfare of mankind" when he kept secret his remedy for "eye soreness." Ever on the lookout for new drugs both at home and abroad, Sloane attempted to procure the secret of Dr. Luke Rugeley's eye balm. Avoiding direct contact with the doctor, Sloane sought out a "very understanding apothecary," a mutual acquaintance of both Rugeley and Sloane—but the apothecary either did not know or was not talking. After Rugeley's death Sloane acquired all his books and manuscripts, including his *materia medica*, but still "all in vain." Finally, a "person" who had made the medicine for Rugeley sold Sloane the secret on the condition that Sloane would not divulge it until after the unnamed person's death (because he or she was still making a living by it). Toward the end of his life Sloane published *An Account of a Most Efficacious Medicine for Soreness . . . of the Eyes*, which he sold for sixpence each.[46]

The great trading companies of the early modern period also guarded their investments through scrupulously protected monopolies. The historian Francisco Guerra has perceptively noted that "medical science has been the heritage of the human race, but the drug trade has been throughout history the object of economic monopolies."[47] The Dutch, for example, held a monopoly on trade with Japan from 1639 to 1854.[48] One of Linnaeus's many students, Charles Thunberg (1743–1828), who traveled to the Cape of Good Hope and on to Java and Japan for the Dutch in the 1770s, told how the Dutch East India Company continued in the eighteenth century to hold a monopoly on the spice trade and opium. "If anyone is caught smuggling," he warned, "it always costs him his life, or at least he is branded with a red hot iron, and imprisoned for life."[49] While many naturalists caught passage on company ships, the East and West India Companies cautioned scholars not to reveal too much in their publications.[50] The French Compagnie des Indes, for example, urged naturalists to limit published information. The French company also blocked British efforts to buy Michel Adanson's papers treating the natural history of Senegal after the British took Saint Louis in 1758. Few French academic reports dealing with this area were published while Senegal was under British control.[51] Commercial exchange—the desire for profits by both individuals and joint-stock companies in this era—countermanded emerging practices of free scientific exchange.

## Drug Prospecting at Home

Beginning in the late seventeenth century and throughout the eighteenth century, academic physicians prospected for drugs inside Europe, using techniques similar to those employed by their colleagues in the colonies. From Sweden, Linnaeus remarked that, "it is the folk whom we must thank for the most efficacious medicines, which they . . . keep secret."[52] In England, Joseph Banks's (1743–1820) interest in botany was kindled when, as a youth in the 1750s, he watched women gathering "simples" for sale to druggists.[53] In encounters strikingly similar to those in the West Indies, European medical men interrogated and cajoled their countrymen and women into disclosing the secrets of their indigenous cures. Sometimes persuasion and at other times the power of the purse yielded the secrets of their indigenous cures. As with colonial plants, physicians began testing on the basis of ethnobotanical clues and published the results. Thomas Sydenham (1624–89) provided a popular rationale for these new practices: any "good citizen," he wrote, in possession of a secret cure was duty-bound "to reveal to the world in general so great a blessing to his race"; medical experiments were to benefit the public good (not only physicians' purses) for, as Sydenham continued, "honors and riches are less in the eyes of good men than virtue and wisdom."[54] Not all would have agreed.

Physicians cast their nets widely in the search for effective cures at home. Published instructions encouraged travelers inside and outside Europe to question and learn from people of all stations and sexes— from statesmen, scholars, and artists as well as from craftsmen, sailors, merchants, peasants, and "wise women."[55] The process of collecting information from old women or particularly successful women healers was strikingly similar to that of prospecting abroad, a kind of internal bio-prospecting. The major strategy was to buy the cure, and often, as noted above, the government put up the money to purchase the secret of a useful remedy. One woman who did well for herself was the spinster Mrs. Joanna Stephens, daughter of a gentleman of good estate and family in Berkshire, England. She was paid five thousand pounds on 17 March 1739 by the king's exchequer for her drug—an eggshell and soap mixture reported to dissolve bladder stones.[56] Her cure for this "painful distemper" was highly prized because the only other option was surgery, known as "cutting for stones." The London surgeon William Cheselden (1688–1752) was able to extract stones in less than one minute (a procedure he perfected on dead bodies); nonetheless, the operation was excruciating and dangerous.[57] Most women, however, whose cures were eventually adopted and published in the various European *Pharmacopoeia*, remained nameless (and probably unpaid), as was true also of the greater part of West Indian indigenes and slaves who offered cures.

As with informants in the West Indies, we have little access to the women's own reactions to their encounters with academic European naturalists. The historian Lisbet Koerner has highlighted an article in a 1769 Stockholm magazine purporting to give voice to "wise women." It was not uncommon in the eighteenth century for articles such as this to be written by men under female pseudonyms. Nonetheless, the wise women's assertions air complaints that appear in other women's writings from this period.[58] The women noted their "joy and pleasure" when a physician, standing by a sick child's bed, judged, "here no other help can be had than that of finding an experienced wise woman." They complained that physicians "exert themselves to both smell and taste our pouches, creams and bandages," attempting to divine the secrets of the medicines. These women, like so many others at the time, ended the article by asking to be admitted to professional training, in this case to the Stockholm Medical College, "for we are after all considered as highly as the gentlemen [physicians] in the homes into which we are called."[59]

The Swedish physicians offered to purchase the wise women's secrets, but, like other folk healers, these women would not sell them because their livelihood, and often that of their descendants, depended on monopolizing the cure. Mr. Ward, for example, who developed Ward's pills, would not divulge the recipe "because it would be worth so much yearly to him and his successors."[60] Thus a conflict arose between irregular healers (and Sloane classed some of his London colleagues in this category), who depended on their secrets for their livings, and the beginnings of academic research medicine with its ethics of sharing results for the "benefit of humankind"—or at least such was the ideology. Patents, developed to safeguard university and company investment in drug development, tempered this ethic.

## Botany and Colonial Expansion

Since Columbus first set foot on the Caribbean islands that came to be known as the West Indies, Europeans scoured these tropical areas for useful and profitable plants of all sorts. Already by the sixteenth century "wonder drugs" such as quinine, jalap, and ipecacuanha had been found and were soon to be introduced into the pharmacopoeia of major European commercial centers. Mercantilists and cameralists all across Europe attempted to staunch the flow of bullion to foreign countries to pay for these new drugs. The French physician and botanist Pierre-Henri-Hippolyte Bodard decried the annual loss of revenue to foreign countries, a loss aggravated by the French upper classes' preference for "objects difficult to obtain."[61] He estimated the loss from Peruvian bark alone at 7,380,000 livres per annum. Other West Indian and South American exotics, such as cacao and jalap, similarly drained European coffers.[62]

Where Europeans could not import and acclimatize these tropical plants inside their own borders, they turned to their colonies to secure such valuable commodities for the metropolis. Bourgeois, speaking from France's richest colony, Saint Domingue, advocated not only fulfilling France's needs by cultivating medicinal plants in the colonies but also the continued search for new and profitable drugs. "Why have recourse to foreign drugs," he asked, "when it is believed that there are so many here [in the Antilles] of use?" There is "a mass" of simples, he continued, in Saint Domingue that need only be examined.[63]

The search for new medicines was fueled by the desire to supply markets back in Europe and also by the need to keep troops, colonists, missionaries, and traders alive in the tropics. Here colonial botanists had to rely on peoples indigenous to these areas to expand their knowledge of the multitude of new diseases they encountered daily. To this end, European-trained physicians and privateers in the West Indies turned to indigenous peoples and slave populations in their efforts to find new and effective drugs.[64]

Historians of science often present the creation of new knowledge as an unencumbered search for truth characterized by open communication of the best results. As we have seen, however, conditions were hard in bio-contact zones. Mortality was high; informants were unwilling. The search for potential life-saving new drugs was mired in relations of conquest, commerce, and slavery. European colonial expansion depended on and fueled the search for new knowledge concerning tropical medicines. At the same time, colonialism bred dynamics of conquest and exploitation that impeded the development of this knowledge.

## Notes

1. Denis Diderot and Jean Le Rond d'Alembert, eds., *Enclopédie: ou, Dictionnaire raisonné des sciences, des arts et des métiers* (Paris, 1751–76), s.v. "Amerique."

2. [Nicolas Bourgeois], *Voyages intéressans dans différentes colonies Françaises, Espagnoles, Anglaises, etc.* (London, 1788), 459.

3. Hans Sloane, *A Voyage to the Islands Madera, Barbadoes, Nieves, St Christophers, and Jamaica; with the Natural History* . . . , 2 vols. (London, 1707–25), vol. 1, preface, n.p.

4. Figures vary from sixty thousand to four million. See Noble David Cook, *Born to Die: Disease and New World Conquest, 1492–1650* (Cambridge: Cambridge University Press, 1998), 23–24.

5. David Watts, *The West Indies: Patterns of Development, Culture and Environmental Change since 1492* (Cambridge: Cambridge University Press, 1987).

6. Jean-Baptiste-René Pouppé-Desportes, *Histoire des Maladies de Saint Domingue*, 3 vols. (Paris, 1770), vol. 3, 59.

7. Londa Schiebinger, *Plants and Empire: Colonial Bioprospecting in the Atlantic World* (Cambridge: Harvard University Press, 2004), chap. 5.

8. Pouppé-Desportes, *Histoire*, vol. 3, 59.

9. Ibid.

10. [Bourgeois], *Voyages*, 67.

11. [Anonymous], *Histoire des désastres de Saint-Domingue* (Paris, 1795), 47.

12. John Stedman, *Stedman's Surinam: Life in Eighteenth-Century Slave Society*, ed. Richard Price and Sally Price (1796; Baltimore: Johns Hopkins University Press, 1992), 63. For the African diaspora of seeds, plant knowledge, and cultivation techniques, see Judith Carney's chapter in this volume.

13. [Bourgeois], *Voyages*, 458, 470.

14. Ibid., 470.

15. Ibid., 468, 470.

16. Galvin de Beer, *Sir Hans Shane and the British Museum* (London: Oxford University Press, 1953), 41–42.

17. Pierre Barrère, *Essai sur l'histoire naturelle de la France Equinoxiale* (Paris, 1741), 204.

18. David de Nassy, *Essai historique sur la colonie de Surinam* (Paramaribo, 1788), 64.

19. Sloane, *Voyage*, vol. 1, xiii–xiv.

20. Charles-Marie de La Condamine, "Sur l'arbre du quinquina" (28 May 1737), in *Histoire mémoires de l'Académie Royale des Sciences* (Amsterdam, 1706–55): 319–46, esp. 330.

21. Pouppé-Desportes, *Histoire,* vol. 3, 81.

22. Edward Long, *The History of Jamaica*, 3 vols. (London, 1774), vol. 2, 381.

23. See also Antonio Barrera, "Local Herbs, Global Medicines: Commerce, Knowledge, and Commodities in Spanish America," in *Merchants and Marvels: Commerce, Science, and Art in Early Modern Europe*, ed. Pamela Smith and Paula Findlen (New York: Routledge, 2002), 163–81, esp. 174.

24. Emma Spary, *Utopia's Garden: French Natural History from the Old Regime to Revolution* (Chicago: University of Chicago Press, 2000), 84.

25. Mary Louise Pratt, *Imperial Eyes: Travel Writing and Transculturation* (London: Routledge, 1992), 6–7.

26. Long called the kidnappers "mantraders" for stealing young men and forcing them into service (*History*, vol. 2, 287); Carl Thunberg called the Dutch East India Company "man-stealers" in *Travels in Europe, Africa and Asia, Performed between the Years 1770 and 1779*, 3 vols. (London, 1795), vol. 1, 73–75.

27. Maria Sibylla Merian, *Metamorphosis insectorum Surinamensium*, ed. Helmut Deckert (1705; Leipzig: Insel Verlag, 1975), commentary to pls. 7, 25, 13.

28. Edward Bancroft, *An Essay on the Natural History of Guiana, in South America* (London, 1769), 3.

29. La Condamine, "Sur l'arbre du quinquina," 340.

30. Charles-Marie de La Condamine, *Relation abrégée d'un voyage* (Paris, 1745), 53–55.

31. Alexander von Humboldt and Aimé Bonpland, *Personal Narrative of Travels*, trans. Helen Williams, 7 vols. (London, 1821), vol. 3, 301–3.

32. Cited in Robin Blackburn, *The Making of New World Slavery: From the Baroque to the Modern: 1492–1800* (London: Verso, 1997), 281.

33. Charles de Rochefort, *Histoire naturelle et morale des Iles Antilles de l'Amerique* (Rotterdam, 1665), 449.

34. Guenter Risse, "Transcending Cultural Barriers: The European Reception of Medicinal Plants from the Americas," in *Botanical Drugs of the Americas in the Old and New Worlds*, ed. Wolfgang-Hagen Hein (Stuttgart: Wissenschaftliche Verlagsgesellschaft, 1984), 31–42, esp. 32.

35. Spary, *Utopia's Garden*, 87. See also Anthony Pagden, *European Encounters with the New World* (New Haven: Yale University Press, 1993), 21.

36. Humboldt (and Bonpland), *Personal Narrative*, vol. 5, 256.

37. Ibid., vol. 5, 132.

38. La Condamine, *Relation abrégée*, 74–75.

39. Sloane, *Voyage*, vol. 1, xviii.

40. [Bourgeois], *Voyages*, 487.

41. Philippe Fermin, *Description generale, historique, geographique et physique de la colonie de Surinam*, 2 vols. (Amsterdam, 1769), vol. 1, preface. In her chapter in this volume Judith Carney emphasizes the contributions of women in other cultures to European pharmacopoeia.

42. [Bourgeois], *Voyages*, 487.

43. Fermin, *Description generale*, vol. 1, 209.

44. Sloane, *Voyage*, vol. 1, liv–v.

45. Jaime Jaramillo-Arango, *The Conquest of Malaria* (London: Heinemann, 1950), 79.

46. Hans Sloane, *An Account of a Most Efficacious Medicine for Soreness, Weakness, and Several Other Distempers of the Eyes* (London, 1745), 1.

47. Francisco Guerra, "Drugs from the Indies and the Political Economy of the Sixteenth Century," in *Analecta Medico-Historia* (Oxford: Pergamon Press, 1966), 29–54, esp. 29.

48. C. R. Boxer, *The Dutch Seaborne Empire: 1600–1800* (New York: Knopf, 1965).

49. Thunberg, *Travels*, vol. 2, 286.

50. Sverker Sörlin, "Ordering the World for Europe: Science as Intelligence and Information as Seen from the Northern Periphery," in *Nature and Empire: Science and the Colonial Enterprise*, ed. Roy MacLeod, special issue of *Osiris* 15 (2000): 51–69, esp. 55.

51. Jean-Paul Nicolas, "Adanson et le mouvement colonial," in *Michel Adanson's "Families des plantes,"* ed. George Lawrence, 2 vols. (Pittsburgh: Carnegie Institute of Technology, 1963), vol. 2, 440.

52. Lisbet Koerner, "Women and Utility in Enlightenment Science," *Configurations* 2 (1995): 233–55, esp. 251.

53. David Mackay, *In the Wake of Cook: Exploration, Science, and Empire, 1780–1801* (London: Croom Helm, 1985), 15.

54. R. G. Lathan, ed., *The Works of Thomas Sydenham*, 2 vols. (London: Sydenham Society, 1848), vol. 1, 82.

55. S. W. Zwicker, *Breviarium apodemicum methodice concinnatum* (Danzig, 1638), cited in Justin Stagl, *A History of Curiosity: The Theory of Travel 1500–1800* (Chur: Harwood Academic Publishers, 1995), 78.

56. Her cure was published in "Mrs. Stephen's Cure for the Stone," *London Gazette*, 16 June 1739, n.p.

57. Zachary Cope, *William Cheselden, 1688–1752* (Edinburgh: Livingstone Ltd., 1953), 24–25.

58. Londa Schiebinger, *The Mind Has No Sex? Women in the Origins of Modern Science* (Cambridge: Harvard University Press, 1989), 237–38.

59. Koerner, "Women and Utility," 250–51.

60. John Douglas, *A Short Account of the State of Midwifery in London, Westminster* (London, 1736), 19.

61. Pierre-Henri-Hippolyte Bodard, *Cours de botanique médicale comparée*, 2 vols. (Paris, 1810), vol. 1, xviii.

62. Ibid., vol. 1, xxx.

63. [Bourgeois], *Voyages*, 460.

64. Pouppé-Desportes, *Histoire*, vol. 3, 59.

LUCILE H. BROCKWAY

# 6. Science and Colonial Expansion

*The Role of the British Royal Botanic Gardens*

KEW GARDENS IS A GREAT scientific research center, staffed by trained botanists who are at the top of their profession. . . . *Science and Colonial Expansion* presents an analysis of the role of Kew Gardens in the expansion of the British Empire. Through its research, its dissemination of scientific information, and its practical activities, which included plant smuggling, Kew Gardens played a major part in the development of several highly profitable and strategically important plant-based industries in the tropical colonies. These new plantation crops complemented Britain's home industries to form a comprehensive system of energy extraction and commodity exchange which for a time, in the nineteenth and early twentieth centuries, made Britain the world's superpower.

This is a historical- anthropological study of the early role of formal scientific institutions in the expansion of empire in recent world history. I propose to demonstrate that such institutions played a critical role in generating and disseminating useful scientific knowledge which facilitated transfers of energy, manpower, and capital on a worldwide basis and on an unprecedented scale. More particularly, I intend to examine the role of Kew Gardens and the colonial botanic gardens—along with other botanical and scientific institutions—in encouraging and facilitating plant transfers which had extraordinary impact in parts of the world subject to western imperial hegemony in the nineteenth and twentieth centuries.

I shall focus on three principal cases, cinchona, rubber, and sisal, tracing their removal under the auspices of Kew Gardens (and in the first two cases, the India office) from their natural habitats in Latin America to their establishment as important commercial crops in the Asian and, to a lesser extent, the African colonies. As important as the physical removal of the plants was their improvement and development by a corps of scientists serving the Royal Botanic Gardens, a network of government botanical stations radiating out of Kew Gardens and stretching from Jamaica to Singapore to Fiji. This new technical knowledge, of improved species and improved methods of cultivation and harvesting, was then transmitted to the colonial planters and was a crucial factor in the success of the new plantation crops and plant-based industries.

In the opening years of the industrial era, before the rise of the chemical industry with its synthetic fibers and pharmaceuticals, botanical knowledge concerning useful plants was a counterpart of today's academic–industrial research. At a time when such research was not yet institutionalized but was the work of semiamateurs, institutions like Kew Gardens were as important in furthering the national welfare as our modern research laboratories today.

I further suggest that the Royal Botanic Gardens at Kew, directed and staffed by eminent figures in the British scientific establishment, served as a control center which regulated the flow of botanical information from the metropolis to the colonial satellites, and disseminated information emanating from them. Much of this botanical information was of great commercial importance, especially in regard to the tropical plantation crops, one of the main sources of wealth of the Empire. Decisions taken at Kew Gardens or implemented with the help of Kew Gardens had far-reaching effects on colonial expansion: if the botanists could suggest where to find a plant that would fill a current demand; how to improve this plant through species selection, hybridization, and new methods of cultivation; where to cultivate this plant with cheap colonial labor; how to process this product for the world market; then the botanists may be said to have had a major role in making a colony a viable and profitable part of the Empire.

The rubber plantations of Malaya, developed from seeds of wild Brazilian rubber, are the best example of the series of events just described, furnishing as they did not only much revenue but a vital strategic resource, whose place in the industrial growth and political hegemony of the West became painfully clear when Southeast Asian sources of natural rubber were cut off in World War II. Cinchona, the Andean fever-bark tree from which quinine is made, underwent a similar development under the leadership of Kew Gardens, and had important demographic and political effects through the control of malaria it afforded, not only in India where the botanical development took place but throughout the tropical world. The colonial penetration of Africa in the late nineteenth century by the European powers was accomplished only after a cheap and reliable source of quinine was available. Sisal, the third case history, illustrates the supranational character of scientific research, which flows easily across national borders to other peoples culturally prepared to use it. When Kew Gardens published the secrets of the carefully guarded Mexican sisal industry, German agronomists were able to find the plants with which to start a modern sisal industry in their East African colony.

Nineteenth-century European colonial expansion was characterized by both competition and cooperation among the powers. The Dutch from their botanic garden on Java engaged in parallel activities of plant transfer and development, especially in the case of cinchona, sometimes competing with the British, sometimes cooperating with them, and in the end, fixing the market through cartel agreements. The French copied British and Dutch plantation methods in their rubber industry in Indochina. In spite of the internal rivalries in Europe which loomed so large at the time and which ultimately instigated two world wars, the industrializing and imperialist nations of the nineteenth century—England, France, Germany, the Netherlands, Belgium (and later the United States and Japan)—shared common interests against the rest of the world. Europe was achieving a global dominance, extracting and mobilizing the energy of the world for its own purposes. In each of my three case studies, a protected plant indigenous to Latin America was transferred by Europeans to Asia or Africa for development as a plantation crop in their colonial possessions. Brazil, Mexico, Colombia, Peru, Ecuador, and Bolivia each lost a native industry as a result of these transfers, but Asia acquired them only in a geographical sense, the real benefits going to Europe.

In this study, we are concerned with the human energy in the form of underpaid labor, and the plant energy extracted by the European core from the tropical peripheries of the world system. Howard S. Irwin, president of the New York Botanical Garden, indicates the way in which energy capture underlies the world political economy:

> Long before the advent of man's dominance of the world, plants and animals living and evolving in systems characterized by give and take, success and failure, struck a balance between the energy available to them in their habitats, and the energy they required to carry out their life processes. Those species unable to adjust to the realities of habitat, including fluctuations in energy availability, passed from the scene. Those able to make the adjustment survived.
>
> Man's use of energy is no different: successful societies grow and expand when energy is abundant; growth slows or ceases altogether when energy is scarce. Industry has expended considerable time and research in order to devise more efficient use of available energy. . . .
>
> From the ecological perspective, dollars simply represent the amount of energy needed to produce a product. [1974:1]

Our study of the energy flow of one particular historical era will be approached from the viewpoint of human ecology, in which the concept of environment includes not only the physical and biotic elements associated with that term but also the human institutions—technological, economic, political—which affect the use of natural resources (Vayda and Rappaport 1968). The institutionalized accretions of past human activities form a large part of the environment in which human populations interact, becoming external factors which limit the strategies open at any one time, or under favorable circumstances, providing for a greater range of alternate choices.

At the other end of the theoretical continuum lies the individual. Societies expand or evolve; individuals choose. Individuals make the shifts from old patterns of behavior that, taken cumulatively, result in social change. . . .

In its broadest aspects, then, our unit of analysis is not any one society or empire, but the network of relations emanating from the West that penetrated all societies, binding colonized to colonizers, and colonizers to each other. The mechanism of the Western botanical expansion that is described in our study is a small, homogeneous scientific elite, an "invisible college" in Diana Crane's phrase (1972),[1] in touch with each other at home and abroad, making and implementing decisions of world-wide implications with the wholehearted support of their government and the commercial establishment. . .

The botanic garden network had three roots: The oldest and deepest being the alliance of botany and medicine, with a Greco-Roman heritage and a revival in the *hortus medicus* of Renaissance universities. The first garden at Kew, started by George III's mother, had been a garden of herbs. The medicinal herbal strain in botany became submerged but never died out, as our cinchona example shows, or the careers of many botanists like Sir David Prain, who went from the Indian Army Medical Corps to the Calcutta Botanic Gardens and on to Kew as director.

Next came the interest in pure science fostered from the late seventeenth century on by the learned societies, which constituted "a series of interrelated information networks," in Hunter Dupree's phrase. These prestigious voluntary associations drew together in social discourse certain powerful nobles, the few professional botanists, numbers of the rising middle class plus wealthy amateurs like Sir Joseph Banks, privy counsellor to the king, president of the Royal Society, benefactor of the Linnean Society, patron of botanists and explorers, who catalyzed the whole scene in the years when he was unofficial director of the royal gardens at Kew.

Most important of all for the future botanical network were the voyages of exploration and territorial acquisition dating from the post-Columbian expansion of the West and continuing through the nineteenth century. To get at the spices of the East, the coffee and the tea, to capture the trade and break the monopolies—whether held by Arab traders, native princes, Portuguese, Spanish, or Dutch trading companies—was a consuming passion of English merchants and trading companies for centuries. The British East India Company's interest in useful plants led it to establish botanical stations and to maintain a department at its headquarters in London devoted to tropical plants under Forbes Royle, who recommended both the tea and the cinchona transfers. In mid-nineteenth century the government assumed to itself all the territorial gains made by the trading companies in India, Ceylon, Penang, etc., and the botanical stations there and at the transit points, for example, the Cape of Good Hope, became part of the national botanical network.

Just as private and governmental enterprise cannot be rigidly separated in colonial expansion, neither can pure and applied science be separated on the voyages of exploration and in Kew's later activities. Captain Cook and Sir Joseph Banks, using a combination of Royal Society money and private money (Banks's), explored, botanized (bringing back 17,000 plant specimens), observed the transit of Venus, visited the Society Islands which Cook named after the Royal Society, and acquired Australia for the Crown, all in one voyage. Captain Bligh, commissioned by the government to bring breadfruit from Tahiti as food for slaves in the West Indies, also brought the "noble canes" which were to improve the sugar yield so significantly, *and* he also brought over 90 other plants to Kew to add to its study collections, as is shown by the *Kew Records Book* of 1793 (King 1976:14). Joseph Hooker collected hundreds of plants in the Himalayas, most of them new to Western botanists, while he was surveying boundaries for the East India Company in Bengal. Charles Darwin benefitted not only from his own observations on the voyage of the government ship, *Beagle*, but from the observations of all the other naturalists who had shipped on these mixed purpose voyages, which enabled him to have a mass of facts about the natural world on which to base his theories. I have dwelt on these exploratory voyages of the eighteenth and nineteenth centuries to which naturalists were attached, of which there were many more than have been mentioned here, because they constitute an almost unrecognized government subsidy to natural science, as important as any more formal government subsidy to science in our own time, both for its contribution to Darwin's theories and through the economically useful plants discovered, studied, and transferred.

These naturalist–explorers also evolved into "Kew collectors" once Kew was established as a government botanical center, and a mission like Captain Bligh's to get breadfruit or Robert Fortune's to collect tea plants in China was a model for Kew's rubber and cinchona seed transfers, and the many other plant transfers Kew directed, the mahogany, cork oak, papyrus, ipecac, varieties of coffee, tobacco, bananas, etc., which were moved into British colonies or from one colony to others.

Kew brought two new factors to the previous ad hoc plant transfers: centralized direction of the information flow, and professional development of the plants to be disbursed. The satellite gardens in the colonies provided the system with multiple centers of research, a point I consider important. The selection of *Hevea* as the most promising rubber source, out of many species, the new methods of cultivating, tapping, and semi-processing Hevea rubber, were developed not only at Kew, or in Ceylon, or in Singapore, or in other places since forgotten, but through the efforts of men in all these places. The size of the botanical network, coupled with the flexibility and informality of its operation, allowed duplication of research effort, which was an asset, as our own experience in multicentered scientific research in the United States confirms.

Knowledge flowed from center to periphery along this network, from periphery back to center where it could be redistributed, and from satellite to satellite through personal contact. This knowledge was stored for later retrieval in the libraries and herbaria, principally at Kew and Edinburgh but also in the larger colonial botanic gardens like Calcutta. It was transmitted, in increasingly formalized fashion as the century unfolded, through the human resources of the botanists trained at Kew, by the many botanical journals, more than can be supported in today's economy and, after 1887, by the *Kew Bulletin*, official organ of the Royal Garden network. The institutionalization of information disbursal made knowledge easily available to international rivals, to the immediate detriment of the British West Indies' sisal industry, although Britain's cumulative strength enabled her to recoup this particular loss by winning back the German East African colony and its sisal industry at the Versailles Treaty. With the colonial scramble over, with the increasing internationalization of knowledge, and with synthetic substitutes for some plant products, it is hard to imagine the critical role played by Kew in information dispersal in the past.

At very little cost to Britain, Kew had searched for economically useful plants, most often in South America, had improved them, and transferred them to Asia, thus participating in extending the plantation system to Asia. Kew also introduced some plantation crops to Africa and tried unsuccessfully to shore up the plantations of the West Indies with new crops to replace sugar, but the Asian plantations were Kew's most significant contribution to empire-building. In mid-nineteenth century the botanic gardens stepped in to help the empire make use of the displaced peasants thrown onto the labor market by the plunder of Asia, especially the immiseration of India's commerce, industry, and agriculture. For the next hundred years, England extracted human energy from Asia in the form of plantation labor which tended the introduced plants. By diverting some of India's excess labor to Ceylon and Malaya, the plantations served as a safety valve against peasant unrest as well as a source of profit and strategic raw materials. One of those raw materials, cinchona bark for quinine, affected social relations between Europeans and the colonized peoples by allowing more Europeans to survive in the tropical colonies as army officers, bureaucrats, plantation managers, and missionaries, and by allowing them to bring their families with them, increasing both social distance and racial snobbery. Quinine also was vital to Europe in wartime, as was rubber.

Thus the activities of Kew were in some measure related to the conditions of human labor in much of the tropical world. We have seen that these conditions could vary, depending on many variables, but above all, on the constraint or lack of constraint imposed by the state on capitalistic excesses, and the degree to which inflowing capital allowed, or more commonly removed, alternative employment opportunities or sources of food from the laborers.

At one extreme, we have found small groups who lived in environments that Europeans wanted to protect for their own enjoyment, for example, the Nilgiri Hills of South India, and who could be useful to Europeans as food producers, so that they had this alternative or supplement to family income. But even there, much of the plantation labor was imported from the plains and was coerced by poverty and lack of alternatives.

Another favored group were the Malay, Chinese, and Javanese smallholders producing rubber in Malaya and Sumatra. Here the heavy demand for rubber and the lack of pressure on land led the colonial governments to adopt a policy of development that was partially extended to subject peoples.

However, the bulk of labor on tropical cash crops was debt-coerced. On the great new plantations of Asia, Tamils, Canarese, Malays, Chinese, Javanese, and others worked the rubber, tea, coffee, and cinchona. The Asian workers were often indentured and thus working away from their home communities, living in barracks, tied to the plantations by debt and by the lack of alternative employment. Only a few escaped into labor-bossing, shopkeeping, or into the cities.

East African sisal workers can be placed in this general category, leaving their home communities to seek white man's wages in order to pay white government head taxes. The tea and coffee plantations in the East African highlands also recruited their labor in this way.

Through debt coercion backed by penal laws and the whole force of the administrative machinery, the North Atlantic powers created a disciplined labor force for the newly developed plant products in their Asian and African colonies.

None of these examples matched the depredations of the wild rubber industry of the Amazon basin where a scarcity of labor, the isolation of the tappers, and weak government supervision led to debt coercion in its most extreme forms, often backed by armed guards. The worst abuses were committed against the Indians, by the private armies of rubber barons who were so powerful in the provincial capitals of Iquitos and Manaus that their genocide of upriver Indian tribes was easily hidden. In the frontier districts, terrorism was used to corral a scarce and unwilling labor force, just as Leopold of Belgium used terrorism in the Congo basin, to satisfy a strong new raw material demand and the greed of unregulated capitalism.

Having had the foresight to remove rubber plants to their Asian possessions, the British could afford to play the role of international policemen and expose these labor brutalities. In their Asian possessions both capital and labor could be manipulated without the inconvenience of dealing with a third party, that is, the independent states of Latin America. As the plant transfers of rubber and cinchona show, when the British had the opportunity, as with plants and animals,[2] both moveable as mineral resources are not, they chose formal empire over informal empire as a source of raw materials. Britain could then concentrate its investments in Latin America in those domains where her financial, managerial, and technical skills were most advanced, for example, finance banking, insurance, shipping, railways, and industrial machinery.

Kew's role was much more than smuggling seeds. Through its scientific development of the plants transferred, Kew converted knowledge to profit and power, for the Empire and for the industrial world system of which Britain was then the leader. Kew gave whole-hearted support to its mission, and shared the nation's spirit of crusading imperialism, as is shown by the words of one of its curators, who wrote a history of Kew at the peak of imperial expansion.

> In the industrial development of British colonies and possessions, the Kew man has always been among the earliest workers. As soon as the *pax Britannica* has been established, and often before, he appears. He founds botanic stations where useful plants are grown for distribution, and he gives demonstrations of the best methods of cultivating them. He fostered the tea industry in India and Ceylon; he also started the cultivation of cinchona there; he helped largely in the regeneration of the West Indian Islands, and at the present time Africa is dotted over with stations he is managing, each one a nucleus of what will probably develop into the most important industries of the continent. Often he suffers the fate common to pioneers: he sows that others may reap. Many a Kew man has laid down his life in the conscientious performance of his duty, as genuine a sacrifice to the cause of empire and humanity as any soldier or missionary has ever made. [Bean 1908:68]

Conversely, Kew received the whole-hearted support of the political and commercial establishment, cognizant of the services it was receiving from the botanical gardens. Kew even enjoyed an independence of action unusual for an institution financed by the government. This can be attributed to the general prestige accorded science in the Victorian era, but more specifically to the personal networks of Kew directors. In spite of the adventure of world travel and exploration, their lifestyles at home, and their class values, were properly conservative. But through personal, family, and intellectual friendships they allied themselves with the great scientific innovators of the times, especially with Charles Darwin. And the Darwinians won the battle of minds. Hooker helped Darwin in gestating, publishing, and defending the *Origin of Species* and Darwin helped Hooker in his contretemps with the scientific bureaucracy, appealing straight to the Prime Minister.

Such a small circle could not have comparable power in a democratic society today, though influential cliques exist in science as in other domains. Nor could one family keep the directorship of a major scientific institution for 65 years, as the Hookers did at Kew. Even in Britain, and much more in the United States, the scientific establishment is just too numerous. In 1975, 9,000 scientists served as government advisors under the umbrella of the National Academy of Sciences and 22,700 other individuals served as consultants on government advisory panels (*Chronicle of Higher Education*, September 20, 1976). Out of such numbers diversity must surface and controversy be aired publicly, so that policy may be made by an informed public.

These numbers give some indication of the ever increasing importance of science and technology in the postindustrial world. It is suggested by Paul Goldstene (1977) that science as a potential countervailing force to the corporate business structure is the dominant political reality of the late twentieth century. It is not simply the organized technical knowledge based on systematic research that is vital to the modern corporation, but scientific decisions of a higher order of complexity. Such power carries responsibility.

Science has never operated in a social vacuum, but it has not always had full social awareness or a consciousness of its role and power. Our examples from the nineteenth century show an unquestioning acceptance of the basic tenets of Victorian philosophy, especially as regards imperialism. In the United States from the start of World War II to the late 1960s debate over Vietnam, American science served the government and a corporate structure intent on extending American political, military, and economic hegemony in a Pax Americana, successor to the pre-1914 Pax

Britannica. In some cases, grantism, careerism, and patriotism—or at least an unreflective support of government policy—overrode independent judgment and ethical considerations. In the social sciences, and with special reference to anthropology, this led to involvements, damaging to the profession and to international trust and good will, in such secret enterprises as Project Camelot, a study financed and organized by the Office of Research and Development, United States Army, although a contract with the Special Operations Research Office of the American University, Washington, D.C. served as a cover (Horowitz 1967). The mission of Project Camelot was to develop a social systems model which would enable the Army "to predict and influence politically significant aspects of social change in the developing nations of the world. . . . [and] to devise procedures for assessing the potential for internal war within national societies [from a document released December 4, 1964 by SORO, American University and quoted in Horowitz, 1967:47]." The first target for Project Camelot was Latin America, where social scientists working for the United States Army were to identify and analyze social unrest so that its expression could be forestalled and contained by counterinsurgency. The exposé of Project Camelot and of similar secret research in the Himalayas (Berreman 1969) and in Thailand raised important ethical and policy issues to professional and public scrutiny.

Salutary lessons were learned from these incidents. Among social scientists, there is now a greater concern for the protection and anonymity of informants and closer cooperation with institutions and colleagues in the host countries; among all scientists there is a harder look at the possible hidden biases of agency-directed research. Today there is a more explicit recognition on the part of the scientific community and the public of the impact of science on social policy, especially in the fields of atomic energy, industrial pollutants, and biomedicine. Questions are being asked and thrashed out in public. The public looks to science to resolve the energy crisis by coming up with alternate forms of energy to supplement fossil fuels, but it does not give carte blanche. Substantial segments of the public and of the scientific community, for example, the Union of Concerned Scientists, have challenged government assertions as to the safety of existing atomic energy plants and modes of disposition of atomic wastes. In biomedicine, science is policing itself. In 1974 at the Asomilar International Conference of Molecular Biologists on Recombinant DNA Molecules, a majority of the biologists convened publicly asked their colleagues to postpone further experiments in gene transplantation on animal viruses, especially tumor viruses, until special laboratories with containment facilities could be built to ensure against the escape into the environment of genetically altered bacteria. Research on recombinant

DNA, with all its promises and its hazards, is now subject to NIH safety guidelines, but is still being fiercely debated in the scientific community and by the public (Bennett and Gurin 1977). The new technology of genetic engineering may revolutionize biology, but the work will not go forward without careful evaluation of its dangers.

In a less controversial field, the botanic gardens, including Kew Gardens, are concerning themselves with the conservation of species, the preservation of the environment, and the ecology of the biosphere. In all these research areas, the public detects its self interest.

But the role of science in converting knowledge into power for the core nations of the industrial world system is not as widely recognized, and its implications are resisted. In the postcolonial era this system continues to drain money, talent, and energy from the undeveloped countries in the form of information monopolies, patents, licenses, fees and other rents on technology, gross inequalities between buyer and seller in the tropical commodities markets, and underpaid labor. Resource exploitation of the Third World continues. Robert Merton's theory of cumulative advantage, which he calls the "Matthew effect" after the passage in the Book of Matthew which runs roughly "to him who has shall be given and from him who has not shall be taken away," applies in the larger ecopolitical sphere as well as in the reward system of social and scientific institutions (Merton 1973:447–459). Citibank, a global institution with headquarters in New York, has 5 percent of its assets invested in Brazil but draws 13 percent of its profits from Brazil (*New York Post*, April 13, 1977). Meanwhile, Brazil spends an estimated $2 billion a year, all told, to buy foreign technology (*Veja*, April 13, 1977), and sees its resources developed according to the needs of foreign corporations and the dictates of foreign markets. "We have not had a real transfer of technology, but only the purchase of instructions, projects, and equipment tending to perpetuate a state of dependency," says the Brazilian physicist Dr. Cerquiera Leite, coordinator of a government sponsored research institute which plans to develop and market a homegrown technology suited to Brazil's needs.

Most developing countries are not in a position to generate their own scientific technology or to take a bargaining position in the face of the information monopolies held by the technically advanced countries. If the Third World is to achieve real development instead of modernization, if it is to alter its relationship to the power centers instead of becoming a more profitable area for them to exploit, or so desperately frustrated as to furnish an occasion for another world war, further questions must be asked about the role of scientific technology in the world power system. It is my hope that this historical study of a period only a century ago, when Britain was marshalling all its institutions and scientific expertise toward the imperial goal, and British botanical science energetically furthered British expropriation of the world's plant resources, will add to the modern perspective on the social and political implications of scientific research.

## Notes

1. Crane uses the phrase as the title of a study of the diffusion of knowledge among modern scientists, but it was originally the name of a scientific club which met at Oxford in the 1640s and later became the Royal Society (Ziman 1976:90).

2. During the unrest of the Napoleonic era, both Britain and the United States had been able to obtain from Spain the very valuable merino sheep. The merino breed became the predominant woolbearer in the dominions of South Africa, Australia, New Zealand, and in Argentina, often called the Sixth Dominion in the nineteenth century. By mid-twentieth century, Australia came to have a monopoly on ever perfected merino breeds. I wonder if the British woolen industry did not find it increasingly rewarding to deal with a development-minded Australia rather than the dependence elite and the *compradores* of Argentina, and if the channeling of scientific research funds did not favor Australia over Argentina, and play a part in the diverging prosperity and stability of the two nations.

## References

Adams, Robert McC. 1974. "Anthropological Perspectives in Ancient Trade." *Current Anthropology* 15:239–258.

Bean, William I. 1908. *The Royal Botanic Gardens, Kew: Historical and Descriptive*. London: Cassell.

Bennett, William, and Joel Gurin. 1977. "Science that Frightens Scientists: The Great Debate over DNA." *Atlantic Monthly* 239:43–62.

Berreman, Gerald D. 1969. "Academic Colonialism: Not so Innocent Abroad." *The Nation*. November 10. Reprinted in Thomas Weaver (ed.), *To See Ourselves: Anthropology and Modern Social Issues*. Glenview, Ill.: Scott, Foresman.

Britan, Gerald, and Bette S. Denich. 1976. "Environment and Choice in Rapid Social Change." *American Ethnologist* 3:55–72.

Crane, Diana. 1972. *Invisible Colleges: Diffusion of Knowledge in Scientific Communities*. Chicago: University of Chicago Press.

Goldstene, Paul N. 1977. *The Collapse of Liberal Empire: Science and Revolution in the Twentieth Century*. New Haven: Yale University.

Horowitz, Irving L. 1967. *The Rise and Fall of Project Camelot: Studies in the Relationship between Social Science and Practical Politics*. Cambridge, Mass.: MIT Press.

Irwin, Howard S. 1974. "Energy and Living Systems." *New York Botanical Newsletter* 9:1.

King, Ronald. 1976. *The World of Kew*. London: Macmillan.

Merton, Robert. 1973. *The Sociology of Science*. Chicago: University of Chicago Press.

Vayda, Andrew P., and Roy Rappaport. 1968. "Ecology, Cultural and Noncultural." In J. Clifton (ed.), *Introduction to Cultural Anthropology*. Boston: Houghton-Mifflin.

Ziman, John. 1976. *The Force of Knowledge: The Scientific Dimension of Society*. Cambridge: Cambridge University Press.

JUDITH CARNEY

# 7. Out of Africa

*Colonial Rice History in the Black Atlantic*

> Wade in de water, Wade in de water, children.
> Wade in de water, God's a goin' to trouble de water.
> See dat band all dress'd in red,
> God's a goin' to trouble de water.
> It looks like de band dat Moses led.
> God's a goin' a trouble de water.

THE AFRICAN AMERICAN SPIRITUAL "Wade in the Water" recalls the passage to freedom that the parting of the Red Sea gave Moses and the enslaved Israelites. It also provides a powerful metaphor for examining rice cultivation and its origins in the Americas. Rice is the only grain that demands copious amounts of water. Its caretakers wade through fields of water for its cultivation. Water is also essential for preparing the cereal for consumption. The children of Africa did not wade, as in the spiritual, but were carried in shackles across the troubled waters of the Atlantic slave trade. Enslaved West Africans brought an indigenous knowledge system that would establish rice as a subsistence and plantation crop over a broad region from South Carolina to tropical South America. With them, rice arrived in the Americas in the holds of slave ships, crossing over the ocean grave of the Middle Passage as provisions for its survivors. The cultivation, processing, and preparation of rice reveal a profound knowledge system brought to the Americas by those enslaved from West African rice-growing societies. The importance of rice cuisine to the African diaspora serves even to this day as recipes of memory and cultural identity throughout the black Atlantic.

The African diaspora was one of plants as well as people. Rice figured significantly among the African plants and agricultural systems that shaped environment, food preferences, economies, and cultural identity throughout the era of plantation slavery. During the eighteenth century rice produced by enslaved labor made colonial South Carolina the wealthiest plantation economy in North America. The foundation for its economic prominence rested with West Africans skilled in growing the crop in diverse environments. This chapter illuminates the African origins of rice history in the western Atlantic, emphasizing the technologies and knowledge systems brought to the Americas by enslaved West Africans, especially those long associated with African women's work.

Of Rice and Slaves

Asia has long been associated with the origin of rice, but the grain was also independently domesticated in West Africa. Two key properties of African rice (*Oryza glaberrima*) illuminate why its history is less well known than that of the Asian species (*Oryza sativa*). *Glaberrima*'s yields are lower than *sativa*'s and when milled by machines the grains of African rice break apart. These crucial factors privileged Asian rice in global economic history and favored the diffusion of sativa elsewhere. Higher-yielding Asian rice, however, failed to dislodge glaberrima from its position as the preferred species in West Africa, where sativa was introduced during the period of Atlantic slavery. Demand for rice grew by leaps and bounds in eighteenth-century Europe. The grain was used for brewing beer, and making paper, and it was increasingly favored among middle-class Catholics to accompany fish on meatless Fridays and during Lent.[1] Slaves accompanied the first Europeans arriving in South Carolina from Barbados in 1670. Rice cultivation was well under way in the colony by the 1690s, with the cereal's transition to a plantation crop completed by the early eighteenth century.

Prior to the Civil War, rice was grown within forty miles of the Atlantic coast along the floodplains of sixteen tidal rivers, from the North Carolina–South Carolina border to the Saint Mary's River, which demarcated Florida from Georgia. On the eve of the Civil War nearly one hundred thousand slaves cultivated some seventy thousand acres of tidal floodplain swamps on about five hundred rice plantations.[2]

Until the 1970s the historiography of the colonial rice economy routinely attributed its origins to the ingenuity of white European planters. Accounts of colonial rice history praised the early planters for discovering how to grow an unfamiliar tropical crop in the swamps found along the coastal corridor.[3] This long-standing view changed with the publication of *Black Majority* by the historian Peter Wood in 1974. He argued that the English and French Huguenot planters, who settled the colony, had no experience growing rice. This was not the case with slaves who originated in West Africa's indigenous rice region—a vast area that extends along the coast from Senegal to the Ivory Coast and inland for a thousand miles to Lake Chad. Wood contended that they alone possessed the requisite knowledge, experience, and skills for developing rice cultivation.[4]

Wood's research resulted in a revised view of the African role in shaping wetland landscapes planted to rice. However, questions remained over whether planters recruited slaves from West Africa's rice region to help them develop a crop whose potential they independently recognized or whether African-born slaves initiated rice cultivation in Carolina swamps through their efforts to grow a favored dietary staple for subsistence. The political-ecological analysis presented here seeks to resolve this issue. It reconsiders the way that historians have conceptualized rice. Instead of treating the cereal solely as a commodity consumed and traded, rice is examined as the product of an indigenous knowledge system whose expression in different environments across geographic space was mediated by cultural traditions and power relations. Shifting the perspective on rice from commodity to knowledge system focuses attention on the environments cultivated to the cereal during the era of transatlantic slavery while facilitating recovery of the crop's cultural origins in the American Atlantic.

Archival and historical materials clearly establish the presence of rice in West Africa prior to European maritime voyages as well as in the environments planted in South Carolina and Brazil during the early colonial period. . . .

European accounts reveal the important role of women in the cultivation, processing, and economy of rice. On food purchases by Dutch traders near the Liberian border with Sierra Leone, Samuel Brun noted in 1624 that "for the rice they wanted only glass corals [beads] for their wives, because rice is the ware of women."[5] During the same decade Richard Jobson detailed the role of females in milling rice along the Gambia River: "I am sure there is no woman can be under more servitude, with such great staves wee call Coole-Staves [pestles], beate and cleanse both Rice, all manner of other graine they eate, which is only womens worke, and very painfull."[6] Writing about Sierra Leone in 1678, Jean Barbot added: "The land abounds in millet or white maize [sorghum] and in rice which they have as their main food. The women pound the rice in slightly hollowed tree-trunks."[7] Francis Moore noted in the 1730s that rice was solely a woman's crop in Gambia. "For every Town almost having 2 common Fields of cleared Ground, one for their Corn [millet and sorghum], and the other for the Rice . . . The Men work the Corn Ground and Women and Girls the Rice Ground."[8] He drew attention to an important distinction between floodplain rice systems. On freshwater floodplains rice was a woman's crop. However, mangrove rice depended on the participation of both males and females. Men built the embankments and canals with long-handled shovels, while women performed hoeing and weeding.[9]

These accounts reveal the existence of a fully elaborated rice culture in West Africa. Rice attracted European attention for the surpluses Africans produced, the cereal's availability for purchase, and the peoples skilled in its cultivation. The Portuguese did not introduce rice culture to the region from their voyages to Asia, as historians would later claim.[10] Rain-fed, inland swamp, and tidal rice cultivation long antedated their arrival.

## Antiquity of African Rice Development

African rice (*Oryza glaberrima*) was domesticated along the inland delta of the middle Niger River in Mali. But broader scientific knowledge of West Africa as an independent center of rice origins dates only to the twentieth century. Early Portuguese accounts revealing the extent and significance of African rice cultivation had faded from memory with the conclusion of four centuries of Atlantic slavery and the imposition of colonialism in the late nineteenth century in Africa. Scholars attributed the presence of rice culture in West Africa to the Portuguese, who introduced the cereal from Asia.[11] These views were reassessed when French botanists, working in the region of glaberrima domestication, argued that the grain's unusual characteristics suggested a separate species. Nineteenth-century botanical collections taken from the Upper Guinea coast also revealed specimens sharing similar features and the same red color. By the second half of the twentieth century scientists agreed that rice was independently domesticated in West Africa. Research now places the domestication of African rice between three thousand and forty-five hundred years ago.[12]

## Rice History in South Carolina

Nowhere in the Americas did rice play such an important economic role as in South Carolina. Within a decade of the colony's settlement in 1670, slaves were growing rice for subsistence.[13] By the 1690s the grain was being cultivated for export. Annual exports from South Carolina exceeded sixty-six million pounds on the eve of the American Revolution, making rice the first globally traded cereal.

About a hundred African slaves accompanied the arrival of Europeans to the Carolina colony. The number brought directly from Africa rapidly increased in a short period of time. By 1708 the black population exceeded that of whites.[14] While slaves cultivated the colony's rice and indigo exports, they also grew the subsistence crops consumed by blacks and whites alike. In a pattern similar to the black settlement of the Cape Verde Islands, slaves from West Africa's rice region began planting preferred food staples. Rice became a subsistence crop in the early Carolina settlement period.

There were multiple introductions of rice seed to the colony between 1685 and the early 1690s, both deliberate and casual. Among the earliest types was the "one called Red Rice in Contradistinction to the White, from the Redness of the inner Husk or Rind [bran] of this Sort, tho' they both clean and become white alike."[15] In the early twentieth century, when the southern historian A. S. Salley researched the types of rice initially grown, French botanical research on the African red rice was not broadly known.[16] However, Salley did not address the commonplace knowledge of Thomas Jefferson and his contemporaries—that rice was grown along the West African coast during the era of Atlantic slavery. In attributing the earliest seed introductions to Asia and ship captains arriving from the Orient, Salley remains silent on what commerce brought such ships to South Carolina in the seventeenth century, what African ports of call were visited en route, and the type of cargo they carried to the colony.[17] In his view, rice culture was an outcome of the age of sail, the result of deliberate exchanges of seed between learned and well-traveled gentlemen.

One record from the colony's early settlement period, however, reveals yet another way that seed rice arrived in Charleston, as leftover provisions from a slave ship: "a *Portuguese* vessel arrived, with slaves from the east, with a considerable quantity of rice, being the ship's provision: this rice the *Carolinians* gladly took in exchange for a supply of their own produce. . . . This unexpected cargo was distributed, which gave new spirit to the undertaking, but was not sufficient to supply the demand of all those that would have procured it to plant."[18] Entering the colony as provender from slave ships, this rice originated, along with its human cargo, in Africa. For reasons detailed below, the rice was likely African glaberrima, the source of the red type planted in the early settlement period. For leftover grains to serve as seed rice, as the quotation indicates, the cereal must have crossed the Middle Passage in the husk, that is, unmilled.

Along the rice-growing Guinea coast slave-ship captains regularly bought rice to provision slave ships. Thomas Phillips purchased some five tons of rice in 1693 for his supplies, while the cereal figured in the provisions procured by James B. Barbot in 1699. In 1750 John Newton purchased nearly eight tons of rice to feed the 200 slaves he carried across the Middle Passage. Another captain, John Matthews, estimated that 700 to 1,000 tons of rice would feed the 3,000 to 3,500 slaves he bought in Sierra Leone. For the 250 slaves he carried to Jamaica in 1793, Samuel Gamble purchased more than eighteen tons of rice.[19] Gamble's journal records a demand for both milled and unmilled rice, but purchases of unhusked rice exceeded those of the white milled product, as it was cheaper and served as inexpensive provision for slave ships. Unmilled rice was also less susceptible to moisture spoilage than was its milled counterpart.[20]

Gamble's journal mentions that the unhusked rice was red, a detail that suggests it was African glaberrima.[21]

"Seed rice" refers to rice grains that have not been milled. Rice in the husk therefore could double as germ plasm for planting, if any remained from a slave voyage (as occurred with the Portuguese slaver mentioned above), and there was interest in growing the cereal. Those forlorn passengers of the Middle Passage, already familiar with planting and processing the cereal, became the agents for pioneering its cultivation in the Americas. Through their conscious efforts to grow a dietary staple, they revealed the cereal's commercial potential to those who held them in bondage.

Additional insight into the development of rice culture in the western Atlantic is gained by examining the cereal's milling. Preparing rice for human consumption requires removal of the indigestible husks or hulls. Until the advent of mechanical devices in the late eighteenth century, rice processing depended on hand-milling the grains in an upright hollowed-out cylinder carved from a tree trunk. It involved striking the grains with a handheld pestle to remove the hulls and minimize grain breakage. Milling, like food preparation more generally, was traditionally the work of African women.[22] Slave ships replicated existing cultural practices by relying on enslaved females to process and cook the food for all the slaves on board.

One journal entry from the slave ship *Mary* (outbound from Senegal), dated Monday, 19 June 1796, noted enslaved females at work in food preparation: "Men [crew] Emp[loye]d tending Slaves and Sundry Necessaries Jobs about the Ship. . . . The Women Cleaning Rice and Grinding Corn for corn cakes."[23] "Cleaning" rice refers to its milling. The reference suggests that the ship purchased unmilled rice that required processing on board. More detail is provided in a doctor's report from 1795 or 1796 of a visit aboard a slave ship recently arrived from West Africa's rice region. Observing what the slaves were eating, Dr. George Pinckard wrote:

> their food is chiefly rice which they prepare by plain and simple boiling. . . .
> We saw several of them employed in beating the red husks off the rice which
> was done by pounding the grain in wooden mortars, with wooden pestles
> sufficiently long to allow them to stand upright while beating in mortars
> placed at their feet. . . . They beat the pestle in time to the song and seemed
> happy; yet nothing of industry marked their toil, for the pounding was per-
> formed by indolently raising the pestle and then leaving it fall by its own
> weight.[24]

Providing another instance of rice as the ship's provision, Pinckard's commentary also reveals the use of mortars and pestles aboard slave ships. Their presence suggests that the rice was purchased in the husk, that slavers relied on their enslaved captives to mill the cereal, and that it was processed with a mortar and pestle—the only way to mill the African species.[25] While gender is not explicit in Pinckard's account of rice processing, the role of enslaved women in beating and pounding rice aboard slave ships becomes clearer in Henry Smeathman's account from the 1770s: "Alas! What a scene of misery and distress is a full slaved ship in the rains. The clanking of chains, the groans of the sick and the stench of the whole is scarce supportable . . . two or three slaves thrown overboard every other day dying of fever, flux, measles, worms all together. All the day the chains rattling or the sound of the armourer rivetting [*sic*] some poor devil just arrived in galling heavy irons. The women slaves in one part beating rice in mortars to cleanse it for cooking."[26]

On slave ships wooden fences or barricades separated women and children from male slaves.[27] Females on a ship were usually placed aft in the quarterdeck near the lodgings of the ship's officers.[28] To minimize the potential for sabotage, slavers relocated the ship's galley from the hold to the deck, typically placing the cast-iron cooking hearth in the area reserved for enslaved females. A door in the barricade facilitated the passage of the crew and food to male slaves at mealtimes when they were allowed on deck. . . .

Wherever blacks were forcibly settled in the Americas, the African mortars and pestles were used to prepare food. The device played a crucial role in the evolution of the Carolina rice economy since prior to the development of mechanical mills in the mid-eighteenth century it was the only means to process the cereal without breakage. A milling technology brought to the Americas by enslaved females thus enabled the Carolina rice economy to realize global prominence.

Although Asian rice eventually captured plantation fields, other types were grown for subsistence. Well into the antebellum period slaves were reported planting "Guinea rice" in their dooryard gardens, along with other African plant domesticates such as "Guinea corn" (sorghum), okra, black-eyed peas, and watermelons.[29] Choice of the toponym "Guinea" for both rice and sorghum suggests their African provenance. The deliberate cultivation of African rice also undoubtedly reflected taste preferences, as one slave-ship captain expressed in 1828: "African rice has more taste and solidity than the Carolina rice, although it is not so white."[30] In cultivating African seeds and food technologies, the enslaved transformed dooryard gardens and provision fields into botanical gardens of the dispossessed. Their efforts paralleled the crop experimentation and seed exchanges carried out by Euro-American scientific societies in the same era. In this manner, Africans and their descendants profoundly shaped the agricultural systems and foodways of the Americas . . .

## Conclusion

Rice cultivation accompanied the forced settlement of African slaves to the western Atlantic throughout the early modern period. In another prominent wetland area of the Americas—near Tabasco along Mexico's Gulf Coast—a Spanish land grantee noted as early as 1579 the cultivation of rice in provision gardens by Africans enslaved for tobacco production.[31] Fugitive slaves also planted the grain in the maroon communities they established in northeastern South America.[32] Rice culture accompanied slavery throughout the Americas—for subsistence in plantation provision fields, export as a plantation crop from South Carolina and Maranhão, and as a food staple in maroon communities. Rice remains the signature cereal of the African diaspora -in the Hoppin' John of the Carolina kitchen, Louisiana gumbo (with African okra), *moros y cristianos* (rice and beans) of Cuba, and *gallo pinto* of Nicaragua.

The history of rice in the Americas reveals an important West African legacy. The grain's arrival in the American Atlantic and role in plantation societies represent a remarkable form of technology transfer from West Africa under conditions of forced labor that are difficult to imagine. We are just beginning to explore the plants and knowledge systems that facilitated survival and cultural identity across the troubled areas of the black Atlantic.

## Notes

I would like to thank Starr Douglas, at Royal Holloway, University of London, for providing the Smeathman quote from her research in the Swedish archives. Additional appreciation is extended to Bruce Mouser, Tony Tibbles, Marlène Elias, Philippe Pétout, Kåre Lauring, and the volume editors.

1. Henry Dethloff, *A History of the American Rice Industry, 1685–1985* (College Station: Texas A&M University Press, 1988).

2. James Clifton, "The Rice Industry in Colonial America," *Agricultural History* 55 (1981): 266–83; Peter Coclanis, *The Shadow of a Dream* (New York: Oxford University Press, 1989).

3. David Doar, *Rice and Rice Planting in the South Carolina Low Country* (1936; Charleston, S.C.: Charleston Museum, 1970).

4. Peter H. Wood, *Black Majority* (New York: Norton, 1974), 57–58.

5. George Brooks, *Landlords and Strangers: Society and Trade in Western Africa, 1000–1630* (Boulder, Colo.: Westview Press, 1993), 318.

6. Richard Jobson, *The Golden Trade* (1623; Devonshire, U.K.: Speight and Walpole, 1904), 68.

7. P. E. H. Hair, Adam Jones, and Robin Law, eds., *Barbot on Guinea: The Writings of Jean Barbot on West Africa, 1678–1712*, 2 vols. (London: The Hakluyt Society, 1992), vol. 1, 186.

8. Francis Moore, *Travels into the Inland Parts of Africa* (London: Edward Cave, 1783), 127.

9. Daniel C. Littlefield, *Rice and Slaves* (Baton Rouge: Louisiana State University Press, 1981).

10. O. Ribeiro, *Aspectos e problemas da Expansão Portuguesa* (Lisbon: Estudos de Ciencias Politicas e Sociais, 1962).

11. Ibid.

12. The sole archaeological excavation on *glaberrima* shows the cereal present between 300 B.C. and 300 A.D. in Mali. Older dates derive from research in botany and historical linguistics. See Susan K. McIntosh, "Paleobotanical and Human Osteological Remains," in *Excavations at Jenne-jeno, Hambarketolo and Kaniana in the Inland Niger Delta (Mali)*, ed. S. K. McIntosh (Berkeley: University of California Press, 1994), 348–53; Roland Portères, "African Cereals: Eleusine, Fonio, Black Fonio, Teff, Brachiaria, Paspalum, Pennisetum, and African Rice," in *Origins of African Plant Domestication*, ed. J. Harlan, J. De Wet, and A. Stemler (The Hague: Mouton, 1976), 409–52; Christopher Ehret, *The Civilizations of Africa: A History to 1800* (Charlottesville: University Press of Virginia, 2002).

13. Wood, *Black Majority*, 57–58.

14. Ibid., 25–26, 36, 62, 143–45.

15. A. S. Salley, "Introduction of Rice into South Carolina," in *Bulletin of the Historical Commission of South Carolina*, vol. 6 (Columbia, S.C.: The State Company, 1919), 11.

16. While *glaberrima* rice is always red in color, some *sativa* rice varieties also are.

17. Judith Carney, *Black Rice: The African Origins of Rice Cultivation in the Americas* (Cambridge: Harvard University Press, 2002), 147–52.

18. P. Collinson, "Of the Introduction of Rice and Tar in Our Colonies," *Gentleman's Magazine* 36 (June 1766): 278–80.

19. Hair et al., *Barbot on Guinea*, vol. 1, 282 n. 13, vol. 2, 681; G. Dow, *Slave Ships and Sailing* (Salem: Marine Research Society, 1927), 57, 73; Boubacar Barry, *Senegambia and the Atlantic Slave Trade* (Cambridge: Cambridge University Press, 1998), 79, 107–8, 117–18; Bruce Mouser, ed., *Slaving Voyage to Africa and Jamaica: The Log of the Sundown, 1793–1794* (Bloomington: Indiana University Press, 2002), 90 n. 295.

20. Bruce L. Mouser, "Who and Where Were the Baga? European Perceptions from 1793 to 1821," *History in Africa* 29 (2002): 337–64, esp. 357 n. 42.

21. Mouser, *Slaving Voyage*, 45 n. 170, 86 n. 282, 99 n. 317; Mouser, "Who and Where," esp. 357 n. 42.

22. Carney, *Black Rice*, 25–27, 48–52.

23. Elizabeth Donnan, *Documents Illustrative of the History of the Slave Trade to America,* 4 vols. (1930–35) (Washington, D.C.: Carnegie Institution, 1932), vol. 3, 121, 376.

24. Dow, *Slave Ships,* xxiii–xxiv.

25. African rice shatters with mechanical milling. See National Research Council, *Lost Crops of Africa* (Washington, D.C.: National Academy of Science, 1996).

26. Smeathman to Drury, Sierra Leone, 10 July 1773, MS D.26, Uppsala University, Sweden.

27. Mouser, "Baga," 8 n. 21; George Howe, "Last Slave Ship," *Scribner's Magazine* 8, no. 1 (July 1890): 113-29; Dow, *Slave Ships,* xxii.

28. Robert Harms, *The Diligent: A Voyage through the Worlds of the Slave Trade* (New Haven: Yale University Press, 2002), 311–12.

29. John Drayton, *A View of South Carolina* (1802; Columbia: University of South Carolina Press, 1972), 125.

30. Theophilus Conneau, *A Slaver's Log Book or 20 Years' Residence in Africa* (Englewood Cliffs, N.J.: Prentice-Hall, 1976), 239.

31. "Relaciones de Yucatan," in R. C. West, N. P. Psuty, and B. G. Thom, *Las Tierras Bajas de Tabaso* (Villahermosa: Gobierno del Estado de Tabasco, 1987), 316.

32. Richard Price and Sally Price, eds., *Stedman's Surinam* (Baltimore: Johns Hopkins University Press, 1992), 208–19.

# II. Other Cultures' Sciences

Columbus's attitude with regard to the Indians is based on his perception of them. We can distinguish here two component parts. . . . Either he conceives the Indians . . . as human beings altogether, having the same rights as himself; but then he sees them not only as equals but also as identical, and this behavior leads to assimilationism, the projection of his own values on the others. Or else he starts from the difference, but the latter is immediately translated into terms of superiority and inferiority. . . . What is denied is the existence of a human substance truly other, something capable of being not merely an imperfect state of oneself. These two elementary figures of the experience of alterity are both grounded in egocentrism, in the identification of our own values with values in general, of our I with the universe—in the conviction that the world is one.
—TZVETAN TODOROV, *The Conquest of America*

The discovery of primitiveness was an ambiguous invention of a history incapable of facing its own double.—V. Y. MUDIMBE, *The Invention of Africa*

In Argentina, when the dominant political system faces dissent, it responds by making the dissidents disappear. The "desparacidos" or the disappeared dissidents share the fate of local knowledge systems throughout the world, which have been conquered through the politics of disappearance, not the politics of debate and dialogue. . . . The "scientific" label assigns a kind of sacredness or social immunity to the western system. By elevating itself above society and other knowledge systems and by simultaneously excluding other knowledge systems from the domain of reliable and systematic knowledge, the dominant system creates its exclusive monopoly. Paradoxically, it is the knowledge systems which are considered most open, that are, in reality, closed to scrutiny and evaluation. —VANDANA SHIVA, *Monocultures of the Mind*

Indigenous post-colonial critique asks: Who asks such questions as "Are indigenous beliefs epistemologically justified?" and "Are indigenous peoples rational?," and why do they ask them? Who bears the burden of proof? Who is in the position to demand self-validation, and from whom is self-validation being demanded? Whose concepts, standards, rules, and criteria govern this exchange? . . . Is epistemology, like history, to be written by the victor?
—JAMES MAFFIE, " 'In the end, we have the Gatling Gun, and they have not': Future Prospects of Indigenous Knowledges"

To see from below is neither easily learned nor unproblematic. . . . The positionings of the subjugated are not exempt from critical reexamination, decoding, deconstruction, and interpretation; that is, from both semiological and hermeneutic modes of critical inquiry. The standpoints of the subjugated are not

"innocent" positions. On the contrary, they are preferred because in principle they are least likely to allow denial of the critical and interpretive core of all knowledge. . . . The subjugated have a decent chance to be on to the god trick and all its dazzling—and, therefore, blinding—illuminations. "Subjugated" standpoints are preferred because they seem to promise more adequate, sustained, objective, transforming accounts of the world.
—DONNA HARAWAY, "Situated Knowledges"

The consideration of other cultures' knowledge systems raises many controversial issues, both for Westerners and also for those other cultures. Who is indigenous and what counts as indigenous knowledge? Is the category itself part of a Eurocentric othering of other cultures? How should Westerners understand the effectiveness of the unfamiliar logics of other cultures' knowledge systems, with their magical thinking, anthropomorphism, and spiritual ontologies as well as ceremonial and ritualistic methodologies? (Maffie 2009). Moreover, are such features really absent from modern European sciences? Why should people who have access to Western sciences' effective ways of interacting with natural and social orders value other cultures' knowledge systems at all? What could and should be the theoretical and practical relations between modern Western sciences and other cultures' knowledge systems?

The assumptions supporting the West's exceptionalist and triumphalist perspectives have come to seem less and less justifiable, as the book introduction pointed out. Moreover, many cultures—as well as international and national governmental and nongovernmental groups around the globe, including in the West—increasingly defend non-Western knowledge systems (see also Sardar, Goonatilake, and Hoppers in part IV). How should Western philosophies of science conceptualize and treat these sciences of other cultures?

## Benefits

Why do other cultures want these sciences that have been so devalued in the West? And why should anyone with access to Western sciences value them? First, such knowledge systems have continually been tested and adapted to changing natural and social environments, enabling those cultures to interact effectively with their environments for their entire histories—that is, for many centuries and even millennia. These sciences tend to be empirically reliable in their relevant natural and social environments (see, e.g., Goodenough, Colin Scott, Appleton et al.). James Maffie (2009) points out that the immense power of Western expansion may have politically defeated the legitimacy of other cultures' technological achievements, but that does not mean that such legitimacy has been disproved.

Moreover, many poor people in other societies do not have access to Western medicine, pharmacology, food production, or manufacturing. For them, traditional knowledge keeps them alive. Their physical survival today depends on being able to take care of their own needs for food, housing, clothing and household supplies, travel, health, medical and pharmacological practices, and protection from often dangerous and increasingly impoverished environments. Few of the more privileged of the globe's inhabitants could survive such environments with only the meager resources that modern scientific rationality and technical expertise would provide, as Goonatilake points out in part IV (Appleton et al., Escobar, Colin Scott).

Equally important to other cultures, such knowledge systems are co-constituted with these cultures' intellectual and ethical legacies, just as are our Western sciences, as the field of science studies has recently documented. They are deeply intertwined with the ways cultures are organized and structured and with their members' cultural identities. It is not just that cultures influence already existing sciences (although that also occurs); rather, social formations and scientific institutions and practices come into mutually supportive existence together (see, e.g., Hess 1995; Shapin and Schaffer 1985; Jasanoff 2004). Indeed, even historical forms of regulative ideals such as objectivity are coconstituted with historically specific kinds of ethical social identities (Daston and Galison 2007). Thus other peoples' defenses of their own knowledge systems also are a valuable political resource for their struggles to protect their cultures from destruction by continued Northern economic, political, and cultural expansion (Escobar, Hayden, Sardar, Colin Scott).

Next, learning about other cultures' knowledge systems exposes Westerners to unfamiliar logics of nature's order and of research. To understand how a Pacific navigator conceptualizes his five-thousand-mile trip in an open canoe from the Carolina Islands to New Zealand and back is not easy (Goodenough). Moreover, grasping such unfamiliar logics reveals the sophisticated abstractions and theoretical frameworks invented by these supposedly primitive societies. Modern scientists are loath to respect the practice of anthropomorphizing geese behavior—of treating geese as fully persons, ones with whom humans reciprocally communicate. Yet Colin Scott shows that doing so provides crucial resources for sustaining the geese and their needed environments as a source of food. Ancient civilizations in the Middle East and Asia also provide many examples of such unfamiliar but effective logics (see, e.g., Needham 1956–2004; Sardar; Selin 2007). Colin Scott points out that the "root metaphors" of our own sciences become implicit and invisible to us, while we immediately identify other cultures' metaphors. It can come as a shock to discover that our own claims that we took to be completely factual statements are part of powerful suppressed metaphors of our own—for example, those of nature as a machine, our planet as the center of the universe ("The sun rose this morning at 7:32"), the earth as a great living body ("The miners have found a new vein of coal"), or of science as a search for religious or moral salvation—that we thought we had left behind (Hesse 1966). Thus, coming to see the world around us through others' interests, values, and practices suggests that it can be rational to think in terms of a world of multiple effective scientific rationalities.

To begin to grasp the benefits of the historical values and interests that have shaped all cultures' sciences may appear to adopt an epistemological relativism that has been avoided only through rigorous commitments to value-neutral objectivity and universal reason "operationalized" through good method. Yet it is clear that such an exploration of our own philosophies of science identifies more of the critical resources that Western sciences have insisted that they value for rigorously examining our own everyday and scientific assumptions and beliefs. In other words, gaining a more reliable understanding of others enables us better to understand ourselves. To put the point another way, this kind of argument strengthens the regulative ideals to which the modern Western philosophy of science legacy is so committed: through recognizing the distinctive strengths and limitations of other cultures' knowledge systems, one can strengthen the objectivity, rationality, and "good method" of modernity's own knowledge systems. And, as we will see in part IV, we can explore the virtues of "polycentric epistemologies" that value "a world of sciences."

In addition, as indicated earlier, Westerners can learn from other cultures' knowledge systems new facts about nature and social relations. Other cultures have asked different kinds of questions about different kinds of environments, within different discursive structures, and with different methods than have been possible for Westerners, as every essay in this part reveals. Many Western institutions have long understood how valuable it is to them that we live in a world of sciences, though they only implicitly acknowledge this. Commercial enterprises, from colonial botany to today's big pharma, have always understood how valuable these other knowledge systems are, and they have actively sought to turn native informants' knowledge into the kind of culturally anonymous information that Western corporations can use, buy, and sell around the globe (Escobar, Hayden, Schiebinger).

Moreover, through studies of other cultures' knowledge systems, one can begin to understand the complex relations between cultural, linguistic, and biological diversity. Protests against the destruction of biodiversity created by agricultural monoculture practices, as well as by the multitude of industrial, military, and other mostly unintentional environmentally destructive practices, have drawn attention to the necessity of biodiversity for human survival. Yet until recently, little attention has focused on the relations between biodiversity, on the one hand, and cultural and linguistic diversity, on the other (Brush; Maffi 2001; Mühlhäusler). This work raises many questions. Is it surprising to discover that the areas of the globe experiencing loss of languages and cultures map closely onto the areas where biodiversity is most rapidly decreasing? What causes this startling relation between language, culture, and environment? Moreover, is it only the indigenous societies that should be assigned the responsibility for reversing such losses? If not, what are appropriate roles in such a project for Western societies? And how should indigenous knowledge be preserved and nourished? For example, gene and data banks take the aspects that interest Westerners out of the cultural contexts that produced them and gave them meaning, thereby eliminating their indigenous character and even putting them out of reach of indigenous peoples. And on top of this, the indigenous societies themselves are rapidly disappearing (Agrawal 1995; Brush, Goonatilake, Hayden, Hoppers, Mühlhäusler).

Finally, learning about other cultures' knowledge systems focuses attention on the need for the West to develop new research protocols for studying other societies. What are the ethics and politics of extracting information from already politically and economically weakened cultures, and of the very focus on "studying others" (Smith 1999; Denzin, Lincoln, and Smith 2008)? On the other hand, it cannot be desirable for Westerners to fail to grasp the astonishing diversity of fruitful human interactions with social and natural environments. Readings in this part and the next two suggest possible transformations of institutions, their cultures and practices, that could begin to move toward ethically and politically more desirable interactions between research agendas and the needs of cultures to protect their material resources, cultural knowledge, and expertise.

## Challenges

There are at least two more challenges in studying other cultures' knowledge systems in addition to those raised earlier and in part I. How should gender issues be addressed in these contexts? (See the book introduction.) In this part, the selection by Appleton and her coauthors identifies the costs when international agencies ignore women's knowledge; but are there other gender issues that other essays also overlook?

Another issue is that fundamentalisms, whether Hindu, Christian, or some other, often legitimize conservative resistances to dominant forms of Western culture through appeal to traditional knowledge systems (see Sardar). If one acknowledges the value of the non-Western knowledge systems defended in these essays, must one also similarly value creationism and intelligent design? Historians and the U.S. courts have balked at recognizing these ways of understanding nature's order as scientific today, while pointing out that in the past they were, indeed, valuable guides to the immensely complicated world within and around us. Before Darwin, every educated European believed that the Christian God created the universe in seven days, just as the Bible said. No one questioned this at all. Moreover, all the early modern scientists thought that one should do the new experimental methods to understand God's mind in even greater detail (Jacob 1988; Marks 2007). The new "natural theologies" proposed that the world is like a great big machine, with phenomena along the lines of levers, pulleys, and clocks making it work as well as it does. So they reasoned that since artifacts such as these machines had a designer and maker, then nature, too, must have one, namely, the Christian God of the Bible.

However, in fact nature is not a machine. While it is *like* a machine in some respects, it is not at all like one in other respects. For example, many of the peculiar regularities of human thinking and feeling cannot be completely understood through such mechanistic models (in spite of the undoubted successes of artificial intelligence scientists at duplicating many aspects of human thinking, such as successful chess moves). Once Darwin's evolutionary theory so spectacularly succeeded in explaining data that remained incomprehensible within the earlier theological metaphors, creationism and intelligent design were left with both inexplicable data and no further record of advancing scientific knowledge. As the historian Jonathan Marks explains it, "It's not that ID is demonstrably wrong—it is just anachronistic" (2007, 94). Indeed, Marks goes on to argue that "there is no scientific theory of intelligent design. It is a legal strategy not a scientific theory" (97).

But what about indigenous knowledge systems today? Insofar as they do explain aspects of nature's regularities and underlying causal tendencies that modern Western sciences have not yet considered, and use conceptual frameworks that reveal more than modern Western ones have so far been able to understand (such as how distinctive cultural practices both preserve biodiversity and enable successful interactions with environments), they already have scientific value, as a number of the essays here argue. Moreover, cultural diversity in knowledge systems appears itself to provide scientific benefits insofar as cultures are reservoirs of new questions about nature and motives for asking them, as well as of innovative methods for exploring and of representations of nature and human interactions with our natural and social surrounds.

## References

Agrawal, Arun. 1995. "Indigenous and Scientific Knowledge: Some Critical Comments." *Development and Change* 26:413–39.

Daston, Lorraine, and Peter Galison. 2007. *Objectivity.* New York: Zone.

Denzin, Norman K., Yvonna S. Lincoln, and Linda Tuhiwai Smith, eds. 2008. *Handbook of Critical and Indigenous Methodologies.* Thousand Oaks, Calif.: Sage.

Haraway, Donna. 1991. "Situated Knowledges: The Science Question in Feminism and the Privilege of Partial Perspectives." In *Simians, Cyborgs, and Women: The Reinvention of Nature.* New York: Routledge.

Harding, Sandra. 1998. *Is Science Multicultural?* Bloomington: Indiana University Press.

Hess, David J. 1995. *Science and Technology in a Multicultural World.* New York: Columbia University Press.

Hesse, Mary. 1966. *Models and Analogies in Science.* Notre Dame: University of Notre Dame Press.

Jacob, Margaret. 1988. *The Cultural Meanings of the Scientific Revolution.* New York: Knopf.

Jasanoff, Sheila, ed. 2004. *States of Knowledge: The Co-production of Science and the Social Order.* New York: Routledge.

Latour, Bruno. 1988. *The Pasteurization of France.* Cambridge: Harvard University Press.

Maffi, Luisa, ed. 2001. *On Biocultural Diversity: Linking Language, Knowledge, and the Environment.* Washington: Smithsonian Institution Press.

Maffie, James. 2003. "To Walk in Balance: An Encounter between Contemporary Western Science and Conquest-Era Nahua Philosophy." In *Science and Other Cultures*, ed. Robert Figueroa and Sandra Harding. New York: Routledge.

———. 2009. " 'In the end, we have the Gatling Gun, and they have not': Future Prospects of Indigenous Knowledges." *Futures* 41:53–65.

Marks, Jonathan. 2007. "Intelligent Design and the Native's Point of View (Assuming the Native Is an Educated Eighteenth-Century European)." In *Darwin and the Bible: The Cultural Confrontation.* Boston: Pearson Education.

Mudimbe, V. Y. 1988. *The Invention of Africa.* Bloomington: Indiana University Press.

Needham, Joseph. 1956–2004. *Science and Civilisation in China.* 7 vols. Cambridge: Cambridge University Press.

———. 1969. "The Laws of Man and the Laws of Nature." In *The Grand Titration: Science and Society in East and West.* Toronto: University of Toronto Press.

Selin, Helaine. 2007. *Encyclopedia of the History of Science, Technology, and Medicine in Non-Western Cultures.* 2nd ed. Dordrecht: Kluwer.

Shapin, Steven, and Simon Schaffer. 1985. *Leviathan and the Air Pump.* Princeton: Princeton University Press.

Shiva, Vandana. 1989. *Staying Alive: Women, Ecology, and Development.* London: Zed.

———. 1993. *Monocultures of the Mind: Perspectives on Biodiversity and Biotechnology.* New York: Zed Books.

Smith, Linda Tuhiwai. 1999. *Decolonizing Methodology: Research and Indigenous Peoples.* New York: St. Martin's Press.

Third World Network. 1993. "Modern Science in Crisis: A Third World Response." In *The "Racial" Economy of Science*, ed. Sandra Harding. Bloomington: Indiana University Press.

Todorov, Tzvetan. 1984. *The Conquest of America: The Question of the Other.* New York: Harper and Row.

Watson-Verran, Helen, and David Turnbull. 1995. "Science and Other Indigenous Knowledge Systems." In *Handbook of Science and Technology Studies*, ed. Sheila Jasanoff et al. Thousand Oaks, Calif.: Sage.

WARD H. GOODENOUGH

# 8. Navigation in the Western Carolines

## A Traditional Science

THE ATOLL DWELLERS OF MICRONESIA are ocean voyagers, unlike the inhabitants of Micronesia's few high islands. In the atolls, knowledge of how to build seaworthy sailing canoes and how to navigate from tiny place to tiny place over fairly long distances of open sea has been actively maintained for centuries as vital to successful life. Somewhat different systems of navigation are used in each of Micronesia's three atoll regions: Kiribati (formerly Gilbert Islands), the Marshall Islands, and the Western Caroline Islands (map 1).

Best known to students of Micronesia's cultures is the system used by navigators in the Western Caroline Islands—the great chain of atolls that lies between Pohnpei (formerly Ponape) in the east and Yap and Belau (formerly Palau) in the west, a distance of over a thousand miles. Throughout this region the same basic system has been in use, with local schools differing only in ways that are not crucial to successful voyaging. Navigators, moreover, all of whom are men, have to observe food taboos that result in their eating separately prepared food in their boathouses. Visiting navigators from other atolls dine with them there, so there is opportunity for exchanges of information and displays of knowledge. Such exchange is facilitated by there being a chain of closely related dialects, neighboring ones being mutually intelligible (or very nearly so) over the entire area. Navigation has been developed here, even in the absence of writing, to a high degree. Like any practical science, its application is an art, requiring both knowledge and skill.[1]

## The Star Structure

Fundamental to the system of navigation is the "star structure" (paafúú), as it is called. Seen near the equator, stars appear to rotate across the heavens from east to west on a north-south axis. Some rise and set farther to the north and others farther to the south, and they do so in succession at different times. The "star structure" divides the great circle of the horizon into thirty-two points. Polaris, just visible, marks north, and the Southern-Cross in upright position marks south. Other points, while conceptually equidistant, are named for the rising and setting of stars, whose actual azimuths of rising and setting are not equally spaced. Frake

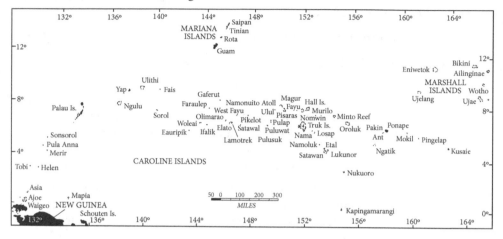

MAP 1  Carolinian navigator's world.

(1994) has observed, therefore, that the stars do not physically mark the points in the "star structure"; rather, they name them.

These thirty-two points, like the points on the European wind rose, form a conceptual compass and serve as the directional points of reference for organizing all directional information about winds, currents, ocean swells, and the relative positions of islands, shoals, reefs, and other seamarks. Every point has another that is conceptually diametrically opposite to it. These diametrical opposites are seen as passing through a point at the center of the compass, and a navigator thinks of himself or any place from which he is determining directions as at this central point, just as western navigators do when using a magnetic compass. Thus whatever point a navigator faces, there is a reciprocal point at his back.

Although navigators represent the thirty-two points of the sidereal compass as equidistant for instructional purposes, there seems to be recognition that the stars that name them are not evenly spaced at their points of rising and setting. Beta and Gamma Aquilae, which are very close to Altair, are omitted from some exercises, the compass being reduced in them to twenty-eight points. To use the stars effectively as directional guides requires either that sailing directions reflect empirical observation or that navigators take account of the difference between the observed position of a star and the position of the point on the compass to which it gives its name. On the basis of the evidence so far available, it seems that the former alternative is more probable. In the absence of writing and accurate maps, the discrepancies between the system as ideally represented and the sailing directions as empirically established are

SETTING                    RISING

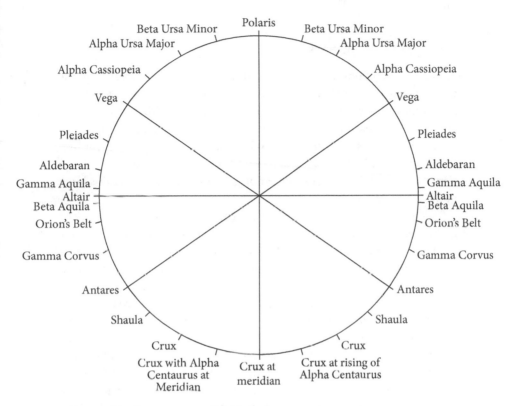

FIGURE 1 The "star structure" (sidereal compass).

not likely to be a matter of concern. The "star structure" is represented in Figure 1 as a compromise between the ideal and the real, showing points as diametrical opposites but bringing them as close as allowed to their actual degrees of azimuth at rising and setting.

A student first learns the compass points. Then he learns their reciprocals. For every pair of reciprocals he learns what other reciprocal pair lies at right angle to it. A compass star on one's beam can thus serve as a guide when the star on which one's course is set is not visible. Navigators develop a feel for the angular distances from one to another of all the points on the compass, just as we develop a feel for the angular distance of numbers on a clock face in describing directions. This feel enables them to maintain a course at the appropriate angle to any visible compass star or other star that "follows the same path" as a compass star. They can do the same thing with reference to any other phenomenon, such as an ocean swell or seasonally prevailing wind, whose compass direction is known.

## Sailing Direction Exercises

All sailing directions are memorized in relation to the sidereal compass or "star structure." So are the relative locations of all places of interest, including such seamarks as reefs, shoals, and regions with distinctive marine life. After learning the "star structure," including its pattern of diametrical opposites, students begin to learn all this information regarding the relative locations of islands and sea marks. Navigators have developed a variety of exercises as aids in memorizing and remembering this large body of information.

The most important set of exercises (figure 2), called "Island Looking" (*woofanú*), takes all the known islands as points of reference and for each one the student goes around the compass naming the nearest places that lie along each radius (or very close to it). Navigators say that "Island Looking" is fundamental in their repertoire of knowledge.

Another exercise requires giving the names of all the "sea roads" (*yelán metaw*) or "sea directions" (*yitimetaw*) between various islands and reefs along with the reciprocal star directions on which they lie; and yet another requires naming the "sea brothers" (*pwiipwiimetaw*), the roads that lie on the same reciprocal star directions. An exercise called "Breadfruit Picker Lashing" (*fééyiyah*) uses as metaphor the breadfruit-picking pole (*yiyah*). In this exercise, one imagines reaching out with the pole along a given star course and picking off in succession all the islands and reefs that lie along it to the end of what is known, then turning and doing the same from that point along another course to its end, and so on until one has picked off all the places, real or imagined, in one's repertoire. Two other exercises, "Reef Hole Probing" (*yaaruwóów*) and "Sea-bass Groping" (*rééyaliy*), involve chasing a fish from island to island, each one cryptically identified by the name of a hole in one of its reefs to which the fish goes. These reef hole names provide a set of esoteric names, known only to navigators, with which they can discuss sailing among themselves without others present knowing what they are talking about.

In all such exercises the navigator follows a course from his home island to the one from which the exercise begins and then proceeds according to a set pattern from one place to another. The pattern may be to box the compass, as in "Island Looking"; to go in a series of zigzags, as in "Reef Hole Probing"; or to follow a main course northward, go east or

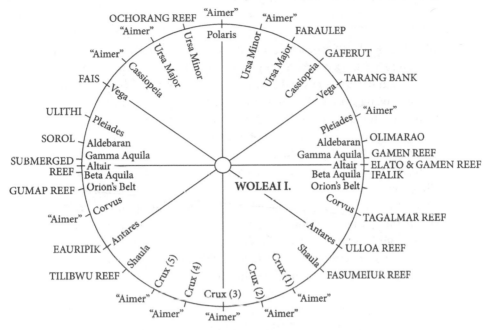

FIGURE 2 "Island looking" exercise, naming places and "aimers" (living sea marks) as one looks out from Woleai Island.

west from it, and then back at a series of points along it, as in an exercise called "Sailing of the Red Snapper" (*herákinimahacca*).

## Living Seamarks

"Aimers" or "aligners" (*yepar*), as the Carolinians call them, are living sea-marks (*pwukof*) associated with particular areas in the vicinity of islands or midway between them. They consist of such things as a tan shark making lazy movements, a ray with a red spot behind the eyes, a lone noisy bird, a swimming swordfish, and so on. Each has its own name and is to be found in a particular drag on a particular star course from its associated island, often on a course along which no island lies. No one sails to find them; rather, one hopes to encounter one of them when one is lost. They serve as a last hope for the navigator who has missed his landfall or lost his bearing, enabling him, if he is lucky enough to encounter one, to align himself once more in the island world. When doing "Island Looking" exercises, advanced students include these "aimers" among the locations to be named in boxing the compass from a given island.

Keeping Track

A major problem for dead reckoning sailors is to estimate distance traveled and keep track of where they are in regard to their course. To do this,
Carolinian navigators use what they call "dragging" or "drags" (*yeták*). It
involves using a place other than one's destination as a point of reference.
If you are traveling from Boston to New York, for example, Albany as
point of reference lies to the west of Boston at the beginning of the journey
and north of New York at the end of it. As you travel from Boston to New
York, Albany moves in relation to where you are through several compass
directions from west, to west northwest, to northwest, to north northwest, to north. The intervals between these changing compass directions
divide the journey into four legs, as we would call them. In just this way,
the Carolinians see the place of reference as being "dragged" through the
intervening directions of their sidereal compass as a voyage progresses.
The number of direction intervals through which the place of reference is
"dragged" comprises the number of "drags" (legs) in the voyage.

Thus the course from Puluwat to Tol in western Truk, 160 miles away, is
almost directly east on the "rising of Altair" (map 2). Pisaras lies 120 miles
northeast of Puluwat on the rising of Vega and a like distance from Tol
on the setting of Vega. During the voyage from Puluwat to Tol, Pisaras
is "dragged" from the rising of Vega through the rising of Cassopeia, the
rising of the main star in Ursa Major, the rising of Kochab in Ursa Minor,
Polaris, and on through the settings of Kochap, Alpha Ursa Major, and
Cassopeia to the setting of Vega, dividing the journey into eight "drags" of
roughly twenty miles each. Estimating the headway he is making, a navigator keeps track of his progress from drag to drag. As sailing conditions
change, he adjusts his reckoning of progress from one drag to the next.
Such reckoning greatly facilitates keeping track of overall progress and
expectation of landfall.

Every course between two islands has an island or seamark of reference
that serves to divide the journey into "drags." Ideally, the end of the first
"drag," called the "drag of visibility" (*etákinikanna*), should come when
the island of departure is no longer visible, and the next drag, the "drag of
birds" (*etákini maan*), should end at the most distant point at which land-
based birds feed at sea. Similarly the next to last begins where birds again
appear and the last when the island of destination becomes visible. These
correlations are understood to be rough, but are useful in that a navigator
knows from his estimation of the number of "drags" traveled when he
should soon be sighting land-based birds and when he should be able to
see his destination. If these signs fail to appear when he has reason thus
to expect them, a navigator knows that he is off course.

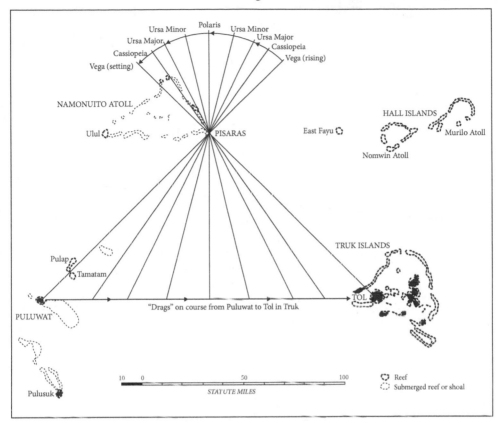

MAP 2 "Drags" on course from Puluwat to Truk.

Imaginary places can serve as points of reference for "dragging" as readily as real ones. All that is required is a convenient set of assumed compass directions to it from islands of departure and destination. For the voyage north from the Carolines to Guam and Saipan, there are no conveniently located islands. Here "ghost islands" and "aimers" are used as reference.

## Schematic Mapping

Without maps or charts, navigators must devise ways of constructing mental equivalents. "Trigger Fish" (*pwuupw*) is the name for one such way of conceiving the geography of the navigator's world. It envisions the locations of five places. Four of them form a diamond to represent the head, tail, and dorsal and ventral fins of the trigger fish. The head is always the eastern point and the tail the western one. The dorsal and ventral

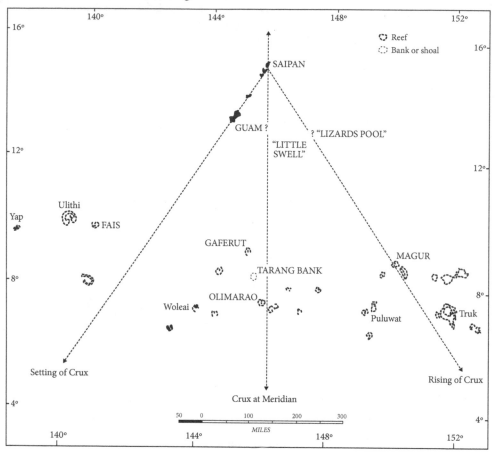

MAP 3  The "great trigger fish" (Fais and Magur tail and head, Gaferut and Olimarao dorsal and ventral fins) with its northern flip (places in capital letters) as located on the map (see fig. 3).

fins can serve either as northern and southern or southern and northern points respectively. The fifth place, at the diamond's center, is the fish's backbone. Any islands (real or imaginary), reefs, shoals, or living sea-marks whose relative locations are suitable can be construed as a trigger fish. On a course between the dorsal and ventral fins, the head can serve as the reference island and the backbone marks midcourse. Trigger fish may be arranged so that the northern point of one is the southern point of another, or the backbone of one is the southern point of another.

"Great Trigger Fish" (*pwuupw lapalap*) are large-scale schematic maps. Of special importance is the one that has Magur and Fais (650 miles

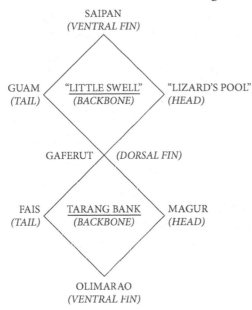

SAIPAN
*(VENTRAL FIN)*

GUAM
*(TAIL)*

"LITTLE SWELL"
*(BACKBONE)*

"LIZARD'S POOL"
*(HEAD)*

GAFERUT  *(DORSAL FIN)*

FAIS
*(TAIL)*

TARANG BANK
*(BACKBONE)*

MAGUR
*(HEAD)*

OLIMARAO
*(VENTRAL FIN)*

FIGURE 3 Carolinian's schematic representation of linked "trigger fish" (in this case the "great trigger fish" and its northern flip; see map 3).

apart) as head and tail, and Gaferut and Olimarao as dorsal and ventral fins (map 3). As one looks south from Saipan, the rising of the Southern Cross, named "Trigger Fish," lies almost directly over Magur, the head, and sets a bit east of Fais, the tail. Sailing south from Saipan or Guam, if a navigator keeps his course within the rising and setting of the Southern Cross, he will end up in the heart of the Caroline Island chain, with its many reefs, shoals, and other seamarks of which he has knowledge.

In one scheme, the northern flip of this great trigger fish has Gaferut as dorsal fin, Saipan as ventral fin, Guam as tail, and the imaginary place "Lizard's Pool" as head (figure 3). A set of lesser overlapping trigger fish involving a series of "ghost" places lying east and west of this north-south course between Saipan and Gaferut provide a series of reference points for dividing it into a convenient number of "drags."

## Predicting the Weather

Prevailing weather conditions are equated with the "months" of a sidereal calendar. Although called "moons" (*maram*), they are not lunar months. In most calendars there are twelve or thirteen months of unequal length, each named for a star. A month begins when its star stands about 45 degrees above the eastern horizon just before dawn, when to look at it one must tilt one's head back to where one feels a roll of skin forming at the back of the neck. The month continues until the next month star reaches the same position.

A weather predicting system, called "storming of stars" (*mworenifúú*), involves the rising of storm stars and the first and last five days of the lunar cycle. Within the span of each month, one or two storm stars make their appearance in the east just before dawn. If there is one such star in the month, there will be stormy weather during the first five days of the next new moon in the west. If there are two such stars in the month, stormy weather will come during the last five days in the lunar cycle after the heliacal rising of the star.

More immediate weather conditions are forecast from the color of the sky at sunrise and sunset and from the shapes of clouds. Large cumulonimbus clouds, "house of wind" (*yimwániyang*), are believed to store up wind. If they are visible at dawn or dusk, a navigator expects the wind to come from their direction.

## Putting the System to Work

It is one thing to learn the "star structure," sailing directions, sea-lane directions, and "aimers" and to become adept in the exercises and drills involving them. It is another thing to put it all to work in actual practice. The stars are not visible by day, the sky may be overcast at night, and the sailing directions in the exercises are, at best, only to the nearest compass point. Conditions vary with the seasons. To use the system, a navigator must rely on what he can actually see. He must also learn how to adjust the sailing directions in the light of his and his fellows' experience.

Ocean swells are a crucial guide in sailing. Navigators recognize up to eight different swells, one from each octant of the compass. Most dependable are those from the north, northeast, and east, associated with the tradewind season (our winter). During our summer, swells come from the southeast and south. The different swells have characteristic intervals. Navigators use opportunities to check the direction of swells against the stars. They then maintain course at the appropriate angle to the swells. When swell systems move across each other, they produce an effect somewhat like that of converging wakes of motorboats, forming an alignment of peaks or "wave nodes" (*pwukonó*) by which to steer. "Wave Nodes" is, indeed, the name used in reference to the whole body of knowledge relating to the interaction of currents, swells, and winds.

Currents reveal themselves by the shape of waves and ripple patterns. These patterns vary according to whether the current is going with the wind, across it, or against it, and according to their set in relation to the direction of swells. They make a significant difference in how one selects a course on the star compass. When setting out from an island, one uses an alignment of landmarks astern (*fótonomwir*) to set one's star course. A navigator should know the configuration of his home island as seen from every compass direction. On reaching the point, called "one tooth," where the island is just visible as a point on the horizon, one sights back along the bow-stern axis of one's vessel. If the island of departure lies directly astern on the axis, no compensation for drift from current is needed. The degree to which it is off, as measured by compass point intervals or their fractions, indicates to a navigator the degree of course adjustment he should make. He can measure the difference by holding out his hand at arm's length. The width of his hand corresponds rather well with the amount of arc on the horizon between adjacent points of a compass of thirty-two points.

In practice a navigator may begin with one star course and change to another at some "drag" point along the way. He adjusts for currents and changing wind conditions. Sailing against the wind may require planning a series of tacks from "drag" to "drag." In doing this, course settings from island of departure and island of destination to reference islands may help to structure the sequence of tacks. Using "drags" as a way to keep mental track of distances covered is crucial in such tacking, as when one plans to cross and recross the line of the direct course (*yallap*) from home to destination at the point where each new drag begins. Again, the set of current is critical in how far one tacks to right or left of one's true course.

Navigators must learn to make all these adjustments of course for the voyages they actually make. Years of sailing experience are necessary to develop skill as a practicing navigator. As with any other skill, not everyone is good at it.

## Navigators as Ritual Specialists

Protective ritual is something else a navigator must learn. He is said to be the "father" of his crew, who depend on him for their safety and welfare. He must know how properly to invoke the patron spirits of navigation; he must carefully observe the associated taboos in regard to sex and food; he must know the spells that will prevent storms and repel sharks; and he must be able to provide his vessel with protective amulets. He and his crew must know the art of righting an overturned sailing canoe. Indeed, they may deliberately overturn it in order to ride out a bad storm without being blown far off course. In addition to all of this, it is useful for a navigator to know enough of the special rhetoric and spells associated with politics and diplomacy (*yitang*) to ensure hospitality and safe conduct for himself and his crew when visiting other islands. Where voyagers have kin or fellow members of their matrilineal clans, they can be relatively sure of hospitality, but otherwise they are liable to be treated with suspicion and hostility. Knowing how to interact properly with a community's official greeter of visiting canoes establishes a navigator as someone to be reckoned with and treated with respect.

Their ritual knowledge and their observance of food and sex taboos set navigators apart. They are perceived to be among the most learned and magically powerful members of Carolinian society. Having demonstrated ability on a test voyage, such as successfully making a 130-mile, direct trip from Puluwat to Satawal, a newly certified navigator is initiated (*pwpwo*) into the ranks of recognized senior colleagues (*palú*). Thereafter, he eats in the canoe house with fellow navigators, whose food must be prepared separately for them. He has achieved the equivalent of a Ph.D. degree or a professional license to practice medicine or law in Euroamerican society.

## Keeping the Knowledge Alive

As should now be clear, sustaining the total body of knowledge in the absence of writing is accomplished by organizing it, making it systematic and schematic. It is taught and learned in this organized form. Indeed, it is overlearned with the use of drills and exercises that build in redundancy and are continually rehearsed.

For memory storage, some of the lore is embedded in chants, whose metric and tonal structures provide aids to recall. These chants are often cryptic in content, requiring commentary in order to understand them. A trainee will learn the words first. Only later will his teacher supply the necessary interpretation. If a teacher should die before passing the commentary on, his pupil must make the best sense he can of the chant. In time, he will develop his own interpretation in the light of his other knowledge and experience.

It is interesting to observe that the new interpretation may be quite different from the earlier one and still be workably consistent with reality. Interesting, too, is the evident elaboration of navigational lore beyond practical requirements, involving sailing directions to many named places that no one ever visits, that lie outside the known world or in the sky world. The living seamarks represent an elaboration beyond what is empirically known, also. Navigators seem to enjoy playing with the possibilities within their system, elaborating on them both for the fun of it and in order to show off superior knowledge to one another. Thus, the practical core of the system that is empirically tested in continual application remains much the same among the competing "schools" of navigation.[2] They differ in their living seamarks, in their chants, in the interpretation of specific chants, in their mythology, and in their magical rituals—in those respects, in short, where difference has little or no effect on successful voyaging and is in regard to what lies beyond the means of ready empirical testing. The navigators' separate mess in the boathouse, where visiting navigators also eat, appears to have played an important role in keeping the common system of navigation shared, at its practical level, over such a wide area.

We should note, finally, that the theoretical assumptions on which the system rests are that the sun and stars revolve around the earth, which remains stationary, and that the star Altair rises due east and sets due west. That these assumptions are false within the framework of modern Western scientific understanding does not deprive the system of practical utility for their purposes nor, indeed, for the purposes of any sailor without instruments and having to navigate by dead reckoning in Carolinian waters.

## Carolinian Navigation as a Practical Science

Several things stand out about Carolinian navigational knowledge. It has all the features of a practical science.[3] It contains a massive amount of discrete information, which, in the absence of writing and reference books, has to be committed to memory. The information is highly organized in a systematic way; the different ways of organizing it provide much redundancy as an aid to recall. It involves highly abstract thinking: the compass as a set of imaginary points at equal intervals around the horizon, named for the stars and abstracted from their perceived motions, but not identical with them; the use of "drags" as imaginary divisions of one's course of travel; the use of imaginary places as points of reference to calculate "drags"; and schematic mapping in the form of "trigger fish." Gladwin (1970) has called attention to the fact that navigators tested low for abstract thinking, and used this discrepancy to question whether the psychological tests in fact were testing for concrete as against abstract thinking at all.

It is also clear that Carolinian lore is based on empirical observation in its practical aspects and becomes fanciful only beyond the bounds of readily verifiable experience or practical application. We should note, moreover, that navigators are ever ready to add to their knowledge. In these respects Carolinian navigational lore is quite similar to Western practical science. Indeed, demonstration of its fundamental soundness was made by Piailug of Satawal, who successfully navigated the Hokule'a, a replica of an ancient Hawai'ian double-hulled voyaging canoe, from

Hawai'i to Tahiti by dead reckoning alone, using only his own knowledge of seamanship and navigation as a Carolinian *palú* (Finney 1979).

There are things about this body of knowledge that have mystified some observers. Gladwin found it strange to base the system of "drags" and its use in tacking on the "concept of a moving island" (1970, 181), seeing the traveller as remaining at the center and the island world moving around him through the sidereal compass as he travels. He saw in this an example of a difference in the way human thinking may be culturally programmed; but I know of no other way to use a compass. If we speak of the direction of places changing in relation to us as we travel while Micronesians speak of places being "dragged" through different points of the compass, the figures of speech may be different, but the underlying understanding of a compass is not.

Lacking other instruments of observation, moreover, Micronesians train themselves to use their own bodies and senses far more than do Euroamericans. Piailug's ability accurately to gauge the direction and force of currents by observing faint patterns in surface ripples amazed Thomas (1987, 32). It takes a lot of practice to be able to assess distance and time of travel accurately enough to use "drags" as reference in tacking upwind. It also requires training to learn to pick up such seamarks as the surface manifestations of the presence of a submerged reef. Yet, obviously, none of these skills are beyond human perceptual abilities.

## Bodies of Knowledge and Cultural Anthropology

Describing the content and organization of the many diverse bodies of knowledge that comprise human understandings is one of the workaday tasks of cultural anthropology. These bodies of knowledge range from how to make fire and catch fish to how to build airplanes and computers, from how to conduct oneself acceptably in one's family relations to how to do so in negotiating a business deal or prosecuting a legal case. Ethnography, as it is called, aims, among other things, to describe what one needs to know in order to engage with a society's members in all their activities in a manner that meets their standards of acceptable performance. Such knowledge is what is technically meant by a people's "culture." Like a language or a game, a culture is something one has to learn in order to describe it. Ethnography is, therefore, an exercise in the systematic learning and presentation of what people are expected to know. Ethnography also tries to describe how people apply and use such knowledge in the affairs of life as that knowledge constructs those affairs, including the skills that application requires and the preferred performance styles. Ethnography is as applicable to what farmers have to know, aeronautical engineers have to know, elementary school educators have to know, or religious specialists have to know to perform acceptably by their respective standards, as it is to what Micronesian navigators have to know to perform acceptably by theirs.

Having said this, I must add that there are relatively few ethnographic accounts of bodies of knowledge that tell us what we need to know to be acceptable practitioners of that knowledge. As I have just said, the only way to record it is for ethnographers to learn it themselves, to pass the test of acceptability, and then, having made a record of what was being learned as it was being learned (the "field notes"), to use that record to describe what they think they now know. In this way they can produce an account whose accuracy can be checked by those already knowledgeable and evaluated by those who try to use it as a guide to becoming knowledgeable themselves. Unfortunately, the customs governing dissertation research proposals in cultural anthropology are such as to steer students away from undertaking to learn and describe other systems of knowledge as a worthy end in itself. Ethnographies rarely describe activities in relation to all that one must know to perform them and all the decision points and criteria for making those decisions in the course of their conduct. But only by doing such things can we acquire good descriptions of the content of human cultures. Only by trying to do this can we confront and solve the methodological challenges it entails. Without such descriptions, however, we lack the information and the understanding that we need in order critically to examine propositions about the relation of culture to human cognition. With this last observation in mind, I ask the reader to consider if Micronesian navigation rests on mental operations of a kind with which we are largely unfamiliar or if, with better understanding of how it works, we find it represents ingenuities of much the same kind that are exemplified in the products of Western thought.

## Notes

1. This chapter is adapted from a paper by Goodenough and Thomas (1987). Here, as in that paper, I try to integrate and summarize material from a number of sources. I have drawn heavily on the work of Gladwin (1970), who provides excellent information on the design of sailing canoes and on the use of "drags"; on that of Riesenberg (1972), who provides details on a number of star lore exercises and the workings of the "Great Trigger Fish"; and on that of Thomas (1987), who has greatly enriched earlier reports with information on exercises, "drags," weather prediction, reading the sea's surface, and the practical application of the formal system of actual sailing. For an earlier account see Damm and Sarfert (1935). See also Lewis (1973) and Turnbull (1991). By and large the same

technical terms are used through all of the Western Carolines, but dialects differ. I have chosen to render them here in the dialect spoken on Puluwat Atoll, where Riesenberg worked. It is very close to that of Satawal, where Thomas studied navigation.

2. Two schools are present in Puluwat; *Wáriyeng,* "Wind Seeing," and *Fáánuur,* "Under the Banana Plant" (Gladwin 1970, 132).

3. In calling this a practical science, I have in mind the kind of knowledge we have traditionally associated with engineering, knowledge that involves empirically tested principles and rules of thumb, organized into a coherent system of ideas, that works well in the achievement of practical objectives. Whether it is science, or craft, or art, or a mix of all three is a matter of how one chooses to fit it into Western intellectual categories about which we Western intellectuals are ourselves in some disagreement.

## References

Damm, H., and E. Sarfert. 1935. *Inseln um Truk: Puluwat, Hok, und Satawal.* Ergebnisse der Sudsee-Expedition 1908–1910, Series 2. B, vol. 6, pt. 2. Hamburg, Germany: Friedrichsen, De Gruyter.

Finney, Ben R. 1979. *Hokulea: The Way to Tahiti.* New York: Dodd, Mead.

Frake, Charles O. 1994. "Dials: A Study in Physical Representations of Cognitive Systems." In *The Ancient Minds: Elements of Cognitive Archaeology,* edited by Colin Renfrew and Ezra Zubrow. Cambridge, U.K.: Cambridge University Press.

Gladwin, Thomas. 1970. *East Is a Big Bird: Navigation and Logic on Puluwat Atoll.* Cambridge, Massachusetts: Harvard University Press.

Goodenough, Ward H., and Stephen D. Thomas. 1987. "Traditional Navigation in the Western Pacific." *Expedition* 29 (3):3–14.

Lewis, David. 1973. *We, the Navigators.* Honolulu: University Press of Hawaii.

Riesenberg, Saul H. 1972. "The Organization of Navigational Knowledge on Puluwat." *The Journal of the Polynesian Society* 81:19–56.

Thomas, Stephen D. 1987. *The Last Navigator.* New York: Henry Holt.

Turnbull, David. 1991. *Mapping the World in the Mind: An Investigation of the Unwritten Knowledge of the Micronesian Navigators.* Geelong, Victoria: Deakin University Press.

# 9. Science for the West, Myth for the Rest?

## *The Case of James Bay Cree Knowledge Construction*

DO CREE HUNTERS PRACTICE SCIENCE? The answer to this question would seem to depend on whether one defines science according to universal features, or culturally specific ones. If one means by science a social activity that draws deductive inferences from first premises, that these inferences are deliberately and systematically verified in relation to experience, and that models of the world are reflexively adjusted to conform to observed regularities in the course of events, then, yes, Cree hunters practice science—as surely all human societies do. At the same time, the paradigms and social contexts of Cree science differ markedly from those of Western science—accustomed as we are in the West to a "root metaphor"[1] of impersonal causal forces that opposes "nature" to "mind," "spirit," and "culture," and conditioned as we also are to view legitimate scientific procedure and production as the prerogative of particular professional and institutionalized elites. While there is no a priori reason to expect that knowledge generated out of non-Western paradigms or social processes should be empirically or predictively less adequate, it has been an effect of Western ethnocentrism to construe non-Western knowledge processes as "pseudoscientific," "protoscientific," or merely "unscientific."[2] Western science, in fostering an ideology of knowledge that supports its own elite status, has assisted the exclusion and disqualification of innumerable "subjugated knowledges" (Foucault 1980).

Indigenous ecological knowledge finds renewed voice, however, in answer to the environmental anxieties of Western industrial societies, as well as aboriginal people's demands to decolonize and to directly manage environmental resources to which they assert primary rights. The Cree of James Bay, Canada, are such a people. Any account of Cree knowledge occupies a context in which jurisdiction for resources is actively contested, in which "science" is invoked both to attack and defend Cree opposition to invasive development projects sponsored by external governments, and in which indigenous knowledge is both advocated and opposed as a basis for deciding development issues.

I do not directly address these political dimensions in this chapter. I focus on the more particular task of exploring how practical, empirical knowledge flows from root metaphors (paradigms) that are not generally associated with "scientific" results in Western thought. The exploration focuses on the way in which root metaphors of pan-species personhood, communication, and reciprocity inform literal models of animal behavior and hunting practice; and how the latter reciprocally transmute the terms of metaphor, as experience is interpreted and actions are formulated.

One conclusion of earlier twentieth-century ethnography, in a line leading from Malinowski through Evans-Pritchard, is that in all societies (including Western civilization) practices dubbed "magical" and "mystical" coexist with rational/ empirical processes. Both anthropologists were alert to

> the danger of double selection by which ("primitive" peoples) are described entirely in terms of their mystical beliefs, ignoring much of their empirical behavior in everyday life, and by which Europeans are described entirely in terms of scientific rational-logical thought, when they too do not inhabit this mental universe all the time. (Tambiah 1990, 92)

Both anthropologists knew that "a person can in a certain context behave mystically, and then switch in another context to a practical empirical everyday frame of mind" (ibid).[3] This legacy poses two problems of immediate concern for anthropology: how do we get beyond the artificial dichotomy that separates Western and non-Western forms of knowledge, simultaneously discrediting and romanticizing the latter; and how are logical/empirical and mystical/magical aspects of thought related, in all traditions?

Perhaps we have begun to see that the distance separating the scientist and the shaman is not so great as was once imagined. But the evolutionary opposition of science for "the West" to myth and magic for "the rest" is far from dissolved; Western self-conception remains profoundly involved with images of rational "self" versus mystical "other." Several trends in late twentieth-century anthropology, to be sure, have continued to erode this dichotomy. Ethnoscientific fieldwork since the 1960s has brought into view empirically elaborate nomenclatures and classification systems from a wide range of "traditional" societies. In the structuralism of Lévi-Strauss (1966), these empirical categories, or "percepts," were signs in the bricolage of mythical thought, which could be seen as the "science of the concrete." The structures of mythical thought can produce scientific results because the mind, perception, and the external world share a common "natural" foundation. For Lévi-Strauss (as for Evans-Pritchard before him) the structures of reason in myth and magic are not fundamentally different from those of science (1966, 1973). Yet in Lévi-Strauss, the science-myth opposition is salvaged via the obscure notion that mythical "signs" are somehow limited and contained by their empirical signifiers, while the "concepts" of science are more free (1966, 18–20).

In other scholarship, ecological adaptation in non-Western societies is seen as systemically reinforced by symbolic structures or cosmologies. Reichel-Dolmatoff (1976) demonstrated the formal compatibility between mythico-ritual structures and ecosystemic principles among Amazonian Tukano. And Rappaport (1968, 1979) showed that supernatural categories—and their ritual entailments—in the "cognized model" of the Maring have homeostatic functions and effects within the analyst's "operational model" of highland New Guinea ecosystems. When the structural or functional connections between abstract cosmology and material ecosystems are the constructs of the analyst, however, it can appear as if the "totalizing" view of Western science has captured what remained unconscious or invisible to native subjects. The intellectual processes involved in framing practical knowledge within cosmological categories, from the actors' point of view, remain largely obscure. The adaptation of native cosmologies to their material-historical environments can then appear to be fortuitously functional, a happy congruence of symbolic and material structures (if functionality and congruence are to be believed); or the outcome of blind selective forces; rather than the outcome of theoretical work and proactive environmental management on the actors' parts.

With the upsurge of multidisciplinary interest in "traditional ecological knowledge," models explicitly held by indigenous people in areas as diverse as forestry, fisheries, and physical geography are being paid increasing attention by western science specialists, who have in some cases established extremely productive long-term dialogues with local experts (see Berkes 1977; Johannes 1981, 1989; Nietschmann 1989). The idea that local experts are often better informed than their scientific peers is at last receiving significant acknowledgment beyond the boundaries of anthropology. Anthropologists may find that we have less knowledge to share with local experts than do our colleagues in biophysical sciences about specific domains of local knowledge, and in this respect we may be at an initial disadvantage in striking up mutually interesting conversations with local experts. On the other hand, anthropology is unique in the degree to which it emphasizes the more inclusive cultural contexts of our local teachers and values ways of translating indigenous knowledge that reflect the symbolic and institutional contexts in which the knowledge is generated. If the sharing of knowledge were to be reduced to a skimming-off by Western specialists of indigenous empirical insights, and their mere insertion into existing Western paradigms, then it would be an impoverished and failed exchange that would ultimately contribute to undermining indigenous societies and cultures.

A number of anthropological studies have addressed the way in which mythico-ritual categories are implicated in actors' modeling of social-environmental practice in situational, strategic discourse about material activity (see Feit 1973, 1978; Nelson 1983; Scott 1989; Brightman 1993). It is this general issue to which the present paper contributes, through consideration of the ethnography of James Bay Cree hunting knowledge. I do not argue that all mythical and ritual symbolism is necessarily directed toward some logic of practical social or environmental knowledge. But I want to highlight that central, recurring propositions within these symbolic discourses are better understood in this way than as mystical precepts; and indeed, our understanding of practical knowledge cannot be adequately formulated without reference to the root metaphors most vividly condensed in myth and ritual.

Here I will proceed in three steps: first, to discuss the significance of ordinary experience within Cree cosmology and epistemology; second, to consider how the "figurative" language of metaphor interacts with the "literal" language of practical/empirical experience (i.e., how paradigms relate to the ordering of empirical experience); and third, to present examples from Cree goose hunting that illustrate the alternation between literal and figurative aspects of knowledge, each providing context and definition to the other.

## Signs in Cree Epistemology

As Hesse (1980), following Black (1962), has argued is true for scientific practice, "literal" or "observation" languages are shaped by the use of metaphor in theory—models are expressions of metaphor; and description, the literal reporting of observed regularities, is not independent or invariant of changes in explanatory models. Observation language "like all natural languages is continually being extended by metaphoric uses" (122). As certain root metaphors become conventionalized, as certain paradigms persist, their presence in observation language becomes less noticeable—they become literally implicit in the empirical description of experience. So it is that we may be largely unconscious of the metaphysical paradigms that underlie our own understandings of the world, while those of other knowledge traditions strike us as exotic, improbable, even "superstitious."

It is only in moments of unusual reflexive insight, for example, that modern Westerners are conscious of the extent to which a (meta-)physics of impersonal forces imposes itself on our perception of "nature." So embedded are the Cartesian myths of the dualities of mind-body, culture-nature, that we tend to privilege models of physical causality, rather than relations of consciousness or significance, in our perception even of sentient nature. It is true that we have begun to culturalize animals in animal communications studies, and to naturalize culture in anthropological ecology. But our conventional attitude is to assume fundamental differences between people and animals, while exploring the nature of their connections. The Cree disposition seems rather the converse: to assume common connections among people, animals, and other entities while exploring the nature of their differences. The connectedness assumed by the Cree reminds me of what Gregory Bateson (1979) has termed the "pattern which connects,"[4] patterns of dancing, interacting parts within larger patterns, the stories "shared by all mind or minds, whether ours or those of redwood forests and sea anemones," the "aesthetic unity" of the world.

In Cree, there is no word corresponding to our term "nature." There is a word *pimaatisiiwin* (life), which includes human as well as animal "persons." The word for "person," *iiyiyuu*, can itself be glossed as "he lives." Humans, animals, spirits, and several geophysical agents are perceived to have qualities of personhood. All persons engage in a reciprocally communicative reality. Human persons are not set over and against a material context of inert nature, but rather are one species of person in a network of reciprocating persons. These reciprocative interactions constitute the events of experience.

Again, there is no Cree category for "culture" that would make it the special province of humans. Cree do, however, have terms that resemble notions of "sign vehicle" (*chischinawaachaawapihtawaawan*) and the "meaning/interpretant" (*iishchiishwaamaakan*) of a sign. Cognates of these terms, incorporating the morpheme *-chis-*, evoke the ubiquity of signs in experience. They include *chischaaimaau* (s/he knows [him/her]); *chischaaitamuun* (information, knowledge); *chischinutihaau* (s/he leads, directs, guides [him/her]); *chischinuwaasinaakusuu* (s/he is used for a sign or s/he gives a sign); and *chischiwaahiicheu* (s/he prophecies) (Mackenzie 1982). Animal actions, particular qualities and features in the bodies of animals, weather, dream images and events, visions, and religious symbols all fall within the Cree notion of "sign," with signs constituting knowledge or guidance for actors. Not only humans, but animals and other nonhuman persons send, interpret, and respond to signs pertinent to various domains of human action: hunting success or failure, birth and death, and, implicit to these, the circumstances of reciprocity between persons in the world.

Signs, then, are part and parcel of action, perception, and experience—of life itself. The term *pimaatisiiwin* (life) was translated by one Cree man as "continuous birth." Consciousness (*umituunaaichikanich*, glossed by the same man as "mind and heart, thought and feeling") is at the threshold of unfolding events, of continuous birth. One consequence of this construction of the world is that an attitude of dogmatic certainty about what one knows is not only untruthful but disrespectful. There are many signs of recurrence and regularity in experience, but interpretations cannot be certain or absolute. To expect a definite future outcome on the basis of signs in the past or present, for example, may presume too much about the cooperation of other persons. Someone (human, animal, or spirit) could even retaliate by frustrating hunters' intentions.

## Relating and Differentiating as Complementary Aspects of Knowledge

Since events are the actions of various mutually responsive persons, human action is subject simultaneously to moral and technical criteria of evaluation. This is a commonly remarked feature of worldview for many egalitarian societies. Roy Wagner (1977, 1981) attributes it to the prominence of figurative ("differentiating") signification in traditional societies, by contrast to the literal ("relational") bent of signification in Western culture. While I think the opposition is misplaced as one of societal types, the theoretical grounds on which it is presented are illuminating for present purposes.

The basic idea is simple enough. In positing relations between things, we depend on some implicit definition of those things; conversely, to distinguish and define the things themselves, we depend on an implicit context of relations among them. The literal and the figurative are complementary and mutually dependent aspects of any knowledge construct. Depending on where we focus our attention when we shape a knowledge construct, we either relate the perceptibly differentiated, or differentiate the perceptibly relational. To choose a common example, for Western science, "natural" objects are implicit in the cause-effect constructions relating them. Objects are combined into total relational patterns that comprise a context (which, at the most inclusive level, we call "nature"). Natural objects and nature at large are experienced as innate; as "naturally" separate from the scientific culture that represents them. Figurative signification, on the other hand, focuses explicitly on defining objects or entities via metaphors that examine the similarities and differences among them; but implicit in those metaphors is a relational context (relations of reciprocity, for example, among many non-Western peoples).

When non-Western peoples focus on the analogies among themselves and other phenomena in the world, they tend to precipitate their own conventional social context (e.g., communicative reciprocity) as the innate character of phenomena in general. Western science tends no less to precipitate its own conventional social context: surely the technical mastery of an objectified nature is metaphorically connected to centralized social hierarchy and control. The separation of culture from nature depends not on a preponderance of literal-mindedness per se, but on which metaphors are used to frame the literal.

The complementary of the literal and the figurative help us to realize that the distinction between myth and science is not structural, but procedural. Myth, in a narrow and derogatory sense, is the dogmatic application of constituent metaphors as literal truths. There is myth, in this sense, in all science. At the same time, no science can embrace the world except through the creative extension of metaphors to emergent experience. We rework our metaphors as our models address particular contexts of experience. Myths in a broader, paradigmatic sense are condensed expressions of root metaphors that reflect the genius of particular knowledge traditions.

Let us return to the point that the interdependence of the figurative and the literal entails the integration of moral and technical aspects of knowledge. The Pacific cultures discussed by Wagner, much like Cree culture, view the world as an innate realization of a conventional social order of reciprocity. For Cree, as we have noted, communicative exchange is extended so ubiquitously to nonhuman domains that it constitutes a root metaphor or paradigm for knowledge in general. Myth, ritual, dreams, and hunting scenarios all express respectful solicitude as the preferred relation among "persons" in the hunter's world. The mental and physical activity of the hunter is directed at maintaining standing in this network, by being generous and respectful to humans and nonhumans, and by ensuring that what is received is in correct proportion. Where moral standards of positive reciprocity are deviated from, its negative corollary ensues: in all contexts, generosity dwindles in response to disrespect and greed.

But the viability of these common premises depends on the rigorous discernment of differences among persons. "It is," as Wagner (1977, 398) has said, "by maintaining a precise awareness of these differences, by differentiating himself and by differentiating the various beings in an appropriate manner, that man precipitates (or from the actor's point of view, invokes) a beneficent relational flow." Here I want to emphasize the creative use of metaphor in interpreting events in specific practical contexts, and to illustrate how permutations of the metaphor of communicative reciprocity vary situationally with the material phenomena that serve as its signifiers.

## Metaphors of Eating and Sexuality:
## Distinguishing Human and Nonhuman

In certain sacred contexts, the identities of hunters and animals are so passionately condensed in the metaphors connecting them, that the aspect of similarity virtually eclipses the aspect of difference. But in most practical contexts, the differences between hunters and animals as reciprocating agents are in the foreground. The definition of these differences flows from quite deliberate relational models that connect hunters and game metonymically—in consumer/consumed complementarity and in cause-effect orders that include other environmental agents such as winds, tides, and topographical features.

Reciprocity among humans is distinguished from reciprocity between humans and animals. In the first instance, biological structures of human reproduction signify a fundamental separation and asymmetry between human community and animal community, in respect of the former consuming the latter. The justification of this asymmetry is no trivial matter. Several Cree myths are concerned with human sexuality as a metaphor for the killing and eating of game, and vice versa. When humans get the terms of their metaphors confused and begin marrying animals or eating other humans (that is, failing to differentiate correctly), the results are impossibly comic or tragic (Preston 1975, 1978). That an animal be available for human consumption is an index of respect and love between hunter and animal; to contemplate the consumption of other humans is horrifying.

The metaphoric juxtaposition/separation of humans and animals is the occasion for much humorous discourse linking the pursuit of sexual partners to the pursuit of game. Hunting and sexuality share a vocabulary: *mitwaaschaau* can mean both "he shoots" and "he ejaculates"; *paaschikan* can refer to both "shotgun" and "penis"; *pukw* to both "gunpowder" and "sperm"; and *spichinaakin* to both "gun sheath" and "condom." But analogy, along with humor, is as much about separation as about similarity. The *atuush*, or "cannibal" figure subverts this separation of human from animal, of sex from food. In one bawdy myth, a cannibal copulates with a woman hunted by his son, before roasting and eating her reproductive organs. In consequence, he consumes his own sperm. He and his son, greatly weakened, are nearly overcome by the superior spiritual power of true human beings.

Another myth illustrates the necessity of killing animals, while respecting certain parameters for doing so. It concerns a supernatural character, Chischihp, who never ate. Chischihp thinks of the food animals whom he loves as his "pets," or "dogs." In this he differs from Cree hunters. Hunters also refer affectionately to certain species as their "pets," but normally it would be the species that an individual hunter is privileged to kill with unusual success. Chischihp would never have begun killing his "pets" had he not met two human sisters on a river journey and desired them for wives. They accept his proposal, but insist that he kill beaver and moose for them. When he objects, they threaten to abandon him. He relents, kills the animals, and eventually, surreptitiously, begins to eat some of the meat himself. However, he goes from excessive abstinence to excessive indulgence, with both the animals and the women, in his imperfect conversion to human status. When he returns to his village, he selfishly hides his wives, preventing their attendance at a public dance. At the dance, he adorns himself with the fatty internal organs and membranes of the moose. These parts are esteemed food delicacies, and their ostentatious display is grossly disrespectful of the animal gift. His wives, for their part, respond with infidelity. Chischihp discovers them sleeping with a lover, whom he promptly murders. He is now classified as a *pwaat*, a subhuman person who lurks at the margins of true human community and who shares some attributes of cannibals. Through treachery, he escapes the wrath of his village, drowns his wives, and is himself transformed into a species of edible waterfowl, the form in which he is known to Cree hunters today.

Human reproduction, then, demands the consumption of animals as positively as it prohibits the consumption of other humans; but there are respectful parameters for both interspecies consumption and intraspecies sexuality that are specific to the form of "reciprocity" in question.

## Knowledge Construction in Cree Hunting

I want to go on now to illustrate how the literal interpretation of animal behavior in the environmental context impels the further figurative differentiation of human and nonhuman persons. Cree hunters continually refer to human and animal capacities as interpretants of one another. The family structure, leadership, memory, and communication processes of animals are all explored as analogs of corresponding human qualities, both individual and social.[5] Here I focus on Canada goose hunting and resource management strategies. Goose hunting is a major ritual and economic event during the exceptionally rich fall and spring migrations along the coasts of Hudson and James Bays. Geese as objects of knowledge are extremely important both ritually and economically—they account for as much as one-quarter of all annual subsistence production for coastal Cree.

There are advantages to the Cree paradigm of a sentient, communicative world that transcends but includes humanity. It has oriented Cree to aspects of animal behavior that Western science, inured by Cartesian metaphors of mechanical nature, has admitted rather belatedly. Lorenz (1979) observes that for the "higher" animals, the expression of emotion involves substantially the same neuromuscular system as in humans. Geese possess some quite "human" affective qualities, including loyalty, jealousy, and grief; furthermore, these qualities are manifest in the context of striking similarities in courtship, mating, and the rearing of young. "The family and social life of wild geese exhibits an enormous number of striking parallels with human behavior," Lorenz observes; "Let no one think it is misleading anthropomorphism to say so" (192).

Lorenz finds a greater gulf separating the rational faculties of geese and humans. Yet here, too, ethologists are finding that animals classify elements of environment with some sophistication:

> Animals create a taxonomy appropriate to their species and ecological niche. Thus predators, for instance, distinguish different categories of prey—by size, appearance, odor, and other signifiers—thus forestalling wastefully indiscriminate attacks. Vice-versa, many potential prey distinguish among different kinds of predators as we observe from their use of sundry warning signs, variations in their flight-distances and flight-reactions. (Sebeok 1975, 93–4)

Sebeok has argued that for animals, as for humans, aesthetics are intimately linked to the extraction and reconstruction of structures from salient environmental features, "even when the process or the product is disunited from its proper biological context" (1975, 61).

The interpretations of Cree hunters suggest that geese are quite apt at learning in what contexts to expect predation, at learning to distinguish predatory from nonpredatory humans, and at communicating appropriate behavioral adaptations to other geese. In other words, there is substantial flexibility for geese to "reinterpret" environmental signs, and this learning is communicated among geese to ramify socially.

It is therefore important for hunters to arrive at precise estimations of goose learning and communication, particularly in relation to themselves as predators. It is heuristically useful but not in itself sufficient to assume, on the basis of the culturally pervasive paradigm, that capacities of intelligence and communication are shared by geese and humans. Hunters need to know more about what is shared and what is different, and in what measure. The more the respective capacities of geese and humans are specifically formulated (i.e., differentiated), the more "literally" they contribute to effective hunting scenarios. The interpretation/modeling of hunting experience is an ongoing refinement of hunters' knowledge of the specific capacities of geese, and the basis for adjustments in hunting practice.

Hunters arrange landscapes that will be attractive and nonthreatening to geese, while exercising caution so that geese will not learn to associate unusual details with the possible presence of hunters. Decoys and goose calls are iconic approximations by hunters of the semiotic landscape of geese. Hunters recognize differences among species of geese. Canada goose decoys must be realistic in profile and must be kept heading into the wind, properly spaced, with decoys appearing in both feeding and alert postures. Generally speaking, greater numbers of decoys are more effective. Snow geese are less sensitive to profile or number of decoys, but respond strongly to color. Two or three white plastic buckets or white rags displayed prominently on a hillside, in conjunction with calling, are sufficient.

The honking of geese is imitated to get the attention of an approaching flock once it is near enough to spot decoys. When the geese have seen the decoys and "made up their minds" to fly over to land, hunters stop calling, or switch to two or three long, low "welcoming" calls at gentle intervals. Calling should be used sparingly—novice hunters must learn both to imitate goose calls accurately, and to know when not to call. If a flock in the air has chosen not to respond to calls but to continue on and away from blinds and decoys, hunters should cease calling, because geese may recognize that such calls are unnatural. When there is less wind, the geese hear calls (and mistakes) more keenly. When a hunter is especially skilled at calling geese, others may prefer to keep silent to reduce the risk of detection.

Neither the semiotic conventions of geese, nor their interpretation and manipulation by hunters, are static. Geese, like hunters, are said to "know the land." Their ability to recontextualize certain perceptual features as signifying the presence of hunters is the potential undoing of the latter, as geese "get wise." For this reason, hunters' precautions to minimize visual and auditory signs of their own presence go well beyond the use of blinds at actual hunting spots. Camps are kept at some distance from concentrations of geese, and are well-hidden in the bush. Snowmobiles and chainsaws are not used near concentrations of geese. Ideally, the only birds on the territory that will be immediately aware of hunters' presence are those from small flocks actually fired on at hunting sites. Shooting on calm days is generally avoided, because the sound of shooting carries over a wide area without a wind to muffle and disperse it. Shooting after dusk is also avoided, because the flame visible at night at the end of a fired shotgun is said to terrify geese. Similarly, the use of lights outdoors at night is restricted.

Hunters' experience is that geese will not return to a hunting spot that has been used too regularly, or where they have been frightened badly. The Wemindji community area along the coast of James Bay is divided into several goose hunting territories, each used by up to a dozen hunters from a number of households linked agnatically, affinally, and by friendship. Hunting activities for each group are under the supervision of a senior "shooting boss." Each territory includes a number of viable hunting spots, and all hunters on a territory are expected to use one and the same spot on any given day, allowing all other spots to "rest." Normally, a new site is chosen each day, so that hunting spots are rotated. In this way, the migrating geese will not learn to expect hunters at any particular location, and will be respectfully permitted to rest and feed undisturbed over the majority of the territory on any given day. At hunting sites, when geese are killed, it should be done accurately and efficiently, to minimize disturbance and to avoid the waste of injured birds that escape.

If these precautions are not taken, geese on the territory will grow increasingly anxious about human presence and adjust their behavior accordingly. Even a fraction of geese too badly frightened communicate their alarm to other geese, which could lead to a reduction in the population staying on the territory, or to incremental avoidance by geese of the spot where the fright occurred. Since the same geese are on the migration route in successive seasons, and since young geese are said to learn their habits from their parents, a hunting spot that has been mishandled can take several seasons to recover.

I will give one example of how changes in goose behavior are effected by hunting activity. In early autumn, when the tide is high and the wind is brisk and onshore from James Bay, the geese fly from the coastal bays to the offshore islands in the bay, early in the morning. They do so to feed on berries there, partly because high tides, onshore wind, and rough waves make it impossible to feed on the eelgrass that grows in shallow water in the coastal bays. The berries, now ripened, are much relished by the geese, besides. This relational model is of key importance in hunters' decisions about where and when to locate themselves to wait for geese. When experience fails to confirm the expected relations, amendments to the model ensue.

It has in fact developed in recent years that geese are less prone than before to fly to hunters waiting on the islands, even when there have been plenty of geese in the coastal bays, plenty of berries on the islands, and

favorable circumstances of wind and tide. The interpretation of this decline by experienced hunters is as follows. In recent years, population
growth and wage and transfer payment income have led to an increase in
the number of and mobility of hunters who have greater access to motor-
ized water craft. This means that any hunting location in the coastal community area is generally accessible within twenty minutes to two hours'
travel from the settlement.

Significantly, these settlement-based hunters do not independently en-
ter the more management-sensitive coastal bays, where concentrations of
geese rest at night and also feed when the tide is low. Only when such a
bay is being hunted as part of a rotational strategy, under the leadership
of a "shooting boss," are hunters from the settlement welcome to join.
But offshore islands can be used with much less risk of disturbance to
the main concentrations of geese in the bays, particularly when there is
enough wind to prevent the sound of shooting from carrying far. It has
developed as a sort of community compromise between full-time hunters
based in camps and wage-earning part-time hunters, who are less flexible
as to the times they can hunt, that the islands can be used without direct
supervision and coordination by shooting bosses. This allows settlement-
based wage-earners to hunt on their days off or after work, even when
conditions are not suitable for hunting in the bays.

However, because this hunting at the islands has become more frequent
and is no longer coordinated in regular rotation, the geese have come increasingly to expect hunters at the islands. Consequently, there have been
more geese flying inland instead of offshore when they leave the bays to
feed. Hunting geese inland is less productive and more difficult, so geese
have learned not to expect hunters there. They therefore fly lower and less
cautiously inland than along the coast and offshore islands.

The kind of interpretation just summarized, and the hunting strategies
entailed, involve years, in some cases generations, of practical empirical
investigation into how much and what kinds of interactions with hunters the geese will tolerate, without withdrawing from strategic locations.
The relational models conventionally signified by this experience help to
define goose communication as different in degrees and respects from
human communication. Literal interpretations precipitate a figurative
complement—the differentiation of human from goose communication,
or of goose communication from that of some other animals.

Some of the attributes of goose communication per se are best contemplated anecdotally. Here, metaphors that evoke human leadership,
speech, and so on, are evident, but the hunting situations referred to are
themselves key interpretants of the appropriate extent and application of
the metaphor.

A friend and I were sitting in our blinds early one spring. A few larger flocks were heading north, but they were too high and were not coming into our calls and decoys. There was a lake nearby, perhaps a half-kilometer back into the bush but within earshot. We could hear the gabbling of numerous geese there. They hardly flew all day, until in mid-afternoon one solitary bird came low over the trees toward us. It wheeled to land among the decoys, an easy shot. We fired half-a-dozen shots, somehow missing, and the goose fled back in the direction of the lake. My companion speculated on the consequences of our poor shooting a few moments later: "Probably that goose told the other ones over there: 'If I don't come back, it's okay to come on over.'" No more geese flew our way from the lake that day. Geese, apparently, could communicate to other geese about phenomena that the latter have not experienced directly. "Scouting" among geese is observed in a variety of contexts, and presumably the behavior of these scouts conveys something about attractive versus dangerous situations. A variety of calls and postures, in flight and on the ground, are distinguished by Cree hunters—from messages of invitation to those of caution and alarm.

On another occasion later the same spring, the same companion and I were waiting for geese on an east-west elevated ridge between two coastal bays. The geese, who were leaving the bay to the south to continue their migration northward, would fly over the ridge. The highest, treeless portion of the ridge was perhaps two hundred meters in length, affording hunters a view of approaching flocks. It happened several times that a flock would cross too far east or too far west of where we sat, so we would be unable to get a shot. This seemed random enough, but then one or more flocks following at intervals of several hundred meters would cross at the same spot as the first flock, again evading us. I wondered if winds might account for this regularity, but rejected that possibility because the flocks had crossed at the same points even when they had approached the ridge from quite different trajectories. We were well-hidden, and the flocks were clearly unable to see us. I was ready to attribute the pattern of evasion to episodes of unfortunate coincidence until my companion remarked: "It always seems to happen like that. I guess they know there's hunters around. They see where the flock ahead of them went over, and they see nothing happens to them, so they think, 'Might as well go over there!'"

This incident was used by my host to instruct me in the attentiveness of individual flocks to the activity of other flocks in selecting a course of action. The safe patterns of a few geese are copied by many, in a context where older geese who are the leaders have learned to expect hunters. It became clear to me how avoidance of certain situations and preference for others could ramify socially among geese, resulting in general behavioral changes for the population as a whole.

At the same time, several aspects of goose awareness and communication remain esoteric. The capacity of animals to anticipate some events is considered superior to, and beyond the ken of many humans. Goose behavior of certain kinds is a predictor of approaching weather; geese begin their preparations before hunters would otherwise be aware of impending changes. Or again, in years when the local berry crop has failed, most geese on the fall migration have been observed to fly, very high up, right on past the James Bay coast. What is outstanding to hunters is that the flocks seem not to have to land to know that the feeding is poor. One interpretation of this phenomenon relates again to scouting behavior, although the precise mechanism of information transfer is ambiguous. In other interpretations, it is supposed that animals experience dream images and corporeal symptoms of the kind that can also alert humans to future or distant events—notable but not particularly unusual premonitions in the Cree world.

This commonalty returns us to the premise of a communicative, reciprocative network that unifies the holistic world. This premise is metaphysically prior to the more particular differentiation of persons in the world—and it is at this level that hunters, animals, geophysical forces, and even God are ultimately of one mind, as it were. Consistent with this premise is the notion that encounters, thoughts, dreams, and rituals involving hunter-animal exchange both index and influence the state of reciprocity that obtains at a given point in time.

## Animal and Human Reciprocators

I have mentioned the effect of metaphors of eating and sexuality in establishing a fundamental difference between human-human and animal-human reciprocity. I'll go on now to illustrate how ritual delineations of reciprocity between humans and animals contain, at the same time, abstract but quite literal constructions of social-ecological principles. The life cycle of the hunter, seasonal cycles, the social roles of men and women are all marked by ritual phases that reflect these principles. A few examples will suffice.

Early in the spring season, the geese killed on the first day of the hunt are cooked and a small share is distributed by a respected woman elder to everyone in a camp. This is done no matter how few the geese or how large the camp, and regardless of who killed the geese. Then, the following geese are saved for a few days until there are enough so that every man, woman, and child in the hunting group receives one, two, or more geese, depending on how many have been killed. Again, the distribution is made by a respected elder of the camp, and without regard to who killed the geese. The group then feasts, with each household roasting some of the geese it has received. This process is considered to be an "invitation" to the geese, since animals are said to come more readily to hunters who share them with others. Only after this feast is it possible for the household to accumulate geese for its own consumption. Significantly, the feast occurs while it is still to be determined whether the migration will linger, ensuring bounty, or pass quickly. Failure to contribute generously to the feast can account for poor luck in hunting later on.

A hunter in his blind often "smokes to the game" (*pwaatikswaau*), or sings goose songs. Tobacco was a traditionally valued item of exchange, and smoke is an appropriate vehicle of exchange with creatures of the air. Hunters' songs express spiritual and aesthetic aspects of the exchange, and include vivid images of the ways in which geese fall to the hunter. When a goose has fallen, the gift is respectfully admired by the hunter and later received as a guest into the lodge by the women of the hunter's household. The women take care to use every part of the goose possible, to avoid spoilage, and to dispose respectfully of the few remains. The cartilage tracheae, including the windpipe and voice organ of the goose, are hung from a tree branch where, poetically, the passing wind carries their call, beckoning geese in future seasons to renew the exchange.[6] When the migration has nearly passed and the last of the geese are departing, the hunter bids them farewell, expressing the hope that, granted continued life, he will be able to see them again on their return.

This ritual complex advances two general propositions about human-animal reciprocity that are of key ecological concern. The first is that respectful activity toward the animals enhances the readiness with which they give themselves, or are given by God, to hunters. The second is that sharing of animal gifts among hunters is an important dimension of respect for the animals. Both propositions are implicit in ritual enactments of the special obligations of hunters toward game, if game animals are to fulfill their own special role in supplying hunters. And both propositions convey literally understood truths about ecological relations.[7]

What is involved in the first of these general propositions? The empirical availability of geese, as we have seen, varies with their treatment by hunters. The specification of "respectful" treatment in day-to-day hunting is as complex as the many situations of interaction, but the general and key notion is that technical efficiency in killing animals must be balanced by restraint, and that only the latter can really guarantee the long-term viability of the former.

A hunter must strive for impeccable technique, both in the interest of his own security and to avoid undue suffering or disturbance to the animals. A hunter who, in spite of his effort, cannot kill animals will be disappointed, but he should keep trying and not be too disappointed because it is wrong to expect more than is freely offered. Perhaps the hunter is not receiving more because the partner is not in the position to give. The wise hunter directs his efforts elsewhere if, after trying hard, it is apparent that a particular species doesn't want to be caught. On the other hand, it is wrong to accept more than one needs, even if it can be taken with the means at one's disposal. The generosity of a partner can be overtaxed.

There is a rather concise set of symbols to summarize and express this balance between efficiency and self-restraint. First, when a hunter who normally kills perhaps fifty geese in a season suddenly kills, let us suppose, three times that number, it is taken as a sign that the hunter has not long to live. It is not surprising, then, that when an individual hunter has accumulated a larger than average kill early in the season, he sometimes stops for the remainder of the season, or lets a younger and less experienced hunter in the household bring home the geese. Collective restraint is also exercised after a particularly abundant daily kill has been made, when all hunters on the territory let the geese rest for a day or two.

There is a second symbol for the perils of excessive killing—the albino Canada goose. Such geese must not be killed under any circumstances or the hunter will find it very difficult to kill geese thereafter. Thirdly, there is the sandhill crane, a relatively uncommon bird on the east coast of James Bay, that is most numerous when there is an unusual abundance of migrating geese feeding and resting in the coastal bays. It is permitted to shoot the crane, but it must be perfectly done. To miss a shot is a sign of the hunter's impending death or that of a near relative. Fourthly, there is a rarely reported Canada goose that is said to bear a luminous collar over its neck and across its breast. The hunter fortunate enough to see this goose must kill it with a single clean shot and retrieve it almost the moment it touches the ground. Otherwise, the collar dissipates, passing to another goose. But the impeccable hunter may reach it in time and retain the collar as a charm. The geese will thereafter fly to such a hunter, even when he is not well-hidden.

To summarize:

(1) Too many geese killed (excessive killing) = Hunter's death
(2) Albino goose killed (excessive killing) = Hunter's poverty (curtailment of gifts)

(3) Crane attempted but not killed impeccably ("insufficient" killing) = Hunter's death

(4) Collar-bearing goose killed impeccably ("sufficient" killing) = Hunter's wealth (abundance of gifts)

This symbolic set signifies that the hunter must practice both excellence and restraint in the killing of game if he is to live well and avoid poverty or death. There is such a thing as "insufficient" killing, expressed in its negative form in the sanction of death when the crane is attempted but failed, and in its positive form in the reward of food wealth when the collar-bearer is impeccably taken. But there is also the possibility of "excessive" killing, expressed in a strong negative form in the sanction of death for killing too much game, and in the milder negative form of poverty for killing an albino. Symbols of the importance of impeccability have "literal" implications where efficiency and excellence in a demanding environment make the difference between prosperity and poverty, life and death. But symbols of the necessity for restraint are equally intelligible in literal terms because the generosity of animals is empirically exhaustible. Geese hunted too noisily or too often in the same place, too much game killed and too little allowed to escape, could lead objectively to dwindling exchange with hunters and to poverty in animal gifts, synonymous in the not too distant past with death.

Let us now turn to the second of the general propositions cited earlier—that positive reciprocity in human society enhances reciprocity with geese. In what literal sense could interspecies reciprocity depend on human reciprocity? Ever more important than sharing food, given households' pride in their autonomy, is sharing the opportunities to kill geese or other game. I have already mentioned that the management of a hunting territory is a delicate matter, requiring a cooperative strategy. Cooperation becomes impossible when generosity is not maintained. For example, if a hunting boss "saves" a rich build-up of geese, then hunts it for the sole benefit of his immediate household without notifying other hunters who would normally be entitled to join in, his reputation as a leader is seriously damaged. Other hunters then feel justified in hunting when and where they choose on such a territory, with the possible result that rotational management becomes impractical. The geese become increasingly wary, or move elsewhere, and hunting productivity on the territory declines. The animal gift, in this literal sense, depends quite literally on human generosity, and it is this social knowledge more than any other factor that accounts for restraint and regimentation in hunting. Social and ecological reciprocity are not just formally interdependent, as inferences from the same root metaphor or paradigm, but interdependent also in human practice.

Both propositions depend on situational elaboration of the reciprocity metaphor; one could say that the empirical contexts of both geese and hunters are assimilated to the terms of the metaphor. It is not as though reciprocity applies "literally" to social relations but only "metaphorically" to relations with animals. Neither set of relations can be said to represent the "primary" meaning—both are part of a reciprocating socio-environmental continuum; but human-animal differences are elaborated and exploited in empirical detail to produce informative permutations of reciprocity across numerous phenomenal domains.

In certain ritual contexts, the identity of the hunter merges radically with that of the animal, a merging accomplished through body-spirit reciprocity. The death of a hunter in a dream is a common omen of an important food animal about to be given in waking life. In a dream, the goose may be a guardian of the hunter's power and essence and may protect him from sorcery. Throughout his life, the hunter receives the gift of geese, and at a hunter's death, it is often a goose that represents his soul on its journey from this life. At the time of a hunter's death, a solitary goose may fly low overhead, or land near the mourning relatives, acting quite unafraid. The hunter's experience of animals as interpretants of his essential self renders all the more poignant the inevitable separation of hunter from prey, and all the more compelling the morality that joins them in the reciprocity of life-giving and life-taking. The transcendence of this tension at significant moments—the death of an animal, a dream encounter, or the death of a hunter—is the experience of the sacred.

## Conclusions

The achievements of indigenous ecological knowledge, as illustrated in the case of Cree hunters, are neither mysterious nor coincidental—they result from intellectual processes not qualitatively different from those of Western science. Western science is distinctive not through any greater logical coherence or empirical fidelity, nor any lesser involvement with metaphysical premises, but through its engagement of particular root metaphors in specific social institutional and socioenvironmental settings. Any number of root metaphors, situationally elaborated in the course of practical engagement with the world, may inform rational explanation and the effective organization of empirical experience. Equally, any number of the same metaphors may obstruct effective knowledge through a dogmatic and misplaced literalism.

Knowledge traditions reflect the morality of the social practices and paradigms in which knowledge is framed. Numerous studies have found that the "anthropomorphic" paradigms of egalitarian hunters and horticulturalists not only generate practical knowledge consistent with the insights of scientific ecology, but simultaneously cultivate an ethic of environmental responsibility that for Western societies has proven elusive.[8] If the inclusion of humans in a figurative world of analogous other-than-human persons promotes environmental responsibility, this depends on the condition of reciprocity in the human society concerned—not on any predominance of figurative versus literal thinking. All societies, whether egalitarian or hierarchical, establish metaphorical connections between the social and the environmental. In all knowledge traditions, literal modeling defines and redefines the relations among objects in the world, relations which in turn are assimilated to the meaning of root metaphors as they are applied in particular situations and contexts. Cree hunters are not less concerned than Western scientists with literal interpretation; nor are Western scientists less involved in figurative invention than Cree hunters. The conventional social context of Western science tends to hierarchy and centralized control, however, and this is the morality that is metaphorically projected onto our own relations with "nature." For this very reason, the historical disqualification and subjugation of indigenous knowledge is intimately linked to Western culture's domination of nature.

## Notes

The ethnography and much of the analysis in this chapter was previously published as Scott (1989). For the publication of *Naked Science*, I was asked to rewrite it with a more general audience in mind. The introduction and conclusions are new, and some technicalities of anthropological semiotics in the body have been clarified or eliminated.

I wish to acknowledge the support of the Social Sciences and Humanities Research Council of Canada for grants supporting this work.

1. Certain metaphors are so pervasive in a knowledge discourse as to constitute what have been termed "paradigms" or "archetypes" (Black 1962). Others have called them "root metaphors" (Pepper 1942; Ortner 1973) or "metaphoric networks" ( Ricoeur 1977). In this chapter, I use the terms "root metaphor" and "paradigm" synonymously.

2. Not to mention the polemical deployment of other adjectives: "magical," "irrational," "superstitious," "traditional," "primitive," etc.

3. For a thoroughgoing historical review of the anthropological analysis of knowledge and belief, see Tambiah (1990).

4. For Bateson, the "totemic analogy" between the social system of which people are the parts and the "larger ecological and biological system in which

the animals and plants and the people are all parts" (1979, 155) was a better analogy than the one that likened people, society, and nature to nineteenth-century machines.

5. One might observe that a consequence of this sort of analogical thinking is to anthropomorphize animals, but that would assume the primacy of the human term in the metaphor. Animal qualities react with perhaps equal force on understandings of humans, so that animal behavior can become a model for human relations. Preston (1978, 152) has suggested that the goose as exemplar of Cree ideals of social coordination, grace, and composure may be "better" than human.

6. The sequence here described in abbreviated fashion is one variation on a general ritual structure for "bringing home animals" (see Tanner 1979).

7. Feit (1973, 1978) has offered seminal discussions of the ecological significance of respect for animal gifts, to which my own analysis owes a great deal. Animals are felt to be given at times and places in which, by virtue of numerical availability and characteristic behavioral traits, they present themselves to the hunter's weapons or traps with maximum efficiency and minimum struggle. When hunters notice animals becoming scarce and difficult to catch, they say it is because the animals are "angry," perhaps because hunters have been taking too many. The respectful response is to stop hunting the species in question until it once again is more freely given.

8. See Rappaport (1968); Reichel-Dolmatoff (1976); Wagner (1977); Feit (1978); Nelson (1983); Bennett (1983); Scott (1983).

## References

Bateson, Gregory. 1979. *Mind and Nature: A Necessary Unity.* Toronto, Canada: Bantam Books.

Bennett, David. 1983. "Some Aspects of Aboriginal and Non-Aboriginal Notions of Responsibility to Non-Human Animals." *Australian Aboriginal Studies* (2):19–24.

Berkes, Fikret. 1977. "Fishery Resource Use in a Subarctic Indian Community." *Human Ecology* 5(4):289–307.

Black, M. 1962. "Metaphor." In *Models and Metaphors*, edited by M. Black. Ithaca, New York: Cornell University Press.

Brightman, Robert. 1993. *Grateful Prey: Rock Cree Human-Animal Relationships.* Berkeley: University of California Press.

Feit, H. 1973. "The Ethno-Ecology of the Waswanipi Cree: Or How Hunters Can Manage Their Resources." In *Cultural Ecology: Readings on Canadian Indians and Eskimos*, edited by Bruce Cox. Toronto, Canada: McClelland and Stewart.

———. 1978. "Waswanipi Realities and Adaptations: Resource Management among Subarctic Hunters." Ph.D. diss. McGill University.

Foucault, Michel. 1980. *Power/Knowledge: Selected Interviews and Other Writings 1972–1977.* New York: Pantheon Books.

Hesse, Mary B. 1980. *Revolutions and Reconstructions in the Philosophy of Science.* Bloomington: Indiana University Press.

Johannes, Robert E. 1981. *Words of the Lagoon—Fishing and Marine Lore in the Palau District of Micronesia.* Berkeley: University of California Press.

———. 1989. "Fishing and Traditional Knowledge." In *Traditional Ecological Knowledge: A Collection of Essays,* edited by R. E. Johannes. Gland, Switzerland: International Union for the Conservation of Nature.

Lévi-Strauss, Claude. 1966. *The Savage Mind.* Chicago: University of Chicago Press.

———. 1973. "Structuralism and Ecology." *Social Science Information* 12 (1): 7–23.

Lorenz, Konrad. 1979. *The Year of the Greylag Goose.* New York: Harcourt Brace Jovanovich.

Mackenzie, Marguerite. 1982. *Cree Dictionary.* Val d'Or, Quebec: Cree School Board.

Nelson, Richard. 1983. *Make Prayers to the Raven: A Koyukon View of the Northern Forest.* Chicago: University of Chicago Press.

Nietschmann, Bernard. 1989. "Traditional Sea Territories, Resources and Rights in Torres Strait." In *A Sea of Small Boats. Cultural Survival Report 26,* edited by John Cordell. Cambridge, Massachusetts: Cultural Survival.

Ortner, Sherry. 1973. "On Key Symbols." *American Anthropologist* 75 (5): 1338–46.

Pepper, Stephen. 1942. *World Hypotheses.* Berkeley: University of California Press.

Preston, Richard. 1975. *Cree Narrative: Expressing the Personal Meanings of Events.* Canadian Ethnology Service Paper No. 30, Mercury Series. Ottawa, Canada: National Museum of Man.

———. 1978. "La Relation Sacrée Entre les Cris et les Oies." *Recherches amerindiennes au Quebec* 8(2):147–52.

Rappaport, Roy. 1968. *Pigs for the Ancestors.* New Haven: Yale University Press.

———. 1979. *Ecology, Meaning and Religion.* Richmond, Virginia: North Atlantic Books.

Reichel-Dolmatoff, Gerardo. 1976. "Cosmology as Ecological Analysis: A View from the Rain Forest." *Man* 11 (3):307–18.

Ricoeur, Paul. 1977. *The Rule of Metaphor: Multidisciplinary Studies in the Creation of Meaning in Language.* Toronto, Canada: University of Toronto Press.

Scott, Colin. 1983. "The Semiotics of Material Life among Wemindji Cree Hunters." Ph.D. diss. McGill University.

———. 1989. "Knowledge Construction among Cree Hunters: Metaphors and Literal Understanding." *Journal de la Société des Américanistes* 75:193–208.

Sebeok, Thomas. 1975. "Zoosemiotics: At the Intersection of Nature and Culture." In *The Tell-Tale Sign: A Survey of Semiotics,* edited by Thomas Sebeok. Lisse, Netherlands: Peter de Ridder Press.

Tambiah, Stanley. 1990. *Magic, Science, Religion and the Scope of Rationality.* Cambridge, U.K.: Cambridge University Press.

Tanner, Adrian. 1979. *Bringing Home Animals: Religious Ideology and Mode of Production of the Mistassini Cree Hunters.* St. John's: Institute of Social and Economic Research, Memorial University.

Wagner, Roy. 1977. "Scientific and Indigenous Papuan Conceptualization of the Innate: A Semiotic Critique of the Ecological Perspective." In *Subsistence and Survival: Rural Ecology in the Pacific,* edited by Timothy P. Bayliss-Smith and Richard G. Feacham. New York: Academic Press.

———. 1981. *The Invention of Culture.* Chicago: University of Chicago Press.

PETER MÜHLHÄUSLER

# 10. Ecolinguistics, Linguistic Diversity, Ecological Diversity

> Coincidences of tribal boundaries to local ecology are not uncommon and imply that a given group of people may achieve stability by becoming the most efficient users of a given area and understanding its potentialities.
> —NORMAN TINDALE, *Aboriginal Tribes of Australia*

ECOLINGUISTICS, AS CHARACTERIZED by Fill (1996), is concerned with two main issues: (a) ecological embeddedness of human communication systems (i.e., language not being a self-contained system but an integral part of a larger ecosystem), and (b) the analysis of environmental discourses (i.e., both how people talk about the local environment, and the discourses of environmentalism). Its key concepts are those of diversity and of functional interrelationships. Its conceptual roots can be traced back to the writings of Humboldt (1836) and Whorf (1956). Both authors address the topic of the function of language diversity and suggest that different languages afford their users different perspectives on the world. The notion of functional interrelationships has a much shorter history in linguistics. It was first raised in the writings of Haugen (1972) with regard to relationships between languages and in my own and others' writings for the relationship between languages and their natural and cultural ecology (Mühlhäusler 1983, 1996a, 1996b; Trampe 1990; Fill 1993; Brockmeier, Harré, and Mühlhäusler 1999). Meanwhile, ecolinguistics has developed rapidly and its results have begun to be taken into account by environmental scholars, though it is too early to speak of a linguistic turn of environmental studies.

## Linguistic Approaches to Diversity

Mainstream modern linguistics has concerned itself—as its other name, general linguistics, suggests—with general principles of *language*, and to be focused on language, not languages, has been its trademark. Significantly, most linguistic pronouncements in recent years have been about principles of formalization or discovery of descriptive devices capable of being applied to a wider and wider range of language data (for a detailed discussion see appendix to Mühlhäusler 1996a).

Those linguists who have concerned themselves with languages, particularly the linguistic picture of largely multilingual areas such as West

Africa, the Americas, or Melanesia, held views that made an ecological understanding difficult. Thus, diversity has tended to be regarded as dysfunctional, the unintended result of language splits, isolation, and lack of cooperation. This view is understandable if one takes into account the mechanistic model of human communication subscribed to by most linguists. That languages, next to transmitting information (their communicative function), also serve a large range of metacommunicative functions such as marking and sustaining group identity has become a topic only fairly recently, begun with the pioneering studies by LePage and Tabouret-Keller (1985).

A particular obstacle to an ecological view has been the working hypothesis that languages are self-contained independent systems and that the boundary between languages and their external environment and between individual languages are categorical. Where linguists have recognized the existence of linguistic diversity, this diversity has been portrayed as an inventory of named entities, related to one another like biological species on a family tree. Moreover, the documentation of human languages has been highly biased toward descriptions of single languages spoken by a group of speakers rather than the speech repertoire of a communication community or the question: What languages are employed when these speakers communicate with outsiders? As a consequence, we have a vast body of descriptions of individual vernaculars and an almost total lack of description of languages of intergroup communication. That the languages of intercultural communication in the Pacific area were perhaps as numerous as its vernaculars was shown only very recently in *Atlas of Languages of Intercultural Communication in the Pacific, Asia, and the Americas* (Wurm, Mühlhäusler, and Tryon 1996).

The independence hypothesis precluded another set of research questions, including questions about adaptation of ways of speaking to specific environmental conditions (see Mühlhäusler 1996b). Such questions are now being raised by ecolinguists who propose that the result of long periods of time in which speakers develop their own interpretation of their environment has been a situation that Whorf (1956, 244) characterized as follows: "Western culture has made, through language, a provisional analysis of reality and, without correctives, holds resolutely to that analysis as final. The only correctives lie in all those other tongues which by aeons of independent evolution have arrived at different, but equally logical provisional analyses." Diversity, in this view, reflects neither regressive compartmentalization nor "progress" in the Western sense, but a large number of progresses (as well as misreadings). Prehistory is full of examples of languages and cultures dying out as a result of misreading their environment.

Diversity of languages in the view I have presented here emerges as a vast repository of accumulated human knowledge and experience, or—to use a term which is becoming fashionable in many branches of knowledge—a memory. By this I mean that in a way comparable to that in which sea currents or layers of ice are "memories" of short- and long-term climatic changes and books are memories of literate cultures, human languages are memories of human inventiveness, adaptation, and survival skills. How to access and read these memories remains an awesome task, but, as Whorf once remarked (1956, 215) with respect to the contribution of American Indian languages to human knowledge: "To exclude the evidence which their languages offer as to what the human mind can do is like expecting botanists to study nothing but food plants and hothouse roses and then tell us what the plant world is like!"

## Linguistic and Biological Diversity

The argument I wish to put forward is that our ability to get on with our environment is a function of our knowledge of it and that by combining specialist knowledge from many languages and by reversing the one-way flow of knowledge dominating the world's education system, solutions to our many environmental problems may be found. In particular, learning from local knowledge, such as learning from the insights and errors of traditional rainforest dwellers or desert nomads, could result in a more informed base for the sustained survival of our species. Such knowledge, I argue, is closely linked to language.

The chance of a productive symbiosis between linguistic-cultural and biological diversity is constrained by two major factors: first, the rapid disappearance of biological species, and, second, the even more rapid disappearance of linguistic diversity. Of the more than 6,000 languages estimated currently to be spoken, as many as 95 percent are believed by some linguists to be on the endangered list, and their rate of extinction appears to be far greater than that of biological species. How big the former loss is to some extent must remain guesswork, as human perception of this loss presupposes knowing the names of the species that are lost. By names I do not mean just labels for single species, nor scientific labels, but native local names, as well as local names for all kinds of ecologies and, very importantly, names of parts of plants and animals of use to human beings. Let me illustrate what I mean with examples selected from the lexicon of Enga, a Papuan language of the New Guinea Highlands. The compiler of this dictionary (Lang 1975) lists a number of tree names for known species, including:

tree—breadfruit (*Ficus dammaropsis*) *kúpí, tokáka, yakáte, yongáte* (T).

—breadfruit (wild) *yokopáti.*

—casuarina (*Casuarina oligodon*) *kupiama, yawále.*

—cedar (*Papuacedrus papuan* [F. Muell]) *ayápa.*

—evergreen (*Podocarpus compactus* Wassch./*P. imbricatus/P. papuanus*) *páu.*

—evergreen (*Podocarpus neriifolius/P. pilgeri*) /*káipu.*

—fig (*Ficus* sp.) *\*peke ítá.*

—mahogany (*Dysoxylum* sp.) *mamá.*

—oak (*Lithocarpus* sp.) *lépa.*

—palm *mulái.*

Lang (1975) also has a long list of tree names not yet described by European botanists, a list that in all likelihood further research would make considerably longer:

> tree—kind of *andaíta, anguana, auki, bóna, gii, káepu, kendu, kipondu, kumú, kúngu, laikiláki, lombá, lyáka, lyakati, lyungúna, matopá, naipí, náka, nápu, opáka, pálá, patepá, péké, pelepéle, pulaka, sángú, sápo, sukú, suú, wayapé, waŋame, wano, yandále, yóké*

It is not inconceivable that the massive logging program currently carried out in Papua New Guinea and recent bushfires triggered by poor forest management will lead to the disappearance of species whose name is known only to the peoples who used to live among them.

From an anthropocentric and utilitarian perspective, the perspective that inescapably drives the human perceptions of nature, most prominent among the Enga names for plants and plant parts are those that this particular culture has identified as being of use as foods, medicines, building materials, and so forth. The Enga dictionary contains a long list of entries naming plants that fall into this category including the following small sample (Lang 1975):

> tree—(bark used as rope) light wood *ángewane* (P), *wanépa.*
>
> —(bark used as string) *enámbó, komau, kotále.*
>
> —(used in leprosy cure) *dílay.*
>
> —(used for throwing stick) *kongéma.*
>
> —(where possums are found) *miná.*
>
> —(seeds eaten) *ámbea mánga, kétá, tapáé, wáima, yombuta.*
>
> —(seeds used for hair-dye) *mílya.*
>
> —(wood used for spears) *mándi.*
>
> —(used for arrows) *mámá, yupi.*
>
> —(used for arrows/bows) black *plam* (?) *kupí, mimá.*

—(used for clubs) *kulepa*.

—(used for drums) *laíyene*.

Knowledge of these plants is under very considerable threat as the Enga are becoming dependent on foods imported in tins and containers, as their children have to attend government schools where they are expected to acquire nontraditional knowledge (which leaves little time or opportunity to acquire the full traditional knowledge), and as the habitat of much of the indigenous fauna and flora is destroyed to make way for coffee plantations and gardens in which introduced food plants are grown, as well as for roads, towns, and airstrips. Studies of many other languages of the New Guinea area point to very much the same development.

## Diversity, Adaptation, Time

Ecologies are functional, adaptive, and dynamic and it is important to focus on these properties rather than static inventories and taxonomies. The question I would like to address is: How do languages adapt to changing environmental conditions?

This question gains importance because of the vastly increased mobility of humans as evidenced in migration, refugee movements, transmigration, tourism, and the like which take speakers of numerous languages into environments where these languages did not develop. Given the magnitude of this question, it is essential to keep the number of variables to a manageable minimum. What I mean is that a study of how Australian Aboriginal peoples adapted their languages to the conditions of a vast continent over 50,000–60,000 years is likely to tell us a great deal less than a study of recent settlements of small populations on small islands.

I have begun a comparative study of a number of such islands (Mühlhäusler 1996b) and arrived at the tentative conclusion that the development of linguistic means to talk about a new environment requires several generations. Let me briefly look at some island situations.

## Pitcairn Island

The story of Pitcairn Island, one of the most isolated islands of Eastern Polynesia (where Tahiti is also found), is well known through several movies dealing with the mutiny on the *Bounty* (see Ball 1973). Pitcairn was settled in 1790 by nine British mutineers, six male and thirteen female Tahitians, with "neither party knowing more than a few words of each other's language and nothing of each other's cultural heritage" (Ross 1964, 57). Pitcairn had once been inhabited:

> The mutineers found many cultivated food-trees and plants already grow-ing in their new home: the coco-nut, bread-fruit, taro, plantain, banana and sugar-cane, as well as the [aute] the paper mulberry from which the native cloth [t^pa'], was made. Other plants, such as yams and sweet potatoes, they brought with them; and for livestock, pigs, goats and chickens. Their food supplies, including sea-birds and fish, were therefore those of Tahiti, while their cultivation methods were a blend between Tahitian tradition and European improvisation—so far as one knows, only Brown, the bota-nist's assistant, had any previous knowledge of horticultural techniques. (Ross 1964, 57–8)

As an uninhabited island, Pitcairn differs from such Indian Ocean is-lands as Mauritius, Reunion, or the Seychelles, which, in the course of European colonial expansion, were settled by Dutch, French, and English plantation owners and their African slaves. Neither group was familiar with the flora and fauna of those islands. In the case of Pitcairn, how-ever, the larger proportion of the settlers of 1790, that is, those of Tahitian extraction, brought with them the knowledge and linguistic expressions needed to talk about their new environment. A limiting factor was that the Tahitians came from the lower strata of society and would not have had extensive knowledge of medicinal plants and life forms associated with the culture of their chiefs. Moreover, Tahitian was soon replaced by Pitkern (an English-based language with some Tahitian elements) and English, and much of the Tahitian linguistic heritage was lost. What re-main are lexical items descriptive of flora, fauna, and topology but not grammatical classification systems or deeper grammar.

Over a few decades the economy changed from a subsistence economy to a market economy in which fruit and vegetables were grown to provide visiting whalers and other vessels. Introduced livestock (specially goats) and excessive use of the scarce timber supply soon caused significant environmental degradation and a number of introduced plant species soon outcompeted local ones. After supplying timber for housing, boat-building, and firewood for over a hundred and seventy years, the supply of indigenous timbers is almost nil today. An introduced plant, *Eugenia jambos*, commonly known as *rose-apple*, covers much of the unused parts of the island and, though generally considered a pest, it supplies nearly all the island's wood; indeed, its presence is the only guarantee of a sufficient supply of firewood. The degradation of Pitcairn was accelerated by the strong vegetarian beliefs its inhabitants held subsequent to their religious conversion, which brought about a decline in hunting of feral animals and a reluctance to use animal manure as fertilizer.

The naming of Pitcairn life forms did profit from the presence of the migrants from nearby Tahiti, although many of the smaller less visible life forms remain unnamed. Pitcairn after all was seen by the mutineers and their companions as a larder from which they could help themselves, not as an ecosystem in need of looking after. The naming of environmental entities bears many similarities with that encountered in other colonial contexts:

> Since the flora and fauna of Pitcairn are so very different from those of England, there must, in almost all these cases, have been transfer; presumably the English settlers applied such names as best they could, guided by real or fancied similarities and, sometimes, no doubt, merely by hazy recollection of the English object. The Tahitians may have indulged in the same sort of linguistic practice, but they perhaps did so to a lesser extent by reason of the similarities between their own flora and fauna and those of the Island. Unfortunately, it is not yet possible to discuss these transfers in general. In the case of Pitcairn, an adequate flora and fauna does not yet exist. In the case of Tahiti, an adequate flora and ichthyology does exist; in the former there has been some coupling of scientific and native names, in the latter this has been rather less. The ornithology of Tahiti seems a virtually untouched field. It is, then, only in one special subject that we can make any useful comment on transfer; this is in the case of the Pitcairnese bird-names which are of English origin. The fact that we are able to do this is due to the work of Mr. Williams. I quote from a letter of his (dated July 27th, 1961). "With regard to names like sparrow, wood-pigeon, snipe, sparrow-hawk and hawk I would say that they had been applied to the nearest equivalents of the English birds. The wood-pigeon is a pigeon but you could not mistake it for the true wood-pigeon; the snipe is a shore bird but could never be mistaken for the British snipe. The sparrow does not look like a house-sparrow—or hedge-sparrow—but it was probably the only bird of the general kind of sparrow on the Island." (Ross 1964, 166)

The very considerable Tahitian knowledge in using the natural environment of Pitcairn is reflected in a number of areas including coconuts and fish:

(a) coconut terminology

| | |
|---|---|
| /a | stuff like gray cheese-cloth faded near the top of the coconut trunk at the base of the fronds |
| etu | sprouting coconut |
| hiwa | failing to reach maturity (of coconuts or bananas) |

| *matapele* | coconut meat extracted whole from split shell |
| *miti coconut* | drinking nut |
| *oʔoʔa* | full ripe coconut |
| *paito* | the baby coconut |
| *palu* | coconut husk |
| *taiʔro* | salt water sauce made by rotting strips of green coconut meat in a bottle full of salt water |

(b) fish names

| *fafaij* | stingray |
| *ihi* | piper or garfish |
| *kuta* | barracuda |
| *iai* | St. Peter's fish |
| *pa* | fish roasted on hot coal |
| *paːluː* | to use ground bait to attract fish |

The growing knowledge of other environmental phenomena is reflected in numerous coinings such as:

| *tuny-nut* | a tree whose nuts have a hole at the top—a note can be produced by blowing across it |
| *whale-bird* | birds that feed on whale offal |
| *trumpet-fish* | fish with tubulous snouts |
| *soap seed* | tree with seeds that lather |
| *shell-in-the-palm* | sea shell found on the dead pandanus |

That some plants and animals could be dangerous is reflected in terms such as:

| *dream fish* | fish whose meat causes nightmares |
| *poison trout* | poisonous trout |
| *tuiluia* | bad-tasting fish (from the adjective *taitai* "tasteless") |

A considerable number of animals and fish are named after the persons who first caught or used them, as in *Austin bird*, *Sandford* (a fish), *Frederick* (a fish), *Bernie flower* (introduced by Bernice from Mangareve), *David shell*, *Allan* (rock submerged at high tide), *Dorcos apple* (pineapple), yet many aspects of Pitcairn nature remain unnamed.

The conditions described by Tindale in the epigram to this chapter do not obtain where humans begin to talk about new environmental conditions, a view which is difficult to square with the misguided egalitarian view of many linguists that all languages are equally capable of expressing what their speakers need to express. I would like to argue, taking up the important points raised by Hymes (1973), that to the contrary languages differ considerably in their ability to do this and that adaptation of any language to a new environment takes several generations of speakers. In the case of Pitcairn, this was not helped by the later relocation of most of the islanders to a very different environment, Norfolk Island.

## Norfolk Island

Norfolk Island lies on the Norfolk Ridge in the South Pacific. One-third the size it was when thrust from the sea bed over two and a half million years ago, the island was discovered by Captain James Cook on 10 October 1774. After circumnavigating the island, Cook landed two small boats on 11 October, finding traces of Polynesian use, but no inhabitants. Noting an abundance of pine and flax (both in great demand in British manufacturing), along with a plethora of unique flora and fauna, Cook is alleged to have recorded in his journal that the island was a "paradise" (e.g., Clarke 1986, 9). An examination of his entries for 11 and 12 October 1774 shows that, while including many positive references to the "products" of Norfolk, they do not contain this epithet. He named the island in honor of the noble family Norfolk, which was one of his patrons.

The island remained uninhabited for fourteen years. On 6 March 1788, Lt. Philip Gidley King landed a party of free settlers and convicts detached from the penal settlement at Botany Bay, Australia, occupying Norfolk Island as a possession of the British Crown. But in 1814, the settlement was abandoned because the pine and flax had been found unsuitable for mast and sail making. Its buildings were either burned or demolished to discourage occupation by other nations. Whilst the government officials and a large proportion of the settlers and convicts were speakers of English, a number of other languages were represented as well (see Wright 1986, 24ff).

In 1825, Norfolk Island was reoccupied as a penal settlement in which punishment was to be as severe as possible. This prison community was closed in 1855 and the island's activities shifted to whaling, trade, and the support of the Pitcairn Islanders, who were given large tracts of land on Norfolk by the British government. The Pitcairnese were transferred to Norfolk Island because of growing food shortages on Pitcairn. Despite early guarantees from the colonial administrator that they would be given free rein in dividing the lands of Norfolk Island, the Pitcairnese actually only gained about one quarter of the total land area, and the administration of land was undertaken from New South Wales, which at the time was under British colonial rule. Some Pitcairnese then returned to their place of origin and established a second colony. With the arrival of the Pitcairn Islanders on Norfolk a tradition of bilingualism began with

English and Pitcairn-Norfolk both being integral parts of the islanders' speech repertoire. (The variant of the informal English-Tahitian language spoken by the Pitcairners on Norfolk gradually changed, and today Pitkern and Norfolk are regarded as distinct languages by their speakers.)

In 1867, the Melanesian mission training school was transferred from Auckland, New Zealand, to Norfolk Island. It was controlled by a small number of English missionaries who educated about two hundred Melanesians at any one time. St. Barnabas chapel and training college were built away from the Pitcairnese settlement, but there were regular and friendly contacts between the two. The mission was built along the lines of a British public school and served the utopian goal of converting Melanesia to Christianity and a British way of life. The college, at its peak, consisted of six dormitories holding thirty boys each, while a much smaller number of girls were divided among the households of married missionaries. The common language of the college was a mission lingua franca, Mota, but in communicating with outsiders English and Melanesian Pidgin English appear to have been used. The mission college was closed down in the early 1920s.

The environment of Norfolk Island thus was named by three separate groups over the last two hundred years, and there has been relatively little continuity in their practices. The inhabitants of the penal colony had left when the Pitcairners arrived in 1856, and the Melanesian members of the Mota-speaking mission community lived in physical as well as almost total social isolation, with only the English missionaries communicating with the Pitcairners. Maiden (1903, 715), with regard to the tree *Exocarpus phyllanthoides*, remarks that this tree, once called cherry tree by the British, became known as Isaac Wood, after Isaac Quintal from Pitcairn, who first pointed it out. "We therefore have an instance of two sets of vernaculars, the pre-Pitcairn and the post-Pitcairn." The Melanesian mission established a large number of Melanesian and English exotics around the boarding school buildings but their names disappeared when the mission left in the 1920s.

Maiden's account of Norfolk Island flora is of ecolinguistic importance on several counts:

(a) It shows that the majority of endemic and native botanical life forms do not have a local name.
(b) Most of the exotics, particularly those introduced by Pitcairners, are named.
(c) When comparing Maiden's recorded names with the most recent Norfolk dictionary (Harrison 1979), one is struck by the fact that a number of local names have disappeared (see table 1).

TABLE 1 Loss of Local Names for Flora in Norfolk:
Maiden's and Harrison's Dictionaries Compared

| Flora | Maiden (1903) | Harrison (1979) |
|---|---|---|
| *waiwai* | "beach" | — |
| *home*[a] *rauti* | *Cordyline terminalis* | — |
| *neh-e* | *Marathia fraxinea* "treefern" | — |

[a]*Home*, the word for "Pitcairn" among the Pitcairn Islanders and their descendants (Maiden 1903:719).

(d) The same local name often refers to plants of quite different species; for example, "sharkwood" is applied to *Coprosma pilosa* (Rubiaceae) and *Sideroxylon costatum* (Sapotadae). The term "maple" is used for two different tree species: *rauti* is recorded as referring to the Norfolk Island breadfruit (*Cordyline australis*) but also to *Lilium modern* and *Dracaena* sp. in Harrison (1979).

(e) Maiden also comments repeatedly on the unwillingness of the Norfolk Islanders to look after their environment, as in the following two excerpts:

> Making every allowance for the islanders, I still feel that they do not make adequate effort to keep the weeds in check. From all that I could gather, the islanders are something fatalists in the matter of weeds. (1903, 768)

> The people have so much land that at present they do not feel the deprivation of these areas which are lost to them through being rendered useless with weeds. (1903, 769)

One concludes that two hundred years have not been enough for the Norfolk language to get fine-tuned to the complex ecology of Norfolk Island with its large proportion of endemic species and that in the absence of the ability to talk about their new environment the islanders unwittingly contributed to the severe environmental degradation of fauna and flora. Today 95 percent of its rainforests are gone and large areas are invaded by feral exotics.

## Conclusions

There is an important aspect to any type of management: one can manage only what one knows; and a corollary: that one knows that for which one has a linguistic expression. My examples were drawn from situations where people found themselves in a new unknown environment and where they had to develop the necessary knowledge and linguistic resources to live in it. There is ample evidence that human colonization brings with it many negative consequences for the environment because of actions and practices that are in conflict with "nature." Profound changes to flora and fauna predate European colonization and occur in "traditional" societies as much as in "modern" ones. A case study such as that of the Marquesas (see Olson 1989) on the one hand reveals enormous destruction in the initial period; on the other, it also denotes that through learning processes over longer periods of time, an approximation between the contours of language and knowledge and the contours of the environment can be achieved and that, in many instances, a sustainable coexistence can be found.

The knowledge we find in older indigenous languages thus emerges as the outcome of many generations' fine-tuning of a language to local conditions, as suggested by Tindale (1974) for Australian Aboriginal societies, where 50,000 years of occupation contrast with 200 on Norfolk. Languages thus are repositories of past experience and once lost, a great deal of effort will be required to recover what has been lost with them.

## References

Ball, I. M. 1973. *Pitcairn: Children of the Bounty*. London: Gollancz.

Brockmeier, J., R. Harré, and P. Mühlhäusler. 1999. *Greenspeak*. Thousand Oaks: Sage.

Clarke, P. 1986. *Hell and Paradise*. Melbourne: Viking.

Fill, A. 1993. *Ökolinguistik*. Tübingen: Narr.

———. 1996. "Ökologie der Linguistik—Linguistik der Ökologie." In *Sprachökologie und Ökolinguistik*, ed. A. Fill. 3–16. Tübingen: Stauffenburg.

Harrison, S. 1979. "Glossary of the Norfolk Island Language." M.A. thesis, Macquarie University, Sydney.

Haugen, E. 1972. *The Ecology of Language*. Selected and introduced by A.S. Dil. Stanford: Stanford University Press.

Humboldt, W. von. 1836. *Über die Verschiedenheit des menschlichen Sprachbaus*. Berlin: Königliche Akademie den Wissenschaften.

Hymes, D. 1973. "Speech and Language: On the Origins and Foundations of Inequality among Speakers." *Daedalus* 102(3):59–85.

Lang, A. 1975. *Enga Dictionary*. Canberra: Pacific Linguistics C-20.

LePage, R. B., and R. Tabouret-Keller. 1985. *Acts of Identity*. Cambridge: Cambridge University Press.

Maiden, J. H. 1903. "The Flora of Norfolk Island." *Proceedings of the Linnean Society of New South Wales* 28:692–785.

Mühlhäusler, P. 1983. "Talking about Environmental Issues." *Language and Communication* 3(1):71–81.

———. 1996a. *Linguistic Ecology*. London: Routledge.

———. 1996b. "Linguistic Adaptation to Changed Environmental Conditions: Some Lessons from the Past." In *Sprachökologie und Ökolinguistik*, ed. A. Fill. 105–130. Tübingen: Stauffenburg.

Olson, S. L. 1989. "Extinction on Islands: Man as a Catastrophe." In *Conservation for the Twenty-First Century*, ed. D. Western and M. C. Peal. 50–53. Oxford: Oxford University Press.

Ross, A. S. C. 1964. *The Pitcairnese Language*. London: Deutsch.

Tindale, N. B. 1974. *Aboriginal Tribes of Australia*. Berkeley: University of California Press.

Trampe, W. 1990. *Ökologische Linguistik*. Opladen: Westfälische Verlag.

Whorf, B. L. 1956. *Language, Thought, and Reality*. Cambridge, Mass.: MIT Press.

Wright, R. 1986. *The Forgotten Generation of Norfolk Island and van Diemen's Land*. Sydney: Library of Australian History.

Wurm, S. A., P. Mühlhäusler, and D. T. Tryon. 1996. *Atlas of Languages of Intercultural Communication in the Pacific, Asia, and the Americas*. Berlin: Mouton de Gruyter.

HELEN APPLETON, MARIA E. FERNANDEZ,

CATHERINE L. M. HILL, AND CONSUELO QUIROZ

## 11. Gender and Indigenous Knowledge

KNOWLEDGE IS GENERATED BY COMMUNITIES, over time, to allow them to understand and cope with their particular agroecological and socioeconomic environment (Brouwers 1993). Such knowledge—referred to as "local," "indigenous," or "traditional"—can be termed science, because it is generated and transformed through a systematic process of observation, experimentation, and adaptation.

Local knowledge systems are geared to dealing with diversity, in both the natural environment and social organization and continue to evolve over time. Like other scientific systems, local knowledge systems develop technology and management practices to improve the quality of life of people. However, local knowledge systems differ fundamentally from those based on modern science and technology (s&t) in that they are managed by the users of the knowledge and they are holistic. Although both "bodies of knowledge"—traditional and modern—are structured by systems of classification, sets of empirical observations about local environments, and systems of self-management that govern resource use (Johnson 1992), they differ in their capacity to deal with local problems and in the degree to which they are accessible to the members of the social group charged with resource management and production.

Because the primary social differentiation among adult, economically active members of a society is based on gender, specific spheres of activity become the domains of different genders as they increase their knowledge and skill over time. As a result, local knowledge and skills held by women often differ from those held by men. For example, in certain parts of the Andes, women have much more knowledge of livestock management practices than men, whereas men know much more about soil classification than women. This specialization is publicly recognized: women are consulted when decisions about health and breeding strategies have to be made; men are consulted when the appropriateness of a particular field for a crop is being weighed (Fernandez 1992). Relations between men and women in a culture will affect hierarchies of access, use, and control resulting in different perceptions and priorities for innovation and their use of technology (Appleton 1993a, b).

Full recognition of local knowledge systems is central to the issue of sustainable and equitable development. Until recently, they have been viewed as "backward," "static," and a "hindrance" to modernization. This negative view has been fostered by a tradition of Western science, which has resulted in today's highly specialized disciplines such as cell biology, molecular biology, and epidemiology (Hill 1994). Although the idea of scientists and technologists working together globally to find solutions to the world's problems is inspiring, the reality is many different units racing independently toward goals that are defined principally in terms of the profit potential (Appleton 1993b).

The view that modern science is capable of providing the solution to "underdevelopment" is also responsible for the depreciative view of indigenous and local knowledge systems. Furthermore, the focus on objectivity, rigor, control, and testing has helped to develop the perception that s&t are value-free, and that they operate outside of the societies in which they are based. Unfortunately, given the tremendous influence of s&t, this attitude has undermined the capacity of local knowledge systems to innovate and has lowered the status of the innovators themselves, especially women whose contribution to technological development has been historically undervalued.

For example, in a study in semi-arid areas of western India, researchers set out to establish the degree of knowledge scientists had about farmers' practices and to highlight the importance of understanding scientists' assumptions about local knowledge (Gupta 1989). The scientists, from various scientific backgrounds, were working for the All India Coordinated Research Project on Dryland Agriculture (aicrpda), Haryana Agricultural University, Hissar, and the University's Dryland Research Station at Bawal. The study concluded (Gupta 1989):

> These scientists have rarely investigated the reasons for the practices they mentioned. Thus the science underlying the rational practices and the myths behind not-so-scientific practices have not been understood. We want to state unambiguously that the mere documentation of peasant practices is not enough. We have to identify the scientific basis of peasant practices and link it with their rationality.

Gupta (1989) also set out to test the validity of biologists' assumptions about women's homestead gardens. At a meeting, the scientists revealed that they believed the homesteaders used space inefficiently; that the vegetation was planted randomly or left to chance; and that trees were grown for only a single purpose, fuel or fruit. The validity of these assumptions was then tested by a team of women scientists working through maps of homesteads with local women. They discovered a complex system of planning, indicating some order in the apparent disorder. It also emerged that responsibilities for the homestead were divided among the men and women, and did not rest solely with the women as had been previously assumed.

The women scientists concluded that greater emphasis had to be placed on women's knowledge and practices: "The role of women in the homestead needed to be understood in terms of their own specialist knowledge and not just by regarding them as exploited workers who contribute to post-harvest chores" (Gupta 1989).

If a productive structure, based on the satisfaction of basic human needs and collective rather than individual consumption, is concomitant with sustainable development, the need for imported technology must be replaced by increased demand for local s&t innovation. However, the development of endogenous s&t capabilities should not necessarily follow the route of s&t in Western industrialized nations (Sagasti 1979). In Sagasti's view of resources, s&t systems focus on control and utilization, whereas local knowledge systems focus on usufruct and management. The generation of s&t is directly linked to centralized control over the distribution of information; information in local knowledge systems is the common property of integrated social groups.

Recognition and reinforcement of local knowledge systems can be the basis for an alternative development model. The capacity of these systems to integrate multiple disciplines, and the resultant synergism, are beginning to demonstrate higher levels of efficiency, effectiveness, adaptability, and sustainability than many conventional technologies. If they are to continue to contribute to sustainable development, however, local knowledge systems must be respected for what they are.

Currently, the United Nations (un) agenda includes two interrelated issues that have to do with the interface of gender, s&t, and respect and recognition of local knowledge systems in their own right:

—Conservation and reproduction of the natural environment for use by future generations; and
—The intellectual property rights (iprs) of local groups who have been responsible over time for the construction and conservation of biodiversity.

These issues directly affect the rights of women and men to manage resources critical to their innovative capacity and, therefore, their ability to contribute to a sustained development from which future generations may benefit.

## Gender, Biodiversity, and New Agrotechnologies

Although women have long been key food producers and "managers" of their environments and play a central role in the sustainable use of biological resources and life support systems, especially in the conservation and enhancement of genetic resources, their work remains relatively unnoticed by researchers and development workers (Shiva and Dankelman

1992, 44). In Dehra Dun, India, for example, local women were able to identify no fewer than 145 species of trees and their uses; forestry "experts" were familiar with only 25 species (Shiva and Dankelman 1992). The stability and sustainability of the intricately interwoven ecosystem of forest, crops, and livestock depended on the practices and knowledge systems of the local women. Their collection of fodder, fuel, and other forest material was vital to the continued flow of resources that maintained the local economy in a sustainable way (Shiva and Dankelman 1992, 46).

The introduction of new agricultural technologies is resulting in women increasingly losing control in areas where they once had considerable control. In India, for example, the shift from subsistence to commercial agriculture has led to a reduction of women's sphere of influence. Women are shown to be increasingly dependent on men for extension services, purchase of seeds, and handling of tools and money (Indian Institute of Management 1992, 47). These problems have been exacerbated by the fact that outside "experts" have tended to interact with men in rural communities. Women, who are often not directly represented in local political decision-making structures, become increasingly disadvantaged, because they lose both their knowledge and their status derived from their control over resources and knowledge. As Shiva and Dankelman (1992) argue, this situation breaks "women's sense of dignity, self-respect and self-determination." There is then the immediate danger that women's ecological knowledge will be "packaged as a product to be collected, owned, and sold in the marketplace of ideas of the scientific community" without them being compensated in any way (Shiva and Dankelman 1992).

Women's knowledge systems tend to be holistic and multidimensional. The introduction of agricultural technologies usually results in "resource fragmentation, undermining the position of women. The flows of biomass resources, that is, plant material for food, fodder, and fuel, as well as animal wastes traditionally maintained by women, are disturbed and the different linkages between the agriculture, forest, and livestock sectors of the system break down" (Shiva and Dankelman 1992, 48). In addition, inputs and outputs become completely dependent on external markets. Within this environment, the "women's role becomes more and more that of a labourer as she loses her control over production and access to resources" (Shiva and Dankelman 1992, 47).

## Gender and Intellectual Property Rights

IPRs refer primarily to international and national legal mechanisms used to protect primarily corporate and individual interests within a profit-motivated S&T system. The term is ineffective when applied to local knowledge as it does not recognize its status as a community responsibility rather than "private" property.

For thousands of years, plant genetic material has been collected, initially by local communities, then by colonizers, and later by botanists, plant breeders, and biotechnologists. Over the last 20 years, germplasm has been systematically collected from and stored in "genebanks." There has been much debate over the "ownership" of these collections as well as the safety of the material, the development of national laws restricting the availability of germplasm, and IPR to new varieties.

Because of the recent practice of "biodiversity prospecting," IPRs focus disproportionately on protecting corporate or individual knowledge in the area of biological products, leaving a whole range of cultural or community knowledge open to exploitation. Genetic resources are often incorrectly referred to as the "raw materials" for biotechnology, whereas in reality they are the products of the intellectual, cultural, and environmental contributions of local innovators, both women and men. Describing them as raw materials allows dominant S&T systems to exploit not only the matter, but also the people, as they are seen to belong to "no one in particular."

An exploitive asymmetry is thus created. When information is collected from Andean women peasants and Amazonian native peoples, for example, scientists consider it to be the "common heritage" of humanity, a public good for which no payment is appropriate or necessary. However, when the information is processed and transformed in the laboratories or factories of so-called "developed" nations, its value is enforced by legal and political mandate.

In the era of biotechnology, all biological "products" and processes could become patentable material, and countries such as the United States could be in a position to act against any country that did not provide exclusive opportunities for their corporations protected by their national laws. As Greaves (1994, ix) argues,

> [Local knowledge] now far more than in the past, is under real or potential assault from those who would gather it up, strip away its honoured meanings, convert it to a product, and sell it. Each time that happens the heritage itself dies a little, and with it its people.

Acquirers of local knowledge have power, technology, "inside" information, and sophisticated economic systems that allow them to take unfair advantage of knowledge innovators, particularly women, who have less access to power structures.

Currently, there are few provisions for the protection of local knowledge systems from outside exploitation. Applying existing patent and

copyright laws to local knowledge is not only impossible, but also impractical for various reasons: there is no identifiable inventor; all traditional
culture is already in the "public domain"; and the protection would, at
best, last only a finite number of years. Furthermore, the present purpose
of patent and copyright protection is to encourage profits for a few, not to
sustain a community and environment as a living system.

A new legal instrument is needed—one that confers ownership and
control of local knowledge on those who create, develop, and enhance it
and that recognizes the differential access of women and men to political
decision-making structures. This instrument would include ownership
of, and control over, knowledge that is commonly held rather than individual. This kind of instrument cannot be developed without the active participation of those who possess the knowledge—both men and
women.

## The Work of Governments, Universities,
## Nongovernmental Organizations, and Local Groups

Few programs have focused specifically on women's indigenous s&t
knowledge. To obtain information in this area, it is necessary to examine
a range of relevant programs and research that fall into three broad categories: s&t programs with women; general women's programs; and programs focused on indigenous knowledge. However, these categories are
self-limiting in terms of adding to knowledge about women's existing s&t
capacities. Also, information derived from an activity-specific approach
encourages a focus on particular areas of work rather than general issues
around women's indigenous technical knowledge. There is little analysis
of how information contributes to a broader understanding of the issues
or, at a strategic level, about the implications of this information for the
design of policies and strategies.

s&t programs generally focus on integration of women into s&t activities. Women are viewed as the recipients of knowledge, rather than the
generators, and the focus is on transfer of technologies to women through
"training" and equipping women with the "necessary skills." This emphasis on delivery *to* women of the necessary opportunities, technologies, and
management skills detracts from examination of existing capacity.

Women's programs tend to focus on improving women's status, access
to resources, education, training, decision-making, and empowerment *in
relation to men*. There is little critical examination of the value of women's
knowledge in relation to identified problems and available resources in
the wider environment, or of the integrity of women's knowledge as a

sphere of knowledge in its own right. The identification of "women" as a group "in need" further militates against recognition of existing strengths, as does the view of s&t as a "male" area of expertise.

Indigenous knowledge programs are not always clear about their approach, either in relation to indigenous knowledge as a system or in relation to the gender-based nature of indigenous knowledge. "Researchers . . . need to be clear in their own minds about whether they aim to legitimize local knowledge solely in the eyes of the scientific community, by picking out the 'tit-bits' of practical information, or whether they are trying to strengthen and maintain its cultural integrity" (Chambers et al. 1989). Knowledge is evaluated in terms of how well it correlates to orthodox scientific and technological thought, rather than in terms of the belief system that supports it (Last and Chavunduka 1986). Even when the system as a whole is examined, differences in type, status, and classification of women's and men's knowledge, which are fundamental to understanding the contributions and priorities of both sexes within a system, are ignored.

In the following sections, we provide examples of research and projects related to women's s&t knowledge. Much existing information is based on work in agriculture or food processing, where the essential contributions of women are finally being recognized; activities and programs in "hard" technology are less evident.

Work designed to strengthen women's indigenous skills is often carried out in teams comprising nongovernmental organizations (NGOs), research institutes, local groups, and universities at local national and international levels. It reflects two main areas of interest: the collection of information about indigenous knowledge systems, that is, their content, validity, and integrity; and the examination and development of suitable participatory research techniques for improving understanding of and working with local knowledge systems. Some universities and academic networks have also attempted to create links between formal research and development (R&D) and local experimentation (see, for example, Chambers et al. 1989, 165).

*Gender and Indigenous Knowledge Systems*

Between 1990 and 1993, offices of the Intermediate Technology Development Group (ITDG) in Asia, Africa, Central and South America, and the United Kingdom carried out research designed to focus on women as technology users, producers, and innovators. The project (IWTC n.d.), called *Do It Herself,* was based on the hypothesis that women's technological capacities are less visible than men's and that a different approach to research would, therefore, be needed. This was achieved by working with researchers (mainly women) from organizations that had established links with women technology users at the community level. Because most of the researchers were relatively inexperienced, they were taught the necessary skills—methods of research and analysis—through a series of group workshops.

The program attempted to build understanding of the existence of women's technical knowledge, and constraints to its recognition, through communication with regional audiences of NGOs, government personnel, and academic networks. After analyzing 22 case studies across a range of technical areas, researchers concluded that the invisibility of women's technology is linked to the domestic nature of their work (which denies its technical content) and the fact that women's techniques tend to focus on processes and organization of production rather than "hardware" and are, therefore, less prestigious and have a lower profile. However, at the community level, it was clear that women's technical skills are critical in survival responses to crises and problems, and that the safety nets created by these responses may be destroyed by insensitive, uninformed policies. The potential contribution that existing skills and knowledge could make to tackling problems is ignored rather than built upon.

For example, in Sudan, as many as 60 fermented food products prepared by women form an important part of people's diet (Dirar 1991). The most complicated, a clear beer called *assaliya*, is the result of a 40-step process, starting with germinated sorghum grain; it takes 3 days to produce.

Fermentation is a complex chemical process that is still not fully understood. Variations in temperature and time during the different stages affect the quality of the final product. Fermentation adds to the nutritional content of food. Using this process, women have been able to produce nutritious food from such substrates as bones, leaves, caterpillars, and cow urine.

Because fermentation increases nutritional value and preservative qualities, the process has played an important role in enabling people to cope with food shortage and famine in the past. Unfortunately, international drought and famine relief operations have been based on supplying imported foods rather than building capacity to produce local foods. This capacity is beginning to diminish as older women die without passing on their knowledge.

The information derived from the *Do It Herself* study has been disseminated to policymakers in NGOs and governments nationally and internationally. However, an important element of research is the feedback of information to the owners of knowledge. Therefore, the program includes the repackaging of information for women technology-users to build up their own knowledge and awareness of the skills and techniques that they are using (see Appleton 1993a,b).

Chambers et al. (1989) document a wealth of evidence of s&t knowledge, innovation, and activities of farmers in the South. The paper sets out "flexible research processes" to facilitate interaction between farmers and scientists and develop or adapt existing methods and technologies. The editors advocate a "complementary relationship" between knowledge possessed by scientists and technical experts and farmers' indigenous s&t knowledge. Although the message is not new, the approach is particularly helpful in bridging the gap between theoretical and abstract literature, and providing actual case material from which practical methods can be developed.

The editors stress the role of women farmers as a group who possess a wealth of often neglected knowledge. For example, the On-Farm Seed Project (OFSP) was a collaborative program of the Peace Corps Senegal, the African Food Systems Initiative, a Senegalese rice agronomist and plant breeder from Institut Sénégalais de Recherches Agricoles, and women rice farmers in the Casamance (southern Senegal). Individual interviews revealed that the women farmers were knowledgeable about the varieties of rice they grow and are using methods best adapted to the local environment. "Rice projects have found it impossible to improve on this indigenous kajando technology" (Chambers et al. 1989, 15).

## Women Promoting Diversity

Several programs have highlighted the importance of recognizing gender issues in the maintenance of diversity. Women and men have different roles and areas of knowledge in relation to seed selection, for example. A further factor is that women depend on environments rich in diversity to ensure household and community survival during periods of crisis.

*Curators of diversity:* A few of the old women farmers in the Quechua communities of the Andes possess rare knowledge of plant breeding, which is probably a legacy of the ancient Inca civilization (Ojeda 1994). Potatoes are normally propagated by asexual reproduction, that is, by planting whole potatoes or sections. The resulting plants are, therefore, clones, genetically identical to the parent plant. However, the wise women gather potato seeds from the fruit of the potato—a practice which has been all but abandoned.

Because potatoes were first cultivated in the Andes, there are countless varieties of the crop, and people have different uses for each type.

Gathering seeds enables the women to breed new varieties with characteristics they prefer. The process includes collecting the fruit, storing it outside until the following season to promote the production of the chemicals that activate the dormant seeds, planting the seeds just before the rains, harvesting tiny tubers and hiding them until the following year, then planting them to produce first generation tubers. The "tuber seed" products of this harvest, "grandchildren" of the original seeds, are sorted by shape, colour, and other characteristics. The various types are usually distributed among the woman's children to be planted to produce food crops. So far younger generations have not taken on this role of "curator of diversity."

*Developers of a new crop:* In 1957, the Tonga of northwestern Zimbabwe were moved to Matabeleland, because their valley was to be flooded during the Kariba hydroelectric scheme (Mpande and Mpofu 1991). Soil conditions at the new site were poor, rainfall low, and hunting was prohibited. People could not produce enough to feed their families and became dependent on government handouts. To survive, Tonga women have invented and adapted food production and processing technologies and have identified new sources of food—47 indigenous plants whose leaves are used for relish and over 100 tree species with a variety of edible parts.

One of these plants is the tamarind, *Tamarindus indica*. Although tamarind is widely used throughout the world for many purposes from medicine to fish preservation, it is relatively unknown in Zimbabwe, except among the Asian population. Women store the fruit for up to 12 months. It has some nutritional value and does not rot, making it especially valuable during famine. Tamarind is processed and used:

—As a flavouring agent in sorghum or millet porridge—the fruit or, in times of shortage, the leaves are soaked and boiled;

—As a substitute for commercial beverages such as tea and coffee, which are expensive or unavailable—ripe or unripe fruits are used, the acidity of the latter being neutralized with ash;

—As a snack—the seeds, which have a high protein content are fried;

—As a substitute for or supplement to scarce maize, sorghum, or millet meals—the seeds are soaked, boiled, pounded, and added to cereals;

—As a medicine—concentrated tamarind juice is used to cure gastrointestinal disorders and may also be added to animal drinking water as it is thought to cure sleeping sickness; and

—As a coagulant—the juice is used to curdle fresh milk.

Tonga women have begun to realize the commercial potential of tamarind and other wild fruits, and are trading the fresh fruit for clothing from agents outside the area. The women are aware of the market, but have not yet developed strategies for dealing with it. They are afraid that large-scale commercialization will cause them to lose control over the source of the fruit, and that it will no longer be available to them as a subsistence crop.

## *The Comparative Advantage of Indigenous Knowledge*

The assumption that technology introduced from the outside is more cost-effective or more productive has begun to be challenged. Various studies highlight the necessity of evaluating existing indigenous technologies, with full understanding of local conditions and local priorities, *before* introducing new ones. Other work demonstrates how interventions can build on the comparative advantages of existing systems, providing an interface between external and internal knowledge.

*Traditional processes versus mechanization:* The aim of one study (Luery et al. 1992) was to "analyse the effect of innovation on rural women, with particular reference to *gari* processing in the Ibaden area of southwestern Nigeria—cassava processed into gari is the most important staple food in most of Nigeria. The study focused initially on obtaining estimates of costs, returns, and amount of labour involved in processing from the 105 women participants. The women were asked to suggest solutions to problems they had identified, and traditional production processes were compared with those used by the "mechanized" cooperative and a nearby factory. The most significant finding was that the traditional gari-processing system is more efficient than mechanized systems, in terms of cost, returns, and relevance to the needs of the village economy.

*Governments using indigenous medicine:* In China, traditional medicine has been practised for about 3,000 years and has developed into a complex system of methods, including acupuncture, herbal medicine, moxibustion, massage, and deep-breathing exercises. Chinese medicine is low cost and accessible. From the early 1950s, the integration of Chinese and Western medicine was encouraged, and the practice was officially recognized in the Chinese constitution of 1982. Today, hospitals and research institutes incorporate both systems. Results have been impressive: major breakthroughs have been achieved in medicines for the treatment of certain types of cancer, hepatitis B, lupus erythematosus, leukemia, bone fracture, acute abdominal disease, and coronary heart disease. Among those who practise integrated medicine, 26 percent are women; women make up 22 percent of doctors working in Chinese medicine and 46 percent of those working in Western medicine.

In its development plan, the Government of Ghana identified as a national priority the need for

> a thorough investigation of the processes and techniques involved in all the important traditional economic activities in farming, processing of agricultural products. . . . This will help evolve and develop the appropriate technology which can help create reasonably self-reliant communities enjoying progressively better standards of living.

With financial support from the Dutch government, a project was launched, under auspices of the International Labour Organisation (ILO), to place special emphasis on improving the status, education, development, and employment of women and to improve their living and working conditions. The specific objectives were to promote the use of appropriate technologies by rural women; arrange the local and indigenous manufacture of the necessary tools and equipment; and strengthen the technological capabilities of indigenous R&D institutions (Ewusi 1987).

For example, soap-making (*alata* and *amonkye*) has been a traditional activity among rural women in Ghana long before the introduction of bar soap (Ewusi 1987). Alata, in particular, is used by women who like its mildness and cosmetic properties. However, the commercial value of light-coloured bar soap has prompted many women to produce it rather than the traditional soaps, even though problems were associated with its production: they were not able to produce enough pale soap to make the enterprise commercially viable; foaming during the boiling stage constituted a health hazard; and caustic soda had to be imported.

At Essam, women use a combination of traditional methods of soap-making and technology developed by the Technical Consultancy Centre (TCC) to overcome these problems. They have also been able to combine palm-oil processing with soap production. The palm oil can be used either for home consumption or in soap-making. The work is carried out by a cooperative. Different members with varying levels of experience are involved at different stages. As a woman gains experience, she takes on different responsibilities; thus, skills are shared within the group.

Overall, the initiative created a valuable contact between individual women's cooperatives and R&D institutions, particularly the TCC, and local manufacturers. The women were able to identify and relay their concerns about the technology and highlight safety, resource, and sociocultural constraints. Because of their experience, they were able to contribute ideas to improve the process. Where the technology proved inadequate or inappropriate, they were able to compensate with traditional methods. The skills that women already possessed contributed to the overall success of the introduction of the technology. Women were also able to suggest enhancements, such as perfumes and alternative oils, to increase the commercial value of the soap.

## References

Appleton, H. 1993a. "Gender, technology, and innovation." *Appropriate Technology* 20 (2): 6–8.

———. 1993b. "Women, science, and technology: looking ahead." *Appropriate Technology* 20 (2): 9–10.

Brouwers, J. H. A. M. 1993. "Rural people's response to soil fertility decline: the Adja case (Benin)." Wageningen Agricultural University, Wageningen, Netherlands. Paper 93–4.

Chambers, R., A. Pacey, and L. A. Thrupp, eds. 1989. *Farmer first: farmer innovations and agricultural research*. Intermediate Technology Publications. London: UK.

Dirar, M. 1991. "Fermented foods in Sudan." *Appropriate Technology* 20 (2): 24.

Ewusi, K. 1987. "Improved appropriate technologies for rural women: a case study of the identification, dissemination, monitoring and evaluation of the adoption of five improved technologies for food processing and home-based industries by rural women in Ghana." Adwinsa Publication. Accra: Ghana.

Fernandez, M. E. 1992. "The social organization of production in community-based agro-pastoralism in the Andes." In McCorkle, C. M., ed., *Plants, animals and people: agropastoral systems research*. Westview Press. Boulder: CO, USA. 99–108.

Greaves, T., ed. 1994. "Intellectual property rights for indigenous peoples: a sourcebook." Society for Applied Anthropology. Oklahoma City: OK, USA.

Gupta, A. K. 1989. "Scientists' views of farmers' practices in India: barriers to effective participation." In R. Chambers, A. Pacey, L. A. Thrupp, eds., *Farmer first: farmer innovations and agricultural research*. Intermediate Technology Publications. London: UK. 24–31.

Hill, C. 1994. "Healthy communities, healthy animals: reconceptualizing health and wellness." In *Indigenous and local community knowledge in animal health and production systems: gender perspectives*. Ottawa: The World Women's Veterinary Association. 4–32.

Indian Institute of Management. 1992. "Regional workshop on the development of microenterprises by women, 4–9 August 1992." Indian Institute of Management. Ahmedabad: India.

Johnson, M., ed. 1992. "Lore: capturing traditional environmental knowledge." International Development Research Centre. Ottawa: ON, Canada.

Last, M., and G. L. Chavunduka, eds. 1986. *The professionalisation of African medicine*. Manchester University Press. Manchester: UK.

Luery, A., M. Bowman, and C. Akinola. 1992. "Technology and women: contradictions in the process of economic and social change—The example of cassava processing, West Africa." Technoserve Inc. Norwalk: CT, USA.

Mpande, R., and N. Mpofu. 1991. "Coping strategies." *Appropriate Technology* 20 (2): 22.

Ojeda, M. 1994. "Potato production in the Andes." In I. Ilkkaracan and H. Appleton, ed., *Women's roles in the development of food cycle technologies.* Intermediate Technology Publications. London: UK.

Sagasti, F. R. 1979. "Towards endogenous science and technology for another development." *Development Dialogue* (1): 68–78.

Shiva, V., and I. Dankelman. 1992. "Women and biological diversity: lessons from the Indian Himalaya." In D. Cooper, R. Vellve, and H. Hobbelink, eds., *Growing diversity: genetic resources and local food security.* Intermediate Technology Publications. London: UK. 44–52.

STEPHEN B. BRUSH

# 12. Whose Knowledge, Whose Genes, Whose Rights?

Intellectual property enables individuals to gain financially from sharing unique and useful knowledge. Compensating indigenous people for sharing their knowledge and resources might both validate and be an equitable reward for indigenous knowledge of biological resources. However, indigenous knowledge, biological resources, intellectual property, compensation, and equity are all ambiguous terms. A contentious debate surrounds these terms and any possible linkage among them. Disagreements exist over the value of different types of knowledge and biological diversity, their rates of loss, and the choice of conservation strategies for each. Likewise, conflicting social perspectives must be addressed concerning the rights of indigenous people, national sovereignty, common heritage, monopolization of biological and knowledge resources, international relations, and responsibility for financing conservation. A goal of this chapter is to review several key terms in this debate: indigenous knowledge, biological resources, compensation, and equity. Another goal is to suggest options that might address both conservation and equity for indigenous people who are stewards of biological resources.

—STEPHEN B. BRUSH AND DOREEN STABINSKY

AMONG PEASANT FARMERS AND TRIBAL inhabitants of the tropical forest are men and women who are versed in the diversity and uses of local plant life. These folk perpetuate legacies of cultural knowledge, and they have few peers as stewards of biological resources. Peasant landscapes and dooryards are often de facto botanical gardens of incredible complexity—stores of biological diversity and natural compounds, providing sources of new hybrids. Yet this stewardship by uneducated farmers is rarely recognized and almost never rewarded. A possible consequence is that the botanical and cultural legacy from the past may be lost as peasant and tribal stewards lose interest in the resources within their fields and forests. When deforestation or technological change impoverishes these landscapes, the world loses biological resources of inestimable value. Threats to biological diversity of tropical forests and traditional crop inventories also imperil cultures that have nurtured and amplified biological resources in the past. The subjects of conservation biology and cultural survival are inexorably bound together, but no field comparable to conservation biology exists for culture and language. Promoting cultural survival has, however, become an important theme, not only to members of endangered groups but also to those interested in the positive contributions of cultural knowledge.

The purpose of this chapter and others in *Valuing Local Knowledge* is to consider a mercantile proposal to promote both cultural survival and biological conservation. The proposal is that cultural or indigenous knowledge be treated as a form of intellectual property in order to increase the economic return from biological resources maintained by peasants and tribal people (Sedjo 1988; Posey 1990). Monetary profit will compensate biological stewardship and encourage conservation. In theory, both stewards and users without property rights over biological resources tend to underinvest in their conservation. When biological resources are public goods, private investment is inadequate to protect them. Turning public goods, such as knowledge and biological resources, into commodities that can be bought and sold could possibly enable tribal herbalists, peasant farmers, or governments to profit from their knowledge and from conserving plant resources.

Four facts suggest the need for indigenous people to control and market their knowledge:

(1) indigenous people control and maintain significant amounts of biological resources,

(2) these resources are useful to industry and to the world community,

(3) both indigenous people and biological resources are threatened, and

(4) intellectual property is an accepted way to encourage the creation and sharing of intellectual goods such as knowledge of plants.

The proposal to create intellectual property for indigenous knowledge is attractive because it links biological resources to cultural knowledge, and it offers a means to connect the users of plant resources to those who maintain them. The links between cultural knowledge and biological diversity are numerous. Cultural knowledge leads to different land-management practices that increase biological diversity—protection of sacred forests, building and maintaining hedgerows, planting a diversity of crops and varieties, and protecting plants in the forest. Intellectual property's positive benefits, however, must be weighed against the possibility that it is unsuited to cultural knowledge or that the cost of implementing it will be prohibitive. Moreover, placing the knowledge and biological resources of farmers and herbalists behind a screen of intellectual property contradicts their historic status as common heritage held in trust for the public good. There is, therefore, a pressing need to weigh the allure of the tools of capitalism against their potential harm in the arena that involves indigenous people, peasants, scientists, and industrial users of biological resources. Intellectual property for indigenous knowledge, commoditizing knowledge and plant life, and biological prospecting are part of a rush to capitalism in times of aversion to common solutions to public problems.

## Diversity Losses Worldwide

Biologists have forcefully argued that biological resources are both very valuable and endangered; the need to conserve these resources is now widely accepted. Biological diversity is important because it is a repository of genetic information gained through long processes of biological evolution. Biological diversity is valued both for its potential use (for instance, as a source of new drugs or crops) and for its aesthetic contribution. The value derived from biological diversity (for instance, cancer-curing drugs and disease-resistant crops) far exceeds the world investment in conservation. Threats to biological diversity are seemingly intractable: population growth, poverty, and commercial interests.

Like biological diversity, language and culture are repositories of information gained through evolution. Linguistic and cultural diversity have incalculable value to present and future generations. Just as human lives are impoverished by the loss of species, so too are they diminished by culture and language loss. Local knowledge about tropical forest plants or different crop varieties is important both to conservation efforts and to identifying useful compounds or genes. Artistic expression, architecture, and cuisine are examples of the aesthetic value that cultural and language diversity bring to the world.

Human cultures and languages are diminishing rapidly, and this loss is as grave as the loss of biological diversity. Culture and language loss depletes the store of information as surely as the loss of biological diversity. Krauss (1992) estimates that 90 percent of the world's languages will either die or become moribund within the next century. At this rate, the loss of languages is an order of magnitude greater than the loss of biological diversity. The loss of biological and linguistic diversity are related; for instance, both tropical forest vegetation and tribal people who live within the forest are reduced by deforestation.

The recently formulated Convention on Biological Diversity is emblematic of the success of biological conservationists in framing a worldwide consensus to respond to the loss of biological diversity. Conserving culture and language is fundamentally different and more problematic than conserving biological resources. While indigenous people, their advocates, and social scientists are acutely aware of the loss of cultural and linguistic diversity, there is no political consensus on how to address this problem or how to conserve cultural knowledge. The dynamics of political and social systems make it far more difficult to design programs of cultural conservation than to lay out biological preserves or to create botanical gardens, zoos, or seed banks. Nevertheless, the value of cultural diversity and its relevance to conserving biological resources warrant an effort to address the loss of cultural knowledge. Cultural knowledge cannot adequately be conserved by setting it aside in a museum, or by recording it on paper or electronically. Like biological diversity, cultural knowledge can only be conserved by keeping it alive and in use. Intellectual property possibly opens a way to harness market forces to this objective.

## Indigenous Knowledge

Indigenous knowledge includes the botanical or pharmacological lexicons of peasants and tribal people, farmers' knowledge of soils, hunters' knowledge of animals, bakers' knowledge of yeast and dough, shamans' ability to read oracle bones, and the rules of football played in schoolyards and sandlots around the world. Two definitions of indigenous knowledge exist. A broader definition refers to popular or folk knowledge that can be contrasted to formal and specialized knowledge that defines scientific, professional, and intellectual elites in both Western and non-Western societies. Broadly defined, indigenous knowledge is the systematic information that remains in the informal sector, usually unwritten and preserved in oral tradition rather than texts. In contrast, formal knowledge is situated in written texts, legal codes, and canonical knowledge. Indigenous knowledge is culture-specific, whereas formal knowledge is decultured.

A narrower definition refers to the knowledge systems of indigenous people and minority cultures. In its current usage, indigenous knowledge is used more often in this narrow sense, to refer to the knowledge of "indigenous groups" rather than to local, popular (folk), or informal knowledge in general. This more narrow usage derives in part from the fact that most research on indigenous knowledge has been done among non-Western cultures and ethnic minorities. Another reason for a more narrow definition is the desire of anthropologists and others to validate the knowledge systems of cultures or languages that are subordinated and often depreciated by the dominant national culture and threatened with extinction.

In 1992, there were only two aged speakers of the Eyak language in Alaska and only five Osage speakers (Krauss 1992). In Australia, fully 90 percent of the 250 aboriginal languages are on the verge of extinction (Krauss 1992). These North American and Australian cases are typical of moribund languages elsewhere. They are spoken by very small cultural groups which are minorities in states dominated by other cultural groups.

These moribund languages are the remnants of cultures that are no longer whole, and their decay is treated as terminal, receiving no state support. In many instances these moribund languages are seen as the last vestiges of primitive or backward cultures that should die. More sensitive terms used to describe these cultural minorities vary from region to region, depending on historical and sociopolitical contexts: "indigenous people," "native people," "ethnic minorities," "aboriginals," and "tribal people."

Anthropologists and linguists have documented the breadth, complexity, regularities, and usefulness of indigenous knowledge (Berlin 1992). Indigenous knowledge shares these attributes with formal and specialized knowledge systems. In fact, such formal systems as Linnaean taxonomy and Western pharmacology originated in indigenous knowledge systems (Atran 1987). Indigenous knowledge (e.g., Mayan folk botanical classification) may be no less systematic than formal or specialized knowledge of Western science (e.g., Linnaean botanical classification), but indigenous knowledge is more commonly shared and accessible (Berlin, Raven, and Breedlove 1974).

"Indigenous people" is a term that is best used in regions with a colonial history that has left a predominant national culture and autochthonous cultures that coexist and compete for limited resources, especially land. This definition is especially suited to the New World, where local cultures (e.g., Hopi, Maya, Quechua, Maupuche, and Yanomamo) survived the European invasion and the establishment of national cultures with distinct but blended cultural identities (e.g., American, Mexican, Peruvian, Chilean, and Brazilian). This definition may be suited to other regions, such as the Himalayas or Central Asia, but it is not suited for large parts of Asia and Africa, where a single hybrid or creole culture (e.g., European–Native) is not dominant. Thus in China, India, or Ethiopia, different ethnic groups are present, but it is not possible to designate some as "indigenous" in contrast to others. Where an indigenous designation is inappropriate, other terms have been used, usually to indicate small size and distinct social organization: tribal, aboriginal, and ethnic minority. These terms often allude to allegedly "simpler" social and political organization.

Although different terms are employed, the common theme is that there are culturally distinct groups who have a minority status within modern nation-states and who are politically and economically subordinate. Ethnic domination, which has resulted in exploitation and ethnocide in different historical and social contexts, shows no signs of abating within modern nation-states. A legacy of domination has been the expropriation of resources without due compensation. This includes the appropriation of knowledge, ideas, and biological resources without payment or acknowledgment. Attribution is often assumed unnecessary. The dominant political status of those who collect indigenous knowledge and biological resources obscures the true origins of the resources.

While the opposition between indigenous and dominant cultures has definite historic and contemporary validity, it is often an overly simplified opposition. This opposition blurs the actual fluidity and permeability of knowledge and cultural boundaries. Indigenous knowledge very often includes information that has been adopted from the dominant culture. Thus, plant taxonomies and crop repertories among Native Americans include cultivars, names, classificatory principles (e.g., "hot" or "cold"), and information brought to the New World by European settlers. Likewise, the culture of the dominant group includes ideas and precepts from minority cultures. The Indian cultures of Mexico divide many elements and plants into hot and cold categories, following European practice brought with Spanish colonizers. The national mestizo culture of Mexico relishes cuisine that is wholly dependent on Indian components: tortillas, chile peppers, and mole.

This give and take among cultures has long been recognized by anthropologists (e.g., Redfield 1962), but the urge persists to reify knowledge systems and set artificial boundaries around culture where none exist in everyday life. In social contexts where domination, expropriation, and exploitation are the defining relationships, boundaries make it easier to define expropriation and exploitation. Ironically, both the dominant classes and the defenders of the oppressed and exploited use the same categories to define cultural differences. Nevertheless, categorical differences between cultures are elusive and usually false. Boundaries that become rigid and impermeable imperil the movement of ideas, thus threatening cultural evolution and survival.

## Biological Resources of Plants

All cultures have systematic knowledge of plants, animals, and natural phenomena. Biological resources are, however, unevenly distributed, so that particular cultures are cognizant of an unusual abundance of biological resources. Indigenous knowledge can be especially elaborate in some regions. Tribal people of the tropical forests of Asia, Africa, and Latin America and peasant cultivators in centers of crop diversity are stewards of large numbers of species and varieties (de Boef et al. 1993; Redford and Padoch 1992). Moreover, tribal and peasant peoples have encouraged and increased the biological diversity of their habitats (e.g., Salick 1992). Tribal people are conspicuous contributors to the wealth of plant resources used by the developed world. Posey (1990) estimates that the annual world market for medicines derived from medicinal plants discovered from indigenous peoples is US $43 billion. Similar estimates are available for the value of germplasm from peasant fields (Fowler and Mooney 1990).

Two types of plant resources and related indigenous knowledge are distinguished in the discussion about intellectual property and conservation: (1) crop germplasm and farmer knowledge of domesticated plants, and (2) natural products derived from wild plants and knowledge about the plants and products. These two types differ by the degree and importance of indigenous management, by the ways they are collected and used in industry, by the appropriate conservation strategy, and by intellectual property protection.

## Crop Germplasm

During the ten millennia in which humans have produced food, farmers have developed the original domesticated plants into complex populations with tremendous phenotypic and genetic diversity. With some important exceptions, crop diversity is usually concentrated in the regions where domestication occurred (Harlan 1992). This concentration derives from the presence of wild ancestors and hybridization, environmental heterogeneity, long-term crop evolution, and human selection. There are four distinct categories of crop germplasm:

(1) wild crop progenitors and relatives,
(2) semidomesticated (weedy) crop relatives,
(3) landraces of ancestral crop species, and
(4) modern crop varieties.

Wild crop relatives, belonging to related genera and species, represent the primary stock genetic material, but until the invention of modern plant breeding methods and genetic recombinant technology, this germplasm was relatively inaccessible. Weedy relatives are a bridge between domesticated and wild plants, often dwelling at the borders of fields and uncultivated land.

Landraces are genetically diverse forms of cultivated plants, a subset of biodiversity at the interface between wild plant species and domesticated species that are manipulated by humans. They are the ancestral populations of modern crop cultivars, comprised of locally named (folk) varieties, most of which are restricted in geographic distribution (Harlan 1992). Farmer knowledge of crop germplasm emphasizes infraspecific variation. These local varieties are often phenotypically and genetically distinct and very diverse.

Landraces are the legacy of hundreds of generations of farming people in cradle areas of domestication, and like other domesticated plants, land-races depend on continued human management to survive. Regions of agricultural diversity are alike in having farming systems that are local-ized in terms of knowledge, the use of few commercial inputs, and an orientation toward subsistence production (Brush 1991). Structural and political conditions such as inequitable land distribution, unfair terms of trade, ethnic domination, and political disenfranchisement have been emphasized as determinants of farming systems where genetic diversity is concentrated (Fowler and Mooney 1990).

Crop germplasm, particularly landraces, has been extensively collected for evaluation, use in breeding programs, and conservation. Hundreds of thousands of accessions gathered from fields and markets are kept in a worldwide network of national and international centers (Plucknett et al. 1987). Landraces provide the bulk of crop germplasm that has been collected and stored in gene banks, and are a primary breeding material in crop-improvement programs. For instance, the International Potato Center in Lima, Peru, has over 6,500 potato varieties in storage, and the International Rice Research Institute in Manila has 78,000 rice accessions in storage, mostly local varieties from farms (Plucknett et al. 1987). These accessions are characterized and evaluated for possible use in breeding programs and are exchanged among scientists. Breeding is done in both public and private programs, but public programs are the more important for food crops. To date, the primary conservation strategy for crop germ-plasm has been *ex situ*, in gene banks and botanical gardens. Farmer-based *in situ* methods have been proposed and implemented in a few instances (Brush 1991; Cooper et al. 1992; Friis-Hansen 1994). A balance of *ex situ* and *in situ* conservation is gaining acceptance as the best way to conserve crop germplasm (Jana 1993; NRC 1993).

## Natural Products from Wild Plants

The second general type of plant resource is natural products derived from wild plants and knowledge about the plants and products. Plants containing pharmacological compounds are especially important to this type of plant resource, but others include plants with industrial com-pounds (e.g., saponins) and with pesticide properties. Unlike the crop genetic resources, which emphasize cultivated landraces, most biologi-cally active plants are wild. Logically, wild plant resources are concen-trated in the tropics, where most plant biodiversity exists. By definition, plants in this category exist without human agency, although they may be protected and encouraged in some instances (e.g., Posey 1985).

Although wild plant resources exist without the helping human hand that maintains landraces, indigenous knowledge of plants has played a critical role in identifying and making useful plants accessible to pharmacology and industry. An extensive literature in ethnobotany and ethnomedicine describes folk identification, classification, and use of medicinal plants (e.g., Schultes and Raffauf 1990). Indigenous knowledge is a valuable tool in the rapid and efficient identification of biologically active plants (Plotkin 1991; Schultes 1991). Farnsworth (1988) reports that at least 25 percent of all drugs prescribed in the United States contain natural compounds from higher plants, and his comparison of the use of plants and plant compounds in medical science and traditional medicine for 119 compounds shows a high correlation between the two uses. Sixty-six percent of the compounds used in the two medical systems were used for the same purpose, and 9 percent are indirectly related. Improved methods to screen for compounds in biodynamic plants are likely to increase the importance of natural compounds to modern pharmacology. Indigenous knowledge is a valuable tool in locating biologically active plants (Plotkin 1991).

The industrial use of indigenous knowledge to find natural compounds differs in several ways from crop breeding with landraces. The manufacture of natural-compound products such as drugs does not rely on public sector research to the same degree as crop breeding. Natural compounds are often very valuable by themselves, while crop germplasm from one landrace or genotype is combined into elaborate hybrids with other landraces. The value of a single natural compound may be identified, but the contribution from a single landrace cannot easily be sorted out from the contributions of others. Natural compounds may first be identified in a particular plant but later synthesized or located in other plants.

Forms of intellectual property rights have been utilized to protect both crop plants and natural compounds obtained from plants. Intellectual property protection for crop germplasm is well established in industrial countries; since the 1930s, industrial countries have granted a limited form of intellectual property protection to whole crop plants. The most common method used is Plant Variety Protection, codified internationally by the Union for the Protection of Plant Varieties (UPOV). In a few countries, such as the United States, utility patents are available for crop plants, but relatively few patents have been issued for this use. However, utility patents are increasingly being used in the United States to protect whole-plant innovations that are the products of biotechnological manipulation. There is no mechanism to secure intellectual property protection for wild plants per se. Rather, specific natural compounds from plants may be patented.

## Compensation

### Equity

While value and scarcity of plant resources are linked, each has led to somewhat different concerns. Attention to value and profit leads to concern about fair treatment for people who nurture and provide plant resources to industrial users (e.g., Mooney 1983; Posey 1990). Attention to the scarcity of biological resources leads to concern for conservation (e.g., Frankel 1970; Plucknett et al. 1987). The inevitable conjoining of value and scarcity gives rise to the idea that compensation for biological resources can address both equity and conservation (Keystone Center 1991; Reid et al. 1993; Rubin and Fish 1994; Sedjo 1988, 1992). Both conservationists and advocates for indigenous people and farmers' interests have proposed that farmers and herbalists with plant resources be compensated for providing industrial users access to those resources (Posey 1990; Rubin and Fish 1994; Sedjo 1992). While drug companies and seed companies in industrial countries can profit by excluding others from using their product, farmers and herbalists have no such legal recourse. Genetic resources in farmers' fields and indigenous people's forests are treated as a public good or common heritage, while genetic resources in industrial laboratories can be treated as private property. Farmers and herbalists lack means of obtaining intellectual property protection over their innovations because of the wide distribution of genetic resources, the existence of public collections with large amounts of germplasm, and ambiguity about the source, uniformity, and novelty of resources in fields and forests.

The existence of patenting and plant breeders' rights in industrial countries and the lack of farmers' or herbalists' rights are particularly salient in the international discourse on the inequity of the existing system. Equity and compensation for indigenous people are important objectives in discussions of the fate of plant resources (e.g., Keystone Center 1991). Equity refers to the quality of being fair or balanced; compensation means to counterbalance, make up for, or make amends. These terms, however, have been used in two different ways in discussions of crop genetic resources. To some, equity and compensation suggest the need for payment to indigenous people and nations who have given their genetic resources in the past. Those who stress the inequity of the current system usually downplay the importance to less developed countries of germplasm and other technology from industrial countries' public research and international development programs (e.g., Kloppenburg and Kleinman 1988). To others, the terms *equity* and *compensation* refer to the basis for future relationships (e.g., NRC 1993; Swanson et al. 1994). Those who stress future needs emphasize the importance of defining mutual interests in developed and less-developed countries.

The Convention on Biological Diversity acknowledges the imbalance between industrial countries, which use genetic resources, and less-developed countries, which provide them. An earlier effort, the Undertaking on Plant Genetic Resources of the Food and Agricultural Organization (FAO), also addresses this imbalance (FAO 1989; Fowler and Mooney 1990). This imbalance might be redressed three ways. First, industrial countries might give up or modify their intellectual property right to monopolize elite germplasm (breeding lines) that uses landraces from farmers' fields. This could represent a return to a universal common heritage system, and it was proposed as part of the initial Undertaking on Plant Genetic Resources (FAO 1989). A variant of this proposal would be to share intellectual property rights or benefits derived from elite germplasm with countries that provided genetic resources used in the elite germplasm. Second, countries with biological diversity may design a new form of intellectual property for landraces and other genetic resources. This might be a variant of plant variety protection or utility patents found in industrial countries. Third, industrial countries might provide compensation to providers of genetic resources in recognition of the global benefit that has resulted from these resources.

The first solution, limiting intellectual property in industrial nations, is untenable. Intellectual property is firmly entrenched in industrial nations and will be increasingly common and uniform following the successful conclusion of the negotiations for General Agreement on Tariffs and Trade (GATT) in 1993. The second solution, creating a new intellectual property system for indigenous people and their genetic resources, is also impractical. The design of sui generis systems of plant protection allowed under GATT will, in all likelihood, follow patterns already established, which do not allow protection for genetic resources per se as intellectual property. Consequently, sui generis systems will be similar to plant variety protection and utility patent protection in industrial nations; such protection is designed to benefit breeders, not farmers. Present intellectual property protection for elite germplasm and natural-product drug compounds is little different from other types of intellectual property. In contrast, there is little chance that landraces or folk plant medicines can satisfy the normal criteria for intellectual property for plants: novelty or uniqueness, the result of nonobvious procedures, uniformity, and stability. Moreover, genetic resources have only recently been defined as national patrimony; their source is often unknown; they are often the result of discovery rather than invention; and collective invention is more frequently important than individual invention.

The third solution to remedy the inequity between suppliers and users of biological resources is to establish a system of compensation whereby industrial users agree to support a system of material recognition of the role of farmers and other stewards of biological resources in less-developed countries. This may be the most feasible of the three methods proposed to redress the inequalities between industrial users and peasant producers of generic resources. This compensation, however, must be seen not only as a way to recognize farmers and tribal herbalists but as a way to provide mutual benefit to both producers and users of genetic resources. In other words, compensation enables the sharing of two distinct benefits: biological resources from peasant fields or tropical forests and economic benefits from using these genetic research laboratories.

## For the Future

Thus, compensation should be defined as payment not for past services but for future options. Compensation, in the form of payment to indigenous people or in some other material transfer from industrial to less-developed countries, may be seen as a way to redress historic inequalities, but this formulation is one-sided. Payment to countries of farmers or tribal herbalists for value gained from plant resources misses the point of balancing of equities. This one-sided theme is, however, echoed in most of the discussions of compensation for genetic resources (e.g., Kloppenburg and Kleinman 1988; Mooney 1983).

Retrospective equity is problematic for several reasons. First, it focuses on only one exchange out of a multiplicity of flows of technology, knowledge, and capital. A balance sheet of benefits that flows among countries is all but impossible to construct. For example, indigenous people in less-developed countries may produce both landraces (genetic resources) and improved varieties that have used germplasm and human capital from industrial countries. Second, biological resources have historically been treated as a nonrivalrous or public good, because collection of seed or plant samples in no way decreases the availability of genetic resources to the farmers who provided it. The origin of a large percentage of biological resources that have been collected and stored is unknown or ambiguous. Third, biological knowledge and resources are usually a product of collective invention, often involving farmers of different regions and nations. Compensation to one farmer, herbalist, community, or nation may arbitrarily ignore the contributions of others. Fourth, payment for past inequities does not necessarily address the future or create lasting conservation incentives for those who provide genetic resources and those who use them. Finally, compensation for past collection does not benefit the particular farmers who originally provided the germplasm. For these reasons and others (Swanson et al. 1994; NRC 1993), retrospective compensation has been largely rejected by the international community as a basis for recognizing the contributions of farmers in maintaining, creating, and providing crop genetic resources.

A more forward-looking approach to equity is grounded in conservation theory and the notion of internalizing the value of genetic resources into the overall budget of agricultural research and industrial production (Cumberland 1991; Norgaard and Howarth 1991; Sedjo 1988). This approach argues that the real balance (equity) is to recognize the future value of genetic resources and the costs of conserving them that are borne by particular individuals, regions, or nations. Because genetic resources are public goods and treated as common property (nonexclusionary, common heritage), farmers who cultivate them cannot benefit directly from their potential value to industrial users. Rather, farmers who have traditional crop genetic resources often have good reasons to abandon them in order to produce more food or income with new technology. Farmers who keep genetic resources may forego the opportunity to increase food production or income, yet there is no current method to offset this opportunity cost and to reward them for conservation. This is the genesis of the problem of genetic erosion. Since genetic resources are public goods, society has a tendency to underinvest in their conservation and to treat them as externalities as long as they remain in farmers' fields.

The imbalance between the future value of genetic resources to the world community and the cost to farmers and less-developed nations for keeping them is measurable and tractable. This perspective replaces compensation based on past contribution with compensation for future use. A balancing of equities in this sense is to compensate farmers for the cost of keeping genetic resources for the future benefit to other consumers and farmers. The equities involved in this formulation are the opportunity costs of conservation on the one hand and the use value of genetic resources in public and private research on the other. Opportunity costs to farmers who maintain genetic resources include the costs of lower productivity or lower income from traditional crop varieties in comparison to modern ones. Equity for conservation requires that these opportunity costs be recompensed by those who benefit from resources that are conserved.

The principle of compensation for conservation does not, however, suggest how to compensate. It is unclear whether compensation can best be accomplished via market or nonmarket means. Little attention has been given to the potential impacts of market methods (intellectual property or contracts) on local farm communities or indigenous groups. The nature of markets for biological resources is poorly understood. Likewise, the efficacy of privatizing genetic resources as a means to conserve them is unknown. On the other hand, nonmarket systems have not been fully planned, implemented, or analyzed.

## Intellectual Property

### Indigenous People's Resources

The advantages of markets for addressing conservation relate to their supposed efficiency and democratic basis (Lesser and Krattiger 1994; Sedjo 1988, 1992). Because environmental degradation is associated with common property, there is little firm basis for setting the level of public investment for conservation. This is an acute problem because spending for conservation competes with funding for other public causes. Defining environmental resources as commodities and internalizing costs of conservation in production budgets allows the market to set the level of spending. This may not only give a better idea of the willingness of the public to invest in conservation, but it may also reduce administrative costs by shifting bureaucratic management of environmental protection to the private sector.

The use of market mechanisms for managing and conserving environmental resources presupposes two things. First, it assumes that environmental resources and conservation costs can be converted to commodities. Second, it assumes that the market will be an effective means for achieving conservation. Both of these assumptions are untested.

Granting intellectual property is a familiar method for converting public goods into private ones (Demsetz 1967). Intellectual property does not directly convey market value to an idea or plant that is protected. Rather, it allows the market to work where it otherwise would not, by permitting a person to exclude others from using his or her ideas or plants, except under license or royalties. The right to exclude effectively becomes the right to profit from selling the idea or plant. Without intellectual property, all ideas are public goods or common property, and no one can be excluded from using another's idea. The right to exercise temporary monopoly power, however, requires that the claimants of the right prove their eligibility. Defining, contesting, and defending this eligibility pose very high costs.

Defining biological resources as eligible for intellectual property protection has a certain superficial appeal. If farmers or herbalists can exclude others from using their genetic resources, then they may benefit from their efforts to create, identify, and preserve genetic resources. The incentive to conserve these resources may be increased if farmers can profit from selling their varieties to seed companies or breeders.

Intellectual property protection over plants generally takes one of two forms, plant-variety protection or utility patents (Jondle 1990). Both plant-variety protection and utility patents for plants are created to protect the rights of specific individuals who have manipulated biological resources to create new crop varieties (Mastenbroek 1988). Plant-variety protection allows breeders to register a variety that meets four criteria: novelty, uniformity, stability, and distinctiveness (Jondle 1990). Protection is given for a limited time (e.g., presently 20 years in the United States), allowing the owner to exclude others from selling the variety during this time. Two major exemptions, however, limit the owners' power to exclude others. The breeders' exemption allows breeders to use registered material to create new varieties without paying royalties to the original owner. The farmers' exemption allows farmers to reproduce the variety for seed and to sell that seed as long as these sales are not the major business of the farm (Jondle 1990). Revisions in 1991 to the international agreement on plant-variety protection, UPOV, allow nations to revoke either of these exemptions. In 1994, the United States eliminated the farmers' exemption from its national plant-variety protection law.

Utility patents are applied to live organisms in a few countries, notably the United States. These patents use different criteria (novelty, utility, and nonobviousness) and are more restrictive than plant-variety protection. Breeders' or farmers' exemptions are not recognized under utility patents. The recently completed Trade Related Intellectual Property Rights (TRIP) agreement of the General Agreement on Tariffs and Trade (GATT) allows nations to design sui generis plant-protection methods that adopt legal elements from industrial countries. While protection for farmer varieties or folk plant medicines is conceivable under a sui generis system of plant protection allowed by GATT, conformity to the prevailing modes of plant protection makes it unlikely that these public goods will be protected by intellectual property.

### As a Conservation Tool

The logic of the intellectual property proposal is to stimulate conservation as part of a profit-making scheme. However, several factors greatly weaken this proposal, at least with existing intellectual-property protection methods. The breeders' exemption under plant-variety protection means that a farmer or group who claims ownership of a local variety may receive little benefit from geneticists who use the local variety to breed a new commercial variety. This impediment to profiting from land-races is compounded by the allowance for derivatives in most intellectual property. Since plant-variety protection allows others to create deriva-tives, this method may be of very little value to farmers hoping for com-pensation from breeders.

A second factor is the limited market potential for genetic resources from landraces or tropical forest plants. Most breeding programs com-bine genetic material from a wide range of sources, making it difficult to designate the contribution of a single source to the final variety or to its commercial value. While unique genes are sometimes found for valu-able traits such as disease or insect resistance, commercial and publicly bred varieties are not usually based on a single gene. This indeterminacy also weakens the position of farmers who might wish to sell landraces to crop breeders. Landraces are likely to have very little commercial value because of breeders' strong preference for genetic material with known agronomic traits rather than exotic, unknown, and unadapted material (Marshall 1989).

The relative abundance of germplasm in public institutions also lessens the possibility that breeders will purchase crop germplasm from farm-ers. Large national gene banks and those at international agricultural research centers of the Consultative Groups on International Agricul-tural Research (CGIAR) system (e.g., International Rice Research Insti-tute (IRRI), CIMMYT, and CIP) control a large percentage of the world's collected germplasm reserves. These institutions have consistently taken a strong stand against commoditizing the germplasm in their collec-tions, through intellectual property or other means (Ayad 1994). Germ-plasm from these collections is more attractive than germplasm from uncollected landraces to breeders in both industrial and nonindustrial countries for several reasons—including seed health, biological identi-fication, and characterization of agronomic traits. Several studies of the uses of crop germplasm (e.g., Marshall 1989; Peeters and Galwey 1988) suggest that current supplies in international and open collections ex-ceed demand. Since most of the use of genetic resources is by the public rather than the private sector, most users are likely to expect a public source for breeding material. The abundance of collected germplasm thus undermines a market based on intellectual property for crop genetic re-sources. There seems little chance that users will pay for unknown germ-plasm when they can obtain it without cost from international and open collections.

Finally, the use of intellectual property as a conservation tool may in-volve high transaction costs. Transaction costs are expenses that are incurred in creating and administering intellectual-property protection—establishing a legal infrastructure, monitoring compliance, and resolving conflicts. Farmers who seek plant-variety or utility-patent protection for landraces will be asked to show that their material is distinctive and not already in the public domain. Enforcing these criteria will be very complex and costly, as other farmers and public programs will want to contest claims. If these criteria are relaxed, the sure result will be an arbitrary benefit to the first farmer, farm group, or nation to file for protection. The conservation benefit that may result from intellectual property, therefore, will be greatly dissipated for farm groups with genetic resources but without financial means for legal assistance.

## Contracts for Biological Prospecting

Contracts between producers of biological resources and private users are a way to avoid the monopoly-related problems associated with intellectual property. Contracts differ from intellectual property in that they do not establish or imply a monopoly over a specific invention. In theory, they are the easiest means to create a market for biological resources because they have fewer transaction costs than intellectual property. Contracts between producers and users of genetic resources might take different forms: e.g., licensing or restrictive-use provisions (Ihnen and Jondle 1990). Contracts are in fact being written by public agencies and private firms for access to biological resources related to plants with potential medicinal properties (Mays et al.; Grifo and Downes; and King et al.—*Valuing Local Knowledge*). The U.S. National Cancer Institute, Merck Pharmaceutical, and Shaman Pharmaceuticals have negotiated contracts with indigenous groups or national institutes for biological prospecting rights (Reid et al. 1993).

The success of using contracts to conserve biological resources depends on the ability of a local group or nation to control and limit the collection and shipment of genetic resources. Also, the group or nation providing biological resources must be able to attract users willing to pay a fee for the right to collect. That fee might either be flat or proportional to commercialization of products derived from the biological resource. The long lag time between collection and use may limit profit sharing for funding immediate conservation programs, but up-front payments can overcome this difficulty to a certain extent (King 1991). The large number of individuals who maintain crop genetic resources greatly reduces the possibility that a single community could control sufficient resources to attract contracts. A cartel among resource-producing communities is possible but dependent on government willingness to enforce limits on collection. Governments, in turn, are likely to expect a share of proceeds from the contract.

The role played by public agencies in maintaining and distributing crop germplasm weakens a market for genetic resources operating through contracts. Users of crop genetic resources usually acquire germplasm from secondary public sources, and most of the users are themselves in the public sector. Disease control and quarantine require public agencies to be part of the genetic-resource supply chain. Efficiencies in storage and screening complement these reasons for the role of public agencies. Because most of the costs of new-product development using genetic resources are borne in the laboratory, collectors and users of genetic resources are apt to be more interested in contracts if they guarantee some kind of exclusive exploration rights. Public agencies, however, are unlikely to act as brokers for, or grant exclusive exploration rights to, one company for a particular region or crop species. The proportion of crop germplasm that has been collected from the total pool of germplasm in landraces is estimated to be high for most crops with large commercial seed markets (Chang 1992). These estimates can be expected to depress the willingness of seed companies or breeders to underwrite costly contracts that permit access to germplasm in regions of crop diversity.

## Conclusion

Economic poverty, exploitation, and biological degradation coexist in areas with the greatest stores of domesticated and wild biological diversity. Indeed, these human and biological conditions are intimately related to one another. Conservation of biological resources in centers of crop origins and in tropical forests is in the public interest of people everywhere, regardless of economic, ethnic, or political status. Effective conservation cannot be planned or accomplished without addressing the issues of poverty, domination, and exploitation. Nevertheless, these problems are centuries old in most places. Understanding them strains the modest theoretical and methodological tools of social science. Solving them is beyond the grasp of the available political tools. Yet the value of human life, cultural diversity, and biological resources is so great that we cannot shirk from the challenge of finding viable conservation methods. The press of poverty and population growth and the urgency of protecting human dignity make this challenge as difficult as any intellectual or political challenge in the modem world.

The allure of simple explanations for poverty and degradation, such as economic exploitation, and simple solutions for conservation, such as privatizing resources, is understandable in this climate of urgency. Yet the problems of reifying cultural boundaries, weighing only one part of the flow of benefits between societies, and assigning monopoly power over resources must be confronted. Unfortunately, the complex problems of poverty and environmental degradation won't disappear with the creation of a new system of property that brings peasants and tribal people into greater contact with and dependence on industrial nation-states. On the other hand, wealthy societies which depend on the ultimate source of biological diversity—in farmers' fields, prairies, and tropical forests—must accept the burden of making conservation an acceptable alternative. Turning public goods into private property is now heavily promoted for conservation purposes. Unfortunately, this is also a high-risk method for societies and cultures that have long been subordinated. Privatization of biological resources could result in greater poverty and exploitation without achieving conservation or equity. An invaluable service to both indigenous people and conservation of biological diversity will be rendered by clarifying the terms of this discourse and the weighing of risks and benefits.

## References

Atran, S. 1987. "Origin of the Species and Genus Concepts: An Anthropological Perspective." *Journal of the History of Biology* 20:195–279.

Ayad, W G. 1994. "The CGIAR and the Convention on Biological Diversity." In *Widening Perspectives on Biodiversity*, edited by A. F. Krattiger et al., 243–54. Gland, Switzerland: IUCN and Geneva: International Academy of the Environment.

Berlin, B. 1992. *Ethnobiological Classification: Principles of Categorization of Plants and Animals in Traditional Societies*. Princeton, NJ: Princeton University Press.

Berlin, B., P. Raven, and D. Breedlove. 1974. *Principles of Tzeltal Plant Classification: An Introduction to Botanical Ethnography of a Mayan-Speaking Community in Highland Chiapas*. New York: Academic Press.

Brush, S. B. 1991. "A Farmer-Based Approach to Conserving Crop Germplasm." *Economic Botany* 45:153–66.

Chang, T. T. 1992. "Availability of Plant Germplasm for Use in Crop Improvement." In *Plant Breeding in the 1990s*, edited by H. T. Stalker and J. P. Murphy, 17–36. Wallingford, UK: C. A. B. International.

Cooper, D., R. Vellvé, and H. Hobbelink (editors). 1992. *Growing Diversity: Genetic Resources and Local Food Security*. London: Intermediate Technology Publications.

Cumberland, J. H. 1991. "Intergenerational Transfers and Ecological Sustainability." In *Ecological Economics: The Science and Management of Sustainability*, edited by R. Costanza, 355–66. New York: Columbia University Press.

de Boef, W., K. Amanor, K. Wellard, and A. Bebbington (editors). 1993. *Cultivating Knowledge: Genetic Diversity, Farmer Experimentation and Crop Research.* London: Intermediate Technology Publications.

Demsetz, H. 1967. "Toward a Theory of Property Rights." *American Economic Review* 57:347–59.

Farnsworth, N. R. 1988. "Screening Plants for New Medicines." In *Biodiversity*, edited by E. O. Wilson, 83–97. Washington, DC: National Academy Press.

Food and Agricultural Organization (FAO). 1989. *Report of the Commission on Plant Genetic Resources.* Rome: FAO.

Fowler, C., and P. Mooney. 1990. *Shattering: Food, Politics and the Loss of Genetic Diversity.* Tucson: University of Arizona Press.

Frankel, O. H. 1970. "Genetic Conservation of Plants Useful to Man." *Biological Conservation* 2:162–9.

Friis-Hansen, E. 1994. "Conceptualizing *In-situ* Conservation of Landraces." In *Widening Perspectives on Biodiversity*, edited by A. F. Krattiger et al., 263–76. Gland, Switzerland: IUCN and Geneva: International Academy of the Environment.

Harlan, J. R. 1992. *Crops and Man.* Madison, WI: American Society of Agronomy and Crop Science Society of America.

Ihnen, J. L., and R. J. Jondle. 1990. "Protecting Plant Germplasm: Alternatives to Patent and Plant Variety Protection." In *Intellectual Property Rights Associated with Plants*, edited by B. E. Caldwell and J. A. Schillinger, 123–44. Madison, WI: Crop Science Society of America, American Society of Agronomy, Soil Science Society of America, ASA Special Publication No. 52.

Jana, S. 1993. "Utilization of Biodiversity from *In-situ* Reserves, with Special Reference to Wild Wheat and Barley." In *Biodiversity and Wheat Improvement*, edited by A. B. Damania, 311–24. Chichester, U.K.: John Wiley & Sons.

Jondle, R. J. 1990. "Overview and Status of Plant Proprietary Rights." In *Intellectual Property Rights Associated with Plants*, edited by B. E. Caldwell and J. A. Schillinger, 5–15. Madison, WI: Crop Science Society of America.

Keystone Center. 1991. *Final Consensus Report of the Keystone International Dialogue Series on Plant Genetic Resources, 3rd Plenary Session.* Oslo, Norway. Keystone, CO: The Keystone Center.

King, S. R. 1991. "The Source of Our Cures." *Cultural Survival Quarterly* 15 (3):19–22.

Kloppenburg, J., Jr., and D. L. Kleinman. 1988. "Seeds of Controversy: National Property Versus Common Heritage." In *Seeds and Sovereignty: The Use and Control of Plant Genetic Resources*, edited by J. Kloppenburg, Jr., 173–203. Durham, NC: Duke University Press.

Krauss, M. 1992. "The World's Languages in Crisis." *Language* 68:4–10.

Lesser, W. H., and A. F. Krattiger. 1994. "Marketing 'Genetic Technologies' in South–North and South–South Exchanges: The Proposed Role of a New Facilitating Organization." In *Widening Perspectives on Biodiversity*, edited by A. F. Krattiger et al., 291–304. Gland, Switzerland: IUCN and Geneva: International Academy of the Environment.

Marshall, D. R. 1989. "Limitations to the Use of Germplasm Collections." In *The Use of Plant Genetic Resources*, edited by A. H. D. Brown, O. H. Frankel, D. R. Marshall, and J. T. Williams, 105–20. Cambridge: Cambridge University Press.

Mastenbroek, C. 1988. "Plant Breeders' Rights, an Equitable Legal System for New Plant Cultivars." *Experimental Agriculture* 24:15–30.

Mooney, P. R. 1983. "The Law of the Seed: Another Development and Plant Genetic Resources." *Development Dialogue* 1983:1–2. Uppsala: Dag Hammarskjöld Foundation.

National Research Council (NRC). 1993. *Managing Global Genetic Resources: Agricultural Crop Issues and Policies*. Washington, DC: National Academy Press.

Norgaard, R. B., and R. B. Howarth. 1991. "Sustainability and Discounting the Future." In *Ecological Economics: The Science and Management of Sustainability*, edited by R. Costanza, 88–101. New York: Columbia University Press.

Peeters, J. P., and N. W. Galwey. 1988. "Germplasm Collections and Breeding Needs in Europe." *Economic Botany* 42:503–21.

Plotkin, M. 1991. "Traditional Knowledge of Medicinal Plants: The Search for New Jungle Medicines." In *The Conservation of Medicinal Plants*, edited by O. Akerele, V. Heywood, and H. Synge, 53–64. Cambridge: Cambridge University Press.

Plucknett, D. L., N. J. H. Smith, J. T. Williams, and N. M. Anishetty. 1987. *Gene Banks and the World's Food*. Princeton, NJ: Princeton University Press.

Posey, D. 1985. "Indigenous Management of Tropical Forest Ecosystems: The Case of the Kayapó Indians of the Brazilian Amazon." *Agroforestry Systems* 3:139–58.

———. 1990. "Intellectual Property Rights and Just Compensation for Indigenous Knowledge." *Anthropology Today* 6:13–16.

Redfield, R. 1962. *The Little Community and Peasant Society and Culture*. Chicago: University of Chicago Press.

Redford, K. H., and C. Padoch (editors). 1992. *Conservation of Neotropical Forests: Working from Traditional Resource Use*. New York: Columbia University Press.

Reid, W. V., S. Laird, C. Meyer, R. Gámez, A. Sittenfeld, D. Janzen, M. Gollin, and C. Juma. 1993. *Biodiversity Prospecting: Using Resources for Sustainable Development*. Washington, DC: World Resources Institute.

Rubin, S. M., and S. C. Fish. 1994. "Biodiversity Prospecting: Using Innovative Contractual Provisions to Foster Ethnobotanical Knowledge, Technology, and Conservation." *Colorado Journal of International Environmental Law and Policy* 5:23–58.

Salick, J. 1992. "Amuesha Forest Use and Management: An Integration of Indigenous Use and Natural Forest Management." In *Conservation of Neotropical*

*Forests: Working from Traditional Resource Use*, edited by K. H. Redford and C. Padoch, 305–32. New York: Columbia University Press.

Schultes, R. E. 1991. "The Reason for Ethnobotanical Conservation." In *The Conservation of Medicinal Plants*, edited by O. Akerele, V. Heywood, and H. Synge, 65–75. Cambridge: Cambridge University Press.

Schultes, R. E., and R. F. Raffauf. 1990. *The Healing Forest: Medicinal and Toxic Plants of the Northwest Amazonia.* Portland, OR: Discordes Press.

Sedjo, R. A. 1988. "Property Rights and the Protection of Plant Genetic Resources." In *Seeds and Sovereignty*, edited by J. R. Kloppenburg Jr., 293–314. Durham, NC: Duke University Press.

———. 1992. "Property Rights, Genetic Resources, and Biotechnological Change." *Journal of Law and Economics* 35:199–213.

Swanson, T. M., D. W. Pearce, and R. Cervigni. 1994. *Appropriation of the Global Benefits of Plant Genetic Resources for Agriculture: An Economic Analysis of Alternative Mechanisms for Biodiversity Conservation.* Report to the Commission on Plant Genetic Resources. Rome: FAO.

## 13. The Role of the Global Network of Indigenous Knowledge Resource Centers in the Conservation of Cultural and Biological Diversity

THE TERMS *indigenous technical knowledge* and *indigenous knowledge* (IK) were first used in publications in 1979 and 1980 (Chambers 1979; Brokensha, Warren, and Werner 1980), but it was the influence of the United Nations Conference on Environment and Development held in June 1992 in Rio de Janeiro that provided a global awareness of the complementary nature of biodiversity and the indigenous knowledge about these natural resources and their uses by human communities. IK refers to the knowledge generated by communities and ethnic groups that usually pass the knowledge from one generation to the next through oral transmission; it is focused on the microenvironment in which it is generated. But from Berlin's comparative study (1992), it appears that there are more commonalities than had been anticipated between IK systems and their counterparts within the global knowledge system generated by the world's network of universities and research laboratories. This has important implications for understanding the universal nature of knowledge systems and how they are generated.

The Agenda 21 documents from the Rio conference refer to IK and the need for its preservation numerous times. One reference states that "governments . . . with the cooperation of intergovernmental organizations . . . should . . . take action to respect, record, protect and promote the wider application of the knowledge, innovations and practices of indigenous and local communities . . . for the conservation of biological diversity and the sustainable use of biological resources" (CGIAR 1993:8). The Global Biodiversity Strategy included as one of its ten principles for conserving biodiversity the principle that "Cultural diversity is closely linked to biodiversity. Humanity's collective knowledge of biodiversity and its use and management rests in cultural diversity; conversely conserving biodiversity often helps strengthen cultural integrity and values" (World Resources Institute, World Conservation Union, and United Nations Environment Programme 1992, 21).

The U.S. National Research Council stated that "development agencies should place greater emphasis on, and assume a stronger role in, systematizing the local knowledge base—indigenous knowledge, 'gray literature,' anecdotal information. A vast heritage of knowledge about species,

ecosystems, and their use exists, but it does not appear in the world literature, being either insufficiently 'scientific' or not 'developmental'" (National Research Council 1992a, 10). "If indigenous knowledge has not been
documented and compiled, doing so should be a research priority of the
highest order. Indigenous knowledge is being lost at an unprecedented
rate, and its preservation, preferably in data base form, must take place
as quickly as possible" (National Research Council 1992a, 45).

This chapter describes the active role that humans and their communities play at the global level with regard to biodiversity and its use by humans, and the role of the growing global network of IK resource centers
in recording this important last frontier of human knowledge.

## IK as Cultural Capital

Much research has been conducted on the role of colonialism and racism
in shaping attitudes and stereotypes that have devalued the contributions
to global knowledge by non-Euro-American communities. Only recently
have scholars and international development agencies begun to recognize
IK as an invaluable resource now regarded as cultural capital (Berkes and
Folke 1992; Hyndman 1994), that is, as important for preservation as biological capital. In recent years, many organizations have recognized the
importance of this overlooked resource, including the Consultative Group
on International Agricultural Research (CGIAR 1993), the International
Labor Organization (Warren 1997a), the United Nations Environmental
Programme (Dowdeswell 1993), the Food and Agricultural Organization
of the United Nations (Herbert 1993; Saouma 1993; den Biggelaar and Hart
1996), the International Board for Plant Genetic Resources (1993), the International Plant Genetic Resources Institute (Guarino 1995; Eyzaguirre
and Iwanaga 1996), the United Nations Development Programme (Rural
Advancement Fund International 1994), the U.K. Department of International Development (Sillitoe 1998), the World Bank (Warren 1991b, 1995a;
Davis 1993), the International Development Research Centre (1993), the
International Center for Living Aquatic Resource Management (Pauly,
Palomares, and Froese 1993), the U.S. National Research Council (1991,
1992a, b, 1993), and UNESCO (1994a, b). Now regarded as intellectual
property, the issue of the rights of communities to their own knowledge
is being actively discussed and debated (Brush and Stabinsky 1996; Posey
and Dutfield 1996).

Of major concern is the rapid loss of the knowledge of many communities as universal formal education is enforced with a curriculum that
usually ignores the contributions of local communities to global knowledge. The loss of knowledge is linked indelibly to language extinction

since language is the major mechanism for preserving and transmitting a community's knowledge from one generation to another (Hunter 1994).

The growing array of recorded knowledge systems related to biological materials include domesticated and nondomesticated plants and animals. A survey of this literature finds case studies for plants, forestry, fisheries, and animals (plants: Altieri and Merrick 1988; Johannes 1989; Juma 1989; Warren 1989, 1995a, b; Oldfield and Alcorn 1991; Berlin 1992; National Research Council 1992a, b, 1993; Gadgil, Berkes, and Folke 1993; Inglis 1993; Morrison, Geraghty, and Crowl 1994; Rajasekaran and Warren 1994; Abbink 1995; Martin 1995; Warren, Slikkerveer, and Brokensha 1995; Aregbeyen 1996; Ferguson and Mkandawire 1996; Setyawati 1996; Adams and Slikkerveer 1997; Innis 1997; Warren and Pinkston 1997; forestry: Posey 1985; Mathias-Mundy et al. 1992; Castro 1995; Ranasinghe 1995; Warren, Slikkerveer, and Brokensha 1995; Hanyani-Mlambo and Hebinck 1996; fisheries: Pinkerton 1989; Hviding and Baines 1992; Pauly, Palomares, and Froese 1993; Ruddle 1994; Warren, Slikkerveer, and Brokensha 1995; and animals: Gunn, Arlooktoo, and Kaomayok 1988; Kohler-Rollefson 1993; Mathias-Mundy and McCorkle 1993; Slaybaugh-Mitchell 1995; McCorkle, Mathias, and Schillhorn van Veen 1996).

It is difficult to discuss biological resources without referring to indigenous approaches to natural resource management of soils and water. There are also numerous resources now available in these areas (Hecht and Posey 1989; Carney 1991; Rajasekaran, Warren, and Babu 1991; Pawluk, Sandor, and Tabor 1992; Quintana 1992; Warren 1992; Davis 1993; Reij 1993; DeWalt 1994; Dialla 1994; Rajasekaran and Warren 1995; Roach 1997).

## Indigenous Experimentation and Biodiversity

All communities have individuals who are regarded as particularly creative in terms of active experimentation and innovation. Often this creativity extends into the realm of biodiversity. Studies among the Yoruba of Nigeria by Warren and Pinkston (1997) and among the Kayapó of Brazil by Posey (1985) demonstrate how biodiversity is actively cultivated, leading to higher indices of biodiversity. We now understand that farmers actively conduct formal on-farm and backyard breeding with similar objectives to those of crop breeders currently working with the new array of biotechnology methodologies (Warren 1996; Bunders, Haverkort, and Hiemstra 1997). There is a rich literature on indigenous approaches to experimentation and innovation (for example, Chambers 1983, 1997; Richards 1986; Ashby, Quiros, and Rivers 1989; Chambers, Pacey, and Thrupp 1989; Juma 1989; Warren 1989, 1991a, 1994, 1996, 1997b; Gamser, Appleton, and Carter 1990; Gupta 1990; den Biggelaar 1991; Haverkort, van der Kamp, and Waters-Bayer 1991; Pretty 1991, 1995; Hiemstra 1992; Atte 1992; Moock and Rhoades 1992; Reijntjes, Haverkort, and Waters-Bayer 1992; Thurston 1992, 1997; de Boef, Amanor, and Wellard 1993; Warren and Rajasekaran 1993, 1994; Haverkort and Millar 1994; McCorkle 1994; Prain and Bagalanon 1994; Rajasekaran 1994; Rural Advancement Fund International 1994; Scoones and Thompson 1994; Ashby et al. 1996; Berg 1996; den Biggelaar and Hart 1996; McCorkle, Mathias, and T. W. Schillhorn van Veen 1996; Selener, Purdy, and Zapata 1996; Innis 1997; Sumberg and Okali 1997; Van Veldhuizen et al. 1997; Warren and Pinkston 1997; Prain, Fujisaka, and Warren 1999; Warren, Slikkerveer, and Brokensha 1995).

## Gendered Knowledge and Biodiversity

The interest in IK resulted in the discovery that such knowledge is often variable depending on a person's gender, age, and occupation. New case studies indicate clearly the important role of women in many communities in preserving and extending knowledge about the biological resources in the area (including Carney 1991; Appleton and Hill 1994; Mishra 1994; Moreno-Black, Somnasang, and Thamthawan 1994; Quiroz 1994; Ulluwishewa 1994; Nazarea-Sandoval 1995; Systemwide . . . 1997; and Zweifel 1997).

## The IK Cycle and the Role of IK in Education

In the past decade the generation of IK in dynamic ways within any given community has been conceived as a cycle. IK provides the starting point in a continual process with the IK serving as the basis for both individual and community decision making. Often the decision making is carried out within indigenous organizations that have development functions such as providing forums for the identification, discussion, and prioritization of community problems as well as the search for solutions to them (Blunt and Warren 1996). The search often involves a variety of indigenous approaches to creativity that include experimentation and innovation, the results of which are evaluated, often through indigenous modes of communication, and new knowledge found to be useful is incorporated into the IK. Case studies of all of these functions within the IK cycle are presented in Warren, Slikkerveer, and Brokensha (1995).

Because IK has not been formally recorded, it has not been convenient to add it into the educational curricula for primary, secondary, and tertiary educational institutions. This has led many persons to believe that the only knowledge worthy of the label comes from the Euro-American scientific tradition. By taking interesting case studies and developing them into teaching modules, one quickly learns that students react very positively to them. There is a growing global network of individuals and institutions engaged in changing educational curricula to provide the visibility and value to IK that it deserves (for example, Kothari 1995; Kroma 1995; Warren, Egunjobi, and Wahab 1996; Kreisler and Semali 1997; Semali 1997).

## Roles of IK Resource Centers in Fostering Biocultural Diversity

The first two IK resource centers, the Center for Indigenous Knowledge for Agriculture and Rural Development (CIKARD) and the Leiden Ethnosystems and Development (LEAD) Programme, were established in 1987. By the end of 1997 there were 33 formally established centers, four universities with formal IK study groups, with another 20 centers in the process of being organized. These centers now exist on every inhabited continent. In September 1992 Canada's International Development Research Centre funded the International Symposium on Indigenous Knowledge and Sustainable Development conducted at the Regional Program for the Promotion of Indigenous Knowledge in Asia (REPPIKA) based at the International Institute of Rural Reconstruction in Silang, the Philippines (Flavier, De Jesus, and Navarro 1995). This was the first opportunity for directors of existing centers as well as individuals working to establish centers to meet personally and think through the role of the centers at the global level (CIRAN 1993a, b).

In half a decade the results of the 1992 meeting have been impressive. Manuals for recording IK have been produced by REPPIKA (IIRR 1996) and the Centre for Traditional Knowledge (1997). Directors and research associates at other centers have developed additional guidelines to facilitate the recording and archiving of IK, for example, REPPIKA (Mathias 1996a, b), the Kenya Resource Centre for Indigenous Knowledge (KENRIK) (Maundu 1995), CIKARD (Warren and McKiernan 1995), and the African Resource Centre for Indigenous Knowledge (ARCIK) (Phillips and Titilola 1995).

National and international policy related to the role of IK in sustainable approaches to natural resource management and development have been discussed by various centers including CIKARD (Rajasekaran, Warren, and Babu 1991) and the Venezuela Resource Center for Indigenous Knowledge (VERCIK) (Quiroz 1996).

The global networking has been facilitated through the publication of the *Indigenous Knowledge and Development Monito*r by the Centre for

International Research and Advisory Networks (CIRAN), published thrice yearly since 1993 and available in hardcopy as well as on the Internet at http://www.nuffic.nl/ciran/ikdm (von Liebenstein, Slikkerveer, and Warren 1995; Warren, von Liebenstein, and Slikkerveer 1993). The *Monitor* now links individuals in more than 130 countries.

Databases of recorded IK have been established at CIKARD (Warren and McKiernan 1995) and LEAD (Slikkerveer 1995). CIKARD now provides a very fast keyword search engine that leads the searcher to citations and abstracts of more than 5,000 documents housed at the CIKARD Library (http://www.iitap.iastate.edu/cikard/cikard.html), as well as IK teaching modules, and full texts of French and Spanish translations of key documents. International Center for Living Aquatic Resource Management has established a global database for IK related to fish (Pauly, Palomares, and Froese 1993).

Several centers and study groups have organized national workshops on the role of IK in sustainable development. Proceedings of several workshops have now been published, for example, the South African Resource Centre for Indigenous Knowledge (SARCIK) (Normann, Snyman, and M. Cohen 1996), the Sri Lanka Resource Centre for Indigenous Knowledge (SLARCIK) (Ulluwishewa and Ranasinghe 1996), Obafemi Awolowo University Indigenous Knowledge Study Group (Warren 1996), and the University of Ibadan Indigenous Knowledge Study Group (Warren, Egunjobi, and Wahab 1996). Other institutes in the process of establishing centers have also carried out workshops, for example, the Center for Integrated Agricultural Development (1994) in Beijing. These conferences and workshops provide opportunities to involve policymakers from government as well as representatives of international donor agencies. They also result in excellent publicity through newspaper, television, and radio coverage.

CIKARD and the Interinstitutional Consortium for Indigenous Knowledge (ICIK), as well as the Indigenous Knowledge Study Group at the University of Ibadan have been instrumental in establishing an agenda for incorporating IK case studies into educational curricula through a growing global network available on the CIKARD home page (Kroma 1995; Warren, Egunjobi, and Wahab 1996; Kreisler and Semali 1997; Semali 1997).

The African Resource Centre for Indigenous Knowledge (ARCIK) has carried out a cost-benefit analysis of using IK in development projects (Titilola 1990). Some centers have provided guidelines for other centers to emulate, for example, ARCIK (Phillips and Titilola 1995), the Philippines Resource Center for Sustainable Development and Indigenous Knowledge (PHIRCSDIK) (Serrano, Labios, and Tung 1993), LEAD

(Slikkerveer and Dechering 1995), SLARCIK (Ulluwishewa 1993), CIRAN (von Liebenstein, Slikkerveer, and Warren 1995), CIKARD (Warren and McKiernan 1995), and REPPIKA (Flavier, De Jesus, and Navarro 1995).

In order to assure easier access to new case studies, three publication series have been established. Bibliographies in Technology and Social Change and Studies in Technology and Social Change are published at CIKARD, while the IT Studies in Indigenous Knowledge and Development book series is published in London by Intermediate Technology Publications.

Guidelines and recommendations for recording IK, the development of methods and manuals for recording IK, the archiving and sharing of IK, the utilization of IK by local groups, extension workers, educators, researchers and policymakers, research on IK, policy issues, and an action plan for the IK centers are all available (CIRAN 1993a, b).

## Conclusions

The relationship between the viability of a language and the knowledge that has been created, preserved, and maintained through that language is inextricable. The recording of IK systems has clearly indicated that they are dynamic, reflecting community reactions to changing sets of circumstances. Many of them are complex and sophisticated. Some of them incorporate exciting discoveries that result from systematic experimentation, results that are being shared with other communities in various parts of the world struggling with similar problems. The use of both the neem tree and vetiver grass and the knowledge generated about them in South Asia has recently been spread globally (National Research Council 1992b, 1993). Communities that live in close contact with the natural environment have extensive knowledge of their natural resources including the biological realm and the soil and water that nurture them. They are the true managers of in situ conservation of biodiversity and the knowledge of their biological realm. With a growing number of committed persons and established IK resource centers around the globe, it is anticipated that the stores of knowledge of communities worldwide will become recorded so they can be recognized as contributions to global knowledge.

## References

Abbink, J. 1995. "Medicinal and Ritual Plants of the Ethiopian Southwest: An Account of Recent Research." *Indigenous Knowledge and Development Monitor* 3 (2):6–8.

Adams, W. V., and L. J. Slikkerveer, eds. 1997. *Indigenous Knowledge and Change in African Agriculture*. Studies in Technology and Social Change no. 26. Ames: CIKARD, Iowa State University.

Altieri, M. A., and L. C. Merrick. 1988. "Agroecology and in situ Conservation of Native Crop Diversity in the Third World." In *Biodiversity*, ed. E. O. Wilson and F. M. Peter. 15–23. Washington, D.C.: National Academy Press.

Appleton, H. E., and C. L. M. Hill. 1994. "Gender and Indigenous Knowledge in Various Organizations." *Indigenous Knowledge and Development Monitor* 2 (3):8–11.

Aregbeyen, J. B. O. 1996. "Traditional Herbal Medicine for Sustainable PHC." *Indigenous Knowledge and Development Monitor* 4 (2):14–15.

Ashby, J. A., C. A. Quiros, and Y. M. Rivers. 1989. "Farmer Participation in Technology Development: Work with Crop Varieties." In *Farmer First: Farmer Innovation and Agricultural Research*, ed. R. Chambers, A. Pacey, and L. A. Thrupp. 115–122. London: Intermediate Technology Publications.

Ashby, J. A., et al. 1996. "Innovation in the Organization of Participatory Plant Breeding." In *Participatory Plant Breeding*, ed. P. Eyzaguirre and M. Iwanaga. 77–97. Rome: International Plant Genetic Resources Institute.

Atte, O. D. 1992. *Indigenous Local Knowledge as a Key to Local Level Development: Possibilities, Constraints, and Planning Issues*. Studies in Technology and Social Change no. 20. Ames: CIKARD, Iowa State University.

Berg, T. 1996. "The Compatibility of Grassroots Breeding and Modern Farming." In *Participatory Plant Breeding*, ed. P. Eyzaguirre and M. Iwanaga. 31–36. Rome: International Plant Genetic Resources Institute.

Berkes, F., and C. Folke. 1992. "A Systems Perspective on the Interrelations between Natural, Human-made, and Cultural Capital." *Ecological Economics* 5:1–8.

Berlin, B. 1992. *Ethnobiological Classification: Principles of Categorization of Plants and Animals in Traditional Societies*. Princeton: Princeton University Press.

Blunt, P., and D. M. Warren, eds. 1996. *Indigenous Organizations and Development*. London: Intermediate Technology Publications.

Brokensha, D. W., D. M. Warren, and O. Werner, eds. 1980. *Indigenous Knowledge Systems and Development*. Lanham, Md.: University Press of America.

Brush, S. B., and D. Stabinsky, eds. 1996. *Valuing Local Knowledge: Indigenous People and Intellectual Property Rights*. Washington, D.C.: Island Press.

Bunders, J., B. Haverkort, and W. Hiemstra, eds. 1997. *Biotechnology: Building on Farmers' Knowledge*. Basingstoke, UK: Macmillan Education.

Carney, J. 1991. "Indigenous Soil and Water Management in Senegambian Rice Farming Systems." *Agriculture and Human Values* 8 (1/2):37–48.

Castro, P. 1995. *Facing Kirinyaga: A Social History of Forest Commons in Southern Mount Kenya*. London: Intermediate Technology Publications.

Center for Integrated Agricultural Development. 1994. *Indigenous Knowledge Systems and Rural Development in China: Proceedings of the Workshop*. Beijing: Beijing Agricultural University.

Centre for Traditional Knowledge. 1997. *Guidelines for Environmental Assessments and Traditional Knowledge*. Ottawa: Centre for Traditional Knowledge.
CGIAR. 1993. "Indigenous Knowledge." In *People and Plants: The Development Agenda*. 8. Rome: Consultative Group on International Agricultural Research.
Chambers, R. 1979. *Rural Development: Whose Knowledge Counts?* Special issue of *IDS Bulletin*, Institute of Development Studies, University of Sussex, 10:2.
———. 1983. *Rural Development: Putting the Last First*. London: Longman.
———. 1997. *Whose Reality Counts? Putting the First Last*. London: Intermediate Technology Publications.
Chambers, R., A. Pacey, and L. A. Thrupp, eds. 1989. *Farmer First: Farmer Innovation and Agricultural Research*. London: Intermediate Technology Publications.
CIRAN. 1993a. "Background to the International Symposium on Indigenous Knowledge and Sustainable Development." *Indigenous Knowledge and Development Monitor* 1 (2):2–5.
———. 1993b. "Recommendations and Action Plan." *Indigenous Knowledge and Development Monitor* 1 (2):24–29.
Davis, S. H., ed. 1993. *Indigenous Views of Land and the Environment*. World Bank Discussion Papers no. 188. Washington, D.C.: World Bank.
De Boef, W., K. Amanor, and K. Wellard. 1993. *Cultivating Knowledge: Genetic Diversity, Farmer Experimentation, and Crop Research*. London: Intermediate Technology Publications.
Den Biggelaar, C. 1991. "Farming Systems Development: Synthesizing Indigenous and Scientific Knowledge Systems." *Agriculture and Human Values* 8 (1/2):25–36.
Den Biggelaar, C., and N. Hart. 1996. *Farmer Experimentation and Innovation: A Case Study of Knowledge Generation Processes in Agroforestry Systems in Rwanda*. Rome: FAO.
DeWalt, B. 1994. "Using Indigenous Knowledge to Improve Agriculture and Natural Resource Management." *Human Organization* 53 (2):123–131.
Dialla, B. E. 1994. "The Adoption of Soil Conservation Practices in Burkina Faso." *Indigenous Knowledge and Development Monitor* 2 (1):10–12.
Dowdeswell, E. 1993. Walking in Two Worlds. Address presented at the Inter-American Indigenous People's Conference, Vancouver, 18 September 1993.
Eyzaguirre, P., and M. Iwanaga, eds. 1996. *Participatory Plant Breeding*. Proceedings of a workshop on Participatory Plant Breeding, 26–29 July 1995, Wageningen, The Netherlands. Rome: International Plant Genetic Resources Institute.
Ferguson, A. E., and R. M. Mkandawire. 1996. "A Crop Diversity Improvement Strategy." *Indigenous Knowledge and Development Monitor* 4 (1):6–7.
Flavier, J. M., A. De Jesus, and C. S. Navarro. 1995. "The Regional Program for the Promotion of Indigenous Knowledge in Asia (REPPIKA). In *The Cultural Dimension of Development: Indigenous Knowledge Systems*, ed.

D. M. Warren, L. J. Slikkerveer, and D. Brokensha. 479–87. London: Intermediate Technology Publications.

Gadgil, M., F. Berkes, and C. Folke. 1993. "Indigenous Knowledge for Biodiversity Conservation." *Ambio* 22 (2/3):151–56.

Gamser, M. S., H. Appleton, and N. Carter, eds. 1990. *Tinker, Tiller, Technical Change.* London: Intermediate Technology Publications.

Guarino, L. 1995. "Secondary Sources on Cultures and Indigenous Knowledge Systems." In *Collecting Plant Genetic Diversity: Technical Guidelines,* ed. L. Guarino, V. Ramanatha Rao, and R. Reid. 195–228. Wallingford, UK: CAB International on behalf of the International Plant Genetic Resources Institute in association with the FAO, IUCN, and UNEP.

Gunn, A., G. Arlooktoo, and D. Kaomayok. 1988. "The Contribution of the Ecological Knowledge of Inuit to Wildlife Management in the Northwest Territories. In *Traditional Knowledge and Renewable Resource Management,* ed. M. M. R. Freeman and L. N. Carbyn. 22–30. Edmonton: Boreal Institute for Northern Studies.

Gupta, A. 1990. *Honey Bee.* [A quarterly journal devoted to indigenous innovations]. Ahmedabad, India: Indian Institute of Management.

Hanyani-Mlambo, B. T., and P. Hebinck. 1996. "Formal and Informal Knowledge Networks in Conservation Forestry in Zimbabwe." *Indigenous Knowledge and Development Monitor* 4 (3):3–6.

Haverkort, B., and D. Millar. 1994. "Constructing Diversity: The Active Role of Rural People in Maintaining and Enhancing Biodiversity." *Etnoecológica* 2 (3):51–64.

Haverkort, B., J. van der Kamp, and A. Waters-Bayer, eds. 1991. *Joining Farmers' Experiments.* London: Intermediate Technology Publications.

Hecht, S. B., and D. A. Posey. 1989. "Preliminary Results on Soil Management Techniques of the Kayapo Indians." *Advance in Economic Botany* 7:174–88.

Herbert, J. 1993. "A Mail-order Catalog of Indigenous Knowledge." *Ceres: The FAO Review* 25 (5):33–37.

Hiemstra, W., with C. Reijntjes, and E. van der Werf. 1992. *Let Farmers Judge: Experiences in Assessing Agriculture Innovations.* London: Intermediate Technology Publications.

Hunter, P. R. 1994. *Language Extinction and the Status of North American Indian Languages.* Studies in Technology and Social Change no. 23. Ames: CIKARD, Iowa State University.

Hviding, E., and G. B. K. Baines. 1992. *Fisheries Management in the Pacific: Tradition and the Challenges of Development in Marovo, Solomon Islands.* Discussion Paper no. 32. Geneva: United Nations Research Institute for Social Development.

Hyndman, D. 1994. "Conservation through Self-determination: Promoting the Interdependence of Cultural and Biological Diversity." *Human Organization* 531 (3):296–302.

IIRR. 1996. *Recording and Using Indigenous Knowledge: A Manual.* Silang, Cavite, Philippines: REPPIKA, International Institute of Rural Reconstruction.

Inglis, J. T., ed. 1993. *Traditional Ecological Knowledge: Concepts and Cases.* Ottawa: International Program on Traditional Ecological Knowledge and International Development Research Centre.

Innis, D. Q. 1997. *Intercropping and the Scientific Basis of Traditional Agriculture.* London: Intermediate Technology Publications.

International Board for Plant Genetic Resources. 1993. "Rural Development and Local Knowledge: The Case of Rice in Sierra Leone." *Geneflow* 1993:12–13.

International Development Research Centre. 1993. Special Issue on Indigenous and Traditional Knowledge. *IDRC Reports* 21 (1).

Johannes, R. E., ed. 1989. *Traditional Ecological Knowledge: A Collection of Essays.* Gland, Switzerland: International Union for the Conservation of Nature.

Juma, C. 1989. *Biological Diversity and Innovation: Conserving and Utilizing Genetic Resources in Kenya.* Nairobi: African Centre for Technology Studies.

Kohler-Rollefson, I. 1993. "Traditional Pastoralists as Guardians of Biological Diversity." *Indigenous Knowledge and Development Monitor* 1 (3):14–16.

Kothari, B. 1995. "From Oral to Written: The Documentation of Knowledge in Ecuador." *Indigenous Knowledge and Development Monitor* 3 (2):9–12.

Kreisler, A., and L. Semali. 1997. "Towards Indigenous Literacy: Science Teachers Learn to Use IK Resources." *Indigenous Knowledge and Development Monitor* 5 (1):13–15.

Kroma, S. 1995. "Popularizing Science Education in Developing Countries through Indigenous Knowledge." *Indigenous Knowledge and Development Monitor* 3 (3):13–15.

Martin, G. J. 1995. *Ethnobotany: A Methods Manual.* London: Chapman and Hall.

Mathias, E. 1996a. "Framework for Enhancing the Use of Indigenous Knowledge." *Indigenous Knowledge and Development Monitor* 3 (2):17–18.

————. 1996b. "How Can Ethnoveterinary Medicine Be Used in Field Projects?" *Indigenous Knowledge and Development Monitor* 4 (2):6–7.

Mathias-Mundy, E., and C. M. McCorkle. 1993. *Ethnoveterinary Medicine: An Annotated Bibliography.* Bibliographies in Technology and Social Change no. 6. Ames: CIKARD Iowa State University.

Mathias-Mundy, E., et al. 1992. *Indigenous Technical Knowledge of Private Tree Management: A Bibliographic Report.* Bibliographies in Technology and Social Change no. 7. Ames: CIKARD, Iowa State University.

Maundu, P. 1995. "Methodology for Collecting and Sharing Indigenous Knowledge: A Case Study." *Indigenous Knowledge and Development Monitor* 3 (2): 3–5.

McCorkle, C. M. 1994. *Farmer Innovation in Niger.* Studies in Technology and Social Change no. 21. Ames: CIKARD, Iowa State University.

McCorkle, C. M., E. Mathias, and T. W. Schillhorn van Veen, eds. 1996. *Ethnoveterinary Research and Development.* London: Intermediate Technology Publications.

Mishra, S. 1994. "Women's Indigenous Knowledge of Forest Management in Orissa (India)." *Indigenous Knowledge and Development Monitor* 2 (3):3–5.

Moock, J. L., and R. E. Rhoades, eds. 1992. *Diversity, Farmer Knowledge, and Sustainability*. Ithaca: Cornell University Press.

Moreno-Black, G., P. Somnasang, and S. Thamthawan. 1994. "Women in Northeastern Thailand: Preservers of Botanical Diversity." *Indigenous Knowledge and Development Monitor* 2 (3):24.

Morrison, J., P. Geraghty, and L. Crowl, eds. 1994. *Science of Pacific Island Peoples*. 4 vols. Suva, Fiji: Institute of Pacific Studies, The University of the South Pacific.

National Research Council. 1991. *Toward Sustainability: A Plan for Collaborative Research on Agriculture and Natural Resource Management*. Washington, D.C.: National Academy Press.

———. 1992a. *Conserving Biodiversity: A Research Agenda for Development Agencies*. Washington, D.C.: National Academy Press.

———. 1992b. *Neem: A Tree for Solving Global Problems*. Washington, D.C.: National Academy Press.

———. 1993. *Vetiver Grass: A Thin Green Line against Erosion*. Washington, D.C.: National Academy Press.

Nazarea-Sandoval, V. D. 1995. "Indigenous Decision-making in Agriculture: A Reflection of Gender and Socioeconomic Status in the Philippines." In *The Cultural Dimension of Development: Indigenous Knowledge Systems*, ed. D. M. Warren, L. J. Slikkerveer, and D. Brokensha. 155–173. London: Intermediate Technology Publications.

Normann, H., I. Snyman, and M. Cohen, eds. 1996. *Indigenous Knowledge and Its Uses in Southern Africa*. Pretoria: Human Sciences Research Council.

Oldfield, M. L., and J. B. Alcorn, eds. 1991. *Biodiversity: Culture, Conservation, and Ecodevelopment*. Boulder: Westview Press.

Pauly, D., M. L. D. Palomares, and R. Froese. 1993. "Some Prose on a Database of Indigenous Knowledge on Fish." *Indigenous Knowledge and Development Monitoring* 1 (1):26–27.

Pawluk, R. R., J. A. Sandor, and J. A. Tabor. 1992. "The Role of Indigenous Soil Knowledge in Agricultural Development." *Journal of Soil and Water Conservation* 47 (4):298–302.

Phillips, A. O., and S. O. Titilola. 1995. "Sustainable Development and Indigenous Knowledge Systems in Nigeria: The Role of the Nigerian Institute of Social and Economic Research (NISER)." In *The Cultural Dimension of Development: Indigenous Knowledge Systems*, ed. D. M. Warren, L. J. Slikkerveer, and D. Brokensha. 475–78. London: Intermediate Technology Publications.

Pinkerton, E., ed. 1989. *Co-operative Management of Local Fisheries: New Directions for Improved Management and Community Development*. Vancouver: University of British Columbia Press.

Posey, D. A. 1985. "Management of Tropical Forest Ecosystems: The Case of the Kayapó Indians of the Brazilian Amazon." *Agroforestry Systems* 3 (2):139–158.

Posey, D. A., and G. Dutfield. 1996. *Beyond Intellectual Property: Toward Traditional Resource Rights for Indigenous Peoples and Local Communities*. Ottawa: IDRC Books.

Prain, G., and C. P. Bagalanon, eds. 1994. *Local Knowledge, Global Science, and Plant Genetic Resources: Towards a Partnership.* Los Baños: UPWARD.

Prain, G., S. Fujisaka, and D. M. Warren, eds. 1999. *Biological and Cultural Diversity: The Role of Indigenous Agricultural Experimentation in Development.* London: Intermediate Technology Publications.

Pretty, J. N. 1991. "Farmers' Extension Practice and Technology Adaptation: Agricultural Revolution in Seventeenth–Nineteenth Century Britain." *Agriculture and Human Values* 8 (1/2):132–148.

———. 1995. *Regenerating Agriculture: Policies and Practice for Sustainability and Self-Reliance.* London: Earthscan.

Quintana, J. 1992. "American Indian Systems for Natural Resource Management." *Akwe:kon Journal* 9 (2):92–97.

Quiroz, C. 1994. "Biodiversity, Indigenous Knowledge, Gender, and Intellectual Property Rights." *Indigenous Knowledge and Development Monitor* 2 (3):12–15.

———. 1996. "Local Knowledge Systems Contribute to Sustainable Development." *Indigenous Knowledge and Development Monitor* 4 (1):3–5.

Rajasekaran, B. 1994. *A Framework for Incorporating Indigenous Knowledge Systems into Agricultural Research, Extension, and NGOs for Sustainable Agricultural Development.* Studies in Technology and Social Change no. 22. Ames: CIKARD, Iowa State University.

Rajasekaran, B., and D. M. Warren. 1994. "IK for Socioeconomic Development and Biodiversity Conservation: The Kolli Hills." *Indigenous Knowledge and Development Monitor* 2 (2):13–17.

———. 1995. "Role of Indigenous Soil Health Care Practices in Improving Soil Fertility: Evidence from South India." *Journal of Soil and Water Conservation* 50 (2):146–49.

Rajasekaran, B., D. M. Warren, and S. C. Babu. 1991. "Indigenous Natural-Resource Management Systems for Sustainable Agricultural Development: A Global Perspective." *Journal of International Development* 3 (4):387–401.

Ranasinghe, H. 1995. "Traditional Tree-Crop Practices in Sri Lanka." *Indigenous Knowledge and Development Monitor* 3 (3):7–9.

Reij, C. 1993. "Improving Indigenous Soil and Water Conservation Techniques: Does It Work?" *Indigenous Knowledge and Development Monitor* 1 (1):11–13.

Reijntjes, C., B. Haverkort, and A. Waters-Bayer. 1992. *Farming for the Future: An Introduction to Low-External-Input and Sustainable Agriculture.* Leusden: ILEIA.

Richards, P. 1986. *Coping with Hunger: Hazard and Experiment in an African Rice-farming System.* London: Allen & Unwin.

Roach, S. A. 1997. *Land Degradation and Indigenous Knowledge in a Swazi Community.* M.A. thesis. Ames: Department of Anthropology, Iowa State University.

Ruddle, K. 1994. *A Guide to the Literature on Traditional Community-Based Fishery Management in the Asia-Pacific Tropics.* FAO Fisheries Circular no. 869. Rome: FAO.

Rural Advancement Fund International. 1994. *Conserving Indigenous Knowledge: Integrating Two Systems of Innovation.* New York: United Nations Development Programme.

Saouma, E. 1993. "Indigenous Knowledge and Biodiversity." In *Harvesting Nature's Diversity.* 4–6. Rome: FAO.

Scoones, I., and J. Thompson, eds. 1994. *Beyond Farmer First: Rural People's Knowledge, Agricultural Research, and Extension Practice.* London: Intermediate Technology Publications.

Selener, D., with C. Purdy and G. Zapata. 1996. *Documenting, Evaluating, and Learning from Our Development Projects: A Participatory Systematization Workbook.* New York: International Institute of Rural Reconstruction.

Semali, L. 1997. "Cultural Identity in African Context: Indigenous Education and Curriculum in East Africa." *Folklore Forum* 28:3–27.

Serrano, R. C., R. V. Labios, and L. Tung. 1993. "Establishing a National IK Resource Centre: The Case of PHIRCSDIK." *Indigenous Knowledge and Development Monitor* 1 (1):5–6.

Setyawati, I. 1996. "Environmental Variability, IK, and the Use of Rice Varieties." *Indigenous Knowledge and Development Monitor* 4 (2):11–13.

Sillitoe, P. 1998. "The Development of Indigenous Knowledge: A New Applied Anthropology." *Current Anthropology* 39 (2):223–252.

Slaybaugh-Mitchell, T. L. 1995. *Indigenous Livestock Production and Husbandry: An Annotated Bibliography.* Bibliographies in Technology and Social Change no. 8. Ames: CIKARD, Iowa State University.

Slikkerveer, L. J. 1995. "INDAKS: A Bibliography and Database on Indigenous Agricultural Knowledge Systems and Sustainable Development in the Tropics." In *The Cultural Dimension of Development: Indigenous Knowledge Systems,* ed. D. M. Warren, L. J. Slikkerveer, and D. Brokensha. 512–16. London: Intermediate Technology Publications.

Slikkerveer, L. J., and W. H. J. C. Dechering. 1995. "LEAD: The Leiden Ethnosystems and Development Programme." In *The Cultural Dimension of Development: Indigenous Knowledge Systems,* ed. D. M. Warren, L. J. Slikkerveer, and D. Brokensha. 435–440. London: Intermediate Technology Publications.

Sumberg, J., and C. Okali. 1997. *Farmers' Experiments: Creating Local Knowledge.* Boulder: Lynne Rienner Publishers.

Systemwide Programme on Participatory Research and Gender Analysis. 1997. *A Global Programme on Participatory Research and Gender Analysis for Technology Development and Organisational Innovation.* AgREN Network Paper no. 72. London: Agricultural Research and Extension Network, UK Overseas Development Administration (ODA).

Thurston, H. D. 1992. *Sustainable Practices for Plant Disease Management in Traditional Farming Systems.* Boulder: Westview Press.

———. 1997. *Slash/Mulch Systems: Sustainable Methods for Tropical Agriculture.* Boulder/London: Westview Press/Intermediate Technology Publications.

Titilola, S. O. 1990. *The Economics of Incorporating Indigenous Knowledge Systems into Agricultural Development: A Model and Analytical Framework.*

Studies in Technology and Social Change no. 17. Ames: CIKARD, Iowa State University.

Ulluwishewa, R. 1993. "Indigenous Knowledge, National IK Resource Centres, and Sustainable Development." *Indigenous Knowledge and Development Monitor* 1 (3):11–13.

———. 1994. "Women's Indigenous Knowledge of Water Management in Sri Lanka." *Indigenous Knowledge and Development Monitor* 2 (3):17–19.

Ulluwishewa, R., and H. Ranasinghe, eds. 1996. *Indigenous Knowledge and Sustainable Development.* Proceedings of the First National Symposium on Indigenous Knowledge and Sustainable Development, Colombo, March 19–20, 1994. Nugegoda: SLARCIK.

UNESCO. 1994a. "Special Issue, Traditional Knowledge in Tropical Environments." *Nature and Resources* 30(1).

———. 1994b. "Special Issue, Traditional Knowledge into the Twenty-First Century." *Nature and Resources* 30(2).

Van Veldhuizen, L., et al., eds. 1997. *Farmers' Experimentation in Practice: Lessons from the Field.* London: Intermediate Technology Publications.

Von Liebenstein, G., L. J. Slikkerveer, and D. M. Warren. 1995. "CIRAN: Networking for Indigenous Knowledge." In *The Cultural Dimension of Development: Indigenous Knowledge Systems,* ed. D. M. Warren, L. J. Slikkerveer, and D. Brokensha. 441–44. London: Intermediate Technology Publications.

Warren, D. M. 1989. "Linking Scientific and Indigenous Agricultural Systems." In *The Transformation of International Agricultural Research and Development,* ed. J. L. Compton. 153–170. Boulder: Lynne Rienner Publishers.

———. 1991a. "The Role of Indigenous Knowledge in Facilitating a Participatory Approach to Agricultural Extension." In *Proceedings of the International Workshop on Agricultural Knowledge Systems and the Role of Extension,* ed. H. J. Tillmann, H. Albrecht, M. A. Salas, M. Dhamotharah, and E. Gottschalk. 161–177. Stuttgart: University of Hohenheim.

———. 1991b. *Using Indigenous Knowledge in Agricultural Development.* World Bank Discussion Papers no. 127. Washington, D.C.: World Bank.

———. 1992. *A Preliminary Analysis of Indigenous Soil Classification and Management Systems in Four Ecozones of Nigeria.* Ibadan: African Resource Centre for Indigenous Knowledge/International Institute of Tropical Agriculture.

———. 1994. "Indigenous Agricultural Knowledge, Technology, and Social Change." In *Sustainable Agriculture in the American Midwest,* ed. G. McIsaac and W. R. Edwards. 35–53. Urbana: University of Illinois Press.

———. 1995a. "Indigenous Knowledge for Agricultural Development." Keynote speech given at the Workshop on Traditional and Modern Approaches to Natural Resource Management in Latin America, World Bank, April 25–26, 1995.

———. 1995b. "Indigenous Knowledge, Biodiversity Conservation, and Development." In *Conservation of Biodiversity in Africa: Local Initiatives and Institutional Roles,* ed. L. A. Bennun, R. A. Aman, and S. A. Crafter. 93–108. Nairobi: Centre for Biodiversity, National Museums of Kenya.

———. 1996. "The Role of Indigenous Knowledge and Biotechnology in Sustainable Agricultural Development." In *Indigenous Knowledge and Biotechnology.* 6–15. Ile-Ife, Nigeria: Indigenous Knowledge Study Group, Obafemi Awolowo University.

———. 1997a. "The Incorporation of Indigenous Knowledge into Project Implementation for the Development of Indigenous Peoples." Address given at the ILO-INDISCO Donor Consultation and Planning Workshop on Employment and Income-Generating for Indigenous and Tribal Peoples: Lessons Learned in Asia, New Delhi, 4–8 November 1997.

———. 1997b. "The Role of Indigenous Knowledge Systems in Facilitating Sustainable Approaches to Development." Proceedings of the International Conference on Nature Knowledge, Istituto Veneto di Scienze, Lettere ed Arti, Venice, 4–6 December 1997.

Warren, D. M., and G. McKiernan. 1995. "CIKARD: A Global Approach to Documenting Indigenous Knowledge for Development." In *The Cultural Dimension of Development: Indigenous Knowledge Systems,* ed. D. M. Warren, L. J. Slikkerveer, and D. Brokensha. 426–434. London: Intermediate Technology Publications.

Warren, D. M., and J. Pinkston. 1997. "Indigenous African Resource Management of a Tropical Rainforest Ecosystem: A Case Study of the Yoruba of Ara, Nigeria." In *Linking Social and Ecological Systems,* ed. F. Berkes and C. Folke. 158–189. Cambridge: Cambridge University Press.

Warren, D. M., and B. Rajasekaran. 1993. "Indigenous Knowledge: Putting Local Knowledge to Good Use." *International Agricultural Development* 13 (4):8–10.

———. 1994. "Using Indigenous Knowledge for Sustainable Dryland Management: A Global Perspective." In *Social Aspects of Sustainable Dryland Management,* ed. D. Stiles. 193–209. New York: John Wiley.

Warren, D. M., L. Egunjobi, and B. Wahab, eds. 1996. *Indigenous Knowledge in Education.* Proceedings of a Regional Workshop on Integration of Indigenous Knowledge into Nigerian Education Curriculum. Ibadan: Indigenous Knowledge Study Group, University of Ibadan.

Warren, D. M., L. J. Slikkerveer, and D. Brokensha, eds. 1995. *The Cultural Dimension of Development: Indigenous Knowledge Systems.* London: Intermediate Technology Publications.

Warren, D. M., G. W. von Liebenstein, and L. J. Slikkerveer. 1993. "Networking for Indigenous Knowledge." *Indigenous Knowledge and Development Monitor* 1 (1):2–4.

World Resources Institute, World Conservation Union, and United Nations Environment Programme. 1992. *Global Biodiversity Strategy: Policy-makers' Guide.* Baltimore: WRI Publications.

Zweifel, H. 1997. "Biodiversity and the Appropriation of Women's Knowledge." *Indigenous Knowledge and Development Monitor* 5 (1):7–9.

# III. Residues and Reinventions

Debates about the nature of science in the Third World are very different from those in Europe and North America. In the industrial capitalist countries science is already deeply entrenched in institutions, intellectual life, public policy, and technology. . . . If the earlier glow of a science linked to liberation has become increasingly tarnished, there is still pride in its achievements and nostalgia for its promise. . . . Science came into the Third World as a rationale for domination with theories of racial superiority, of "progress," and of its own intellectual superiority. . . . After the troops depart, the investments remain; after direct ownership is removed, managerial skills, patents, textbooks, and journals remain, repeating the message that only by adopting their ways can we progress, only by going to their universities can we learn, only by emulating their universities can we teach.
—RICHARD LEVINS AND RICHARD LEWONTIN, "Applied Biology in the Third World: The Struggle for Revolutionary Science"

Nature is part of history and culture, not the other way around. Sociologists and historians of science tend to know that. Most scientists do not. Because I was trained as a scientist, it has taken me many years to understand that "in science, just as in art and life, only that which is true to culture is true to nature."
—RUTH HUBBARD, *The Politics of Women's Biology*

Certainly big science requires an audience . . . If one gazes at [the audience of practitioners] for only a moment, one does not see a collection of equal scientists. Instead, they are ranked, not only individually, but also by nationality. They are nearer and farther from "the action," nearer and farther from stage center. Scientists say that the differential right to judge (or act) in their arena is determined solely by merit; they also argue that scientists are much more likely to be meritorious if they are from Europe or North America (and male). It is this discourse on center and margin, core and periphery, among scientists that concerns me. . . . I am interested in how this "colonialist discourse" might be reenacted in the organization of large laboratories and research groups, in the daily production of scientists and scientific knowledge. The discourse of dominant and subordinate nations and peoples, I find, is reinscribed in scientific discourse by all its speakers, even the included others, those scientists from subordinated or marginalized nations, regions, and peoples.—SHARON TRAWEEK, "Big Science and Colonialist Discourse: Building High-Energy Physics in Japan"

WITH THE END OF WORLD WAR II and the beginning of the end of the remaining formal European rule around the globe, European and North American governments argued that the road to perpetual peace required global prosperity. Third World poverty and "backwardness" could be

eradicated only by increased economic productivity through the dissemination of scientific rationality and technological expertise from the First
World to the Third World. The new United Nations would promote such
goals, as would institutions and agencies established by the European and
North American governments. The World Bank and U.S. Agency for International Development were just two of the most important of such
institutions. Only through the loans and supervision of such institutions
could the standard of living in Third World societies rise to that of the
West as these societies became markets for Western goods and services
and producers of goods and services to be sold in the West. This was the
rhetoric of the West's so-called development policies for Third World societies. The "underdeveloped societies" apparently welcomed the vision
proposed by such institutions.

But why did so many societies agree that they were impoverished and
backward? asks Arturo Escobar. By the 1970s, the development dream
showed signs that it was turning into a nightmare. Small middle classes
did emerge around the globe. Yet the conditions of two-thirds or more
of the globe's peoples in fact worsened. Traditional social bonds were
disrupted and communities destroyed. This occurred as men were recruited to often only low-level technical educations and manufacturing
jobs in urban centers and in mining and plantation enterprises distant
from their families and communities. Ecological destruction intensified
as transnational agribusiness corporations appropriated the "empty land"
on which peasant and tribal peoples lived. The corporations removed
forests, prairies, and local agricultural systems. They substituted mono-
crops that required commercial pesticides and fertilizers that toxified soil
and water. It became clear that not enough natural resources existed for a
world of affluent Western-style consuming societies.

By the 1980s, debt crises plagued Third World nations as Northern banks
demanded repayment of their development loans, to be accomplished
through World Bank–designed "structural readjustment" programs. Such
programs ended what social gains had been made as education, health,
childcare, and social welfare agencies and institutions were disbanded
as luxuries that the Western-installed market economies could not
afford. The employees of these agencies—disproportionately women—
were sent home to provide for free the social services they had been providing in the now-defunct agencies (Sparr 1994; Visvanathan et al. 1997).
It became clear that poor societies, already weakened by colonial occupation, were being destroyed even further by the supposed help provided
by the development programs. These societies were *made* impoverished
and backward in ways they had not been before. In his influential ac-

count, Escobar argues that this de-development and maldevelopment of the world's poorest societies was not accidental; it was the mostly foreseen "unfortunate" cost of the modernization programs—that is, of the continuation of the West's economic and political expansion.

Moreover, feminists began to point out that a significant contributor to the worsening of poor people's conditions in the Third World was the fact that—intentionally or not—development policies systematically ensured the destruction of resources for poor women and their dependents (Appleton et al., Catherine Scott; Shiva 1989; Mies 1986). However, this was not how the population control agencies (founded in the 1950s) saw the matter. They proclaimed that it was women and their reproductive practices that should be blamed for excessive population expansion and consequent environmental destruction and widespread poverty. It has been difficult to detach the field of demography from such an assumption. Indeed, many highly educated people today still assume that poverty is caused by too many births. Yet for several decades now, economists have known that the causality runs the other way. Poor people need more children to work in the kinds of low-pay, labor-intensive work that is available to them (such as agriculture and home manufacturing), as well as to provide all the social services—domestic work, childcare, support and care for the sick and for elderly kin—that middle-class people in the West can count on from the state, private agencies, their employers, or their own salaries. Betsy Hartmann and the Committee on Women, Population, and the Environment point out how sexist assumptions have conveniently allowed policymakers and administrators in militaries, international relations, and supposedly progressive modernization planning and environmental movements to escape critical scrutiny for their roles in worsening the lot of poor people and environments around the globe.

By the 1980s, *development* itself had become a "toxic word" (Escobar). Yet it remains difficult in the First World as well as the Third World to think outside the conceptual framework established by the development institutions and their practices. This framework makes several false and dangerous assumptions in addition to the ones already mentioned. One is that natural resources are infinite and thus everyone can consume at Western levels. Another holds that women's time and energy are infinite, and their needs—beyond those required to work in export industries—should not concern modernization planners. Employers can demand that women devote 100 percent of their nonsleeping time and energy to creating profit for the employers through export production (at wages too low to support them), even though they must also find the time, energy, and resources to provide daily subsistence for themselves and their dependents. And this is so for societies with little infrastructure; declining quantity and quality of water, air, and usable land; and few, if any, government-provided social security networks such as healthcare, childcare, retirement allowances, and senior care.

Third, development policies assume that Westerners are simply rational, and certainly not greedy, to prioritize their consumer desires over everyone else's subsistence requirements. There is no need for the North to scale back its use of Third World material resources and labor to assist the modernization projects for the developing world. Finally, the development discourses assume that the West has provided the one and only rational model of modernization. This assumption persists in spite of the fact that in the period when the West modernized, it did not have to compete for natural resources, economic support, and labor with greedy superpowers like the ones with which the Third World has had to contend. For supporters of the development discourses, there is and can be no other desirable model of modernization or therefore of development.

Well-intentioned modifications to development principles and assumptions introduced since the 1970s have not provided more than stopgap short-term relief, at best. Skepticism about development's assumptions and practices has slowly been building, though transnational corporations and their neoliberal governmental sponsors remain mostly resistant to criticism. Meanwhile new disasters continually appear, to which development assumptions and practices often seem to have contributed: monocrop failures, famines, worsening health conditions, increasing ecological destruction, escalating spread of drug, sex-work, and armament industries, and huge human costs—largely avoidable—in the natural disasters created by earthquakes, mudslides, typhoons, and floods. Then there is the current global financial crisis of a previously unimagined size. Escobar focuses on a question asked throughout Latin America and echoed around the globe: what would be desirable pro-democratic alternatives to development, not just modifications of its existing problematic assumptions and principles? He offers some valuable suggestions, and we return to explore this question further in part IV.

There are other aspects of how colonialism and imperialism "by other means" persist and are also reinvented today. Catherine V. Scott argues that both the dominant modernization theory and its major critics, the Marxian dependency and world systems theories, are grounded in gender and racial stereotypes. These gender and racial subtexts seem to leave women out of national and international plans for social progress. Worse, they define standards for progress in terms of their distance from women, household, family, and local communities while in fact appropriating the resources produced in such "traditional" ways of life to support modernization projects. In these policies and practices, modern sciences and technologies are seen as the methods—the motors—for producing social progress. However, as discussed in the book introduction, simply adding attention to women's issues to lists of "to do" projects, while always a good start, will not in itself succeed in eliminating gender and imperial-racial inequities on local or global scales. Rather, a direct program of explicitly exorcising the deep fears of the racial-gender-cultural other at the heart of the West's social and scientific projects, and designing projects that address diverse women's needs and desires, is required. Only in such ways can social progress for the increasingly immiserated vast majority of the globe's citizens be advanced (Harding 2008).

In the case of the Human Genome Diversity Project, well-intentioned population geneticists and evolutionary biologists set out to "sample and archive the world's human genetic diversity" as a way of advancing an antiracist agenda. Among the distinguished leaders of the project were biologists who had long fought against racism and had promoted human rights. Yet soon "over a hundred groups advocating for the rights of tribes in the United States and indigenous groups worldwide had signed declarations condemning the project as inventing a new form of colonialism" (Reardon). The problem was that conventional assumptions about the actual and desirable relationships between science and power blocked the ability of these scientists to raise the kinds of questions required by the genomic revolution of which the project would be a significant part. The scientists never foresaw that any attempt today to order human diversity with the purpose of advancing scientific research would appear to provide resources for still ongoing histories of racist social programs. Today there is no way to separate supposedly pure science from the political consequences of even doing such research.

Cori Hayden identifies "the 'interests' that local knowledge is expected to bear, represent, or animate" in bioprospecting agreements discussed in the preceding part of the book. She identifies the unforeseen consequences of conceptualizing the plants that Western sciences develop and then patent as arriving "with the innovative labor of 'local' or 'indigenous' people *already embedded*." This reconceptualization was intended to enable indigenous people to "have the right to stake a claim should these plants prove to be the key to, for example, the ever elusive 'cure for cancer.' " Yet corporations' perceptions of the political and financial difficulty of successfully working out benefit-sharing agreements with indigenous groups is leading them "away from Mexican plants and the people who come with them, and towards collecting sites that seem to promise more bioactivity and less social-political complexity," such as the international zones of oceans.

What is to be done? We have seen some proposals for transformation in the readings in these first three parts of the book. The book's final part focuses on a variety of responses to this question.

## References

Harding, Sandra. 2008. *Sciences from Below: Feminisms, Postcolonialities, and Modernities*. Durham: Duke University Press.

Hubbard, Ruth. 1990. *The Politics of Women's Biology*. New Brunswick, N.J.: Rutgers University Press.

Levins, Richard, and Richard Lewontin. 1988. "Applied Biology in the Third World: The Struggle for Revolutionary Science." In their *The Dialectical Biologist*, 225–37. Cambridge: Harvard University Press.

McClintock, Anne. 1992. "The Angel of Progress: Pitfalls of the Term 'Postcolonialism.'" *Social Text* 31–32.

Mies, Maria. 1986. *Patriarchy and Accumulation on a World Scale: Women in the International Division of Labor*. Atlantic Highlands, N.J.: Zed.

Shiva, Vandana. 1989. *Staying Alive: Women, Ecology, and Development*. London: Zed.

Sparr, Pamela, ed. 1994. *Mortgaging Women's Lives: Feminist Critiques of Structural Adjustment*. London: Zed.

Traweek, Sharon. 1992. "Big Science and Colonialist Discourse: Building High-Energy Physics in Japan." In *Big Science*, ed. Peter Galison and Bruce Hevly, 100–28. Stanford: Stanford University Press.

Visvanathan, Nalini, et al., eds. 1997. *The Women, Gender, and Development Reader*. London: Zed.

ARTURO ESCOBAR

# 14. Development and
# the Anthropology of Modernity

> There is a sense in which rapid economic progress is impossible without painful adjustments. Ancient philosophies have to be scrapped; old social institutions have to disintegrate; bonds of caste, creed and race have to burst; and large numbers of persons who cannot keep up with progress have to have their expectations of a comfortable life frustrated. Very few communities are willing to pay the full price of economic progress.
> —United Nations, Department of Social and Economic Affairs,
> *Measures for the Economic Development of Underdeveloped Countries,* 1951

IN IIIS INAUGURAL ADDRESS as president of the United States on January 20, 1949, Harry Truman announced his concept of a "fair deal" for the entire world. An essential component of this concept was his appeal to the United States and the world to solve the problems of the "underdeveloped areas" of the globe.

> More than half the people of the world are living in conditions approaching misery. Their food is inadequate, they are victims of disease. Their economic life is primitive and stagnant. Their poverty is a handicap and a threat both to them and to more prosperous areas. For the first time in history humanity possesses the knowledge and the skill to relieve the suffering of these people. . . . I believe that we should make available to peace-loving peoples the benefits of our store of technical knowledge in order to help them realize their aspirations for a better life. . . . What we envisage is a program of development based on the concepts of democractic fair dealing. . . . Greater production is the key to prosperity and peace. And the key to greater production is a wider and more vigorous application of modern scientific and technical knowledge. (Truman [1949] 1964)

The Truman doctrine initiated a new era in the understanding and management of world affairs, particularly those concerning the less economically accomplished countries of the world. The intent was quite ambitious: to bring about the conditions necessary to replicating the world over the features that characterized the "advanced" societies of the time— high levels of industrialization and urbanization, technicalization of agriculture, rapid growth of material production and living standards, and the widespread adoption of modern education and cultural values. In

Truman's vision, capital, science, and technology were the main ingredients that would make this massive revolution possible. Only in this way could the American dream of peace and abundance be extended to all the peoples of the planet.

This dream was not solely the creation of the United States but the result of the specific historical conjuncture at the end of the Second World War. Within a few years, the dream was universally embraced by those in power. The dream was not seen as an easy process, however; predictably perhaps, the obstacles perceived ahead contributed to consolidating the mission. One of the most influential documents of the period, prepared by a group of experts convened by the United Nations with the objective of designing concrete policies and measures "for the economic development of underdeveloped countries," put it thus:

> There is a sense in which rapid economic progress is impossible without painful adjustments. Ancient philosophies have to be scrapped; old social institutions have to disintegrate; bonds of caste, creed and race have to burst; and large numbers of persons who cannot keep up with progress have to have their expectations of a comfortable life frustrated. Very few communities are willing to pay the full price of economic progress. (United Nations, Department of Social and Economic Affairs [1951], 15).[1]

The report suggested no less than a total restructuring of "underdeveloped" societies. The statement quoted earlier might seem to us today amazingly ethnocentric and arrogant, at best naive; yet what has to be explained is precisely the fact that it was uttered and that it made perfect sense. The statement exemplified a growing will to transform drastically two-thirds of the world in the pursuit of the goal of material prosperity and economic progress. By the early 1950s, such a will had become hegemonic at the level of the circles of power.

*Encountering Development* tells, in part, the story of this dream and how it progressively turned into a nightmare. For instead of the kingdom of abundance promised by theorists and politicians in the 1950s, the discourse and strategy of development produced its opposite: massive underdevelopment and impoverishment, untold exploitation and oppression. The debt crisis, the Sahelian famine, increasing poverty, malnutrition, and violence are only the most pathetic signs of the failure of forty years of development. In this way, the book can be read as the history of the loss of an illusion, in which many genuinely believed. Above all, however, it is about how the "Third World" has been produced by the discourses and practices of development since their inception in the early post–World War II period.

Orientalism, Africanism, and Developmentalism

Until the late 1970s, the central stake in discussions on Asia, Africa, and Latin America was the nature of development. As we will see, from the economic development theories of the 1950s to the "basic human needs approach" of the 1970s—which emphasized not only economic growth per se as in earlier decades but also the distribution of the benefits of growth—the main preoccupation of theorists and politicians was the kinds of development that needed to be pursued to solve the social and economic problems of these parts of the world. Even those who opposed the prevailing capitalist strategies were obliged to couch their critique in terms of the need for development, through concepts such as "another development," "participatory development," "socialist development," and the like. In short, one could criticize a given approach and propose modifications or improvements accordingly, but the fact of development itself, and the need for it, could not be doubted. Development had achieved the status of a certainty in the social imaginary.

Indeed, it seemed impossible to conceptualize social reality in other terms. Wherever one looked, one found the repetitive and omnipresent reality of development: governments designing and implementing ambitious development plans, institutions carrying out development programs in city and countryside alike, experts of all kinds studying underdevelopment and producing theories ad nauseam. The fact that most people's conditions not only did not improve but deteriorated with the passing of time did not seem to bother most experts. Reality, in sum, had been colonized by the development discourse, and those who were dissatisfied with this state of affairs had to struggle for bits and pieces of freedom within it, in the hope that in the process a different reality could be constructed.[2]

More recently, however, the development of new tools of analysis, in gestation since the late 1960s but the application of which became widespread only during the 1980s, has made possible analyses of this type of "colonization of reality" which seek to account for this very fact: how certain representations become dominant and shape indelibly the ways in which reality is imagined and acted upon. Foucault's work on the dynamics of discourse and power in the representation of social reality, in particular, has been instrumental in unveiling the mechanisms by which a certain order of discourse produces permissible modes of being and thinking while disqualifying and even making others impossible. Extensions of Foucault's insights to colonial and postcolonial situations by authors such as Edward Said, V. Y. Mudimbe, Chandra Mohanty, and Homi

Bhabha, among others, have opened up new ways of thinking about representations of the Third World. Anthropology's self-critique and renewal during the 1980s have also been important in this regard.

Thinking of development in terms of discourse makes it possible to maintain the focus on domination—as earlier Marxist analyses, for instance, did—and at the same time to explore more fruitfully the conditions of possibility and the most pervasive effects of development. Discourse analysis creates the possibility of "stand[ing] detached from [the development discourse], bracketing its familiarity, in order to analyze the theoretical and practical context with which it has been associated" (Foucault 1986, 3). It gives us the possibility of singling out "development" as an encompassing cultural space and at the same time of separating ourselves from it by perceiving it in a totally new form. This is the task that I set out to accomplish.

To see development as a historically produced discourse entails an examination of why so many countries started to see themselves as underdeveloped in the early post–World War II period, how "to develop" became a fundamental problem for them, and how, finally, they embarked upon the task of "un-underdeveloping" themselves by subjecting their societies to increasingly systematic, detailed, and comprehensive interventions. As Western experts and politicians started to see certain conditions in Asia, Africa, and Latin America as a problem—mostly what was perceived as poverty and backwardness—a new domain of thought and experience, namely, development, came into being, resulting in a new strategy for dealing with the alleged problems. Initiated in the United States and Western Europe, this strategy became in a few years a powerful force in the Third World.

The study of development as discourse is akin to Said's study of the discourses on the Orient. "Orientalism," writes Said,

> can be discussed and analyzed as the corporate institution for dealing with the Orient—dealing with it by making statements about it, authorizing views of it, describing it, by teaching it, settling it, ruling over it: in short, Orientalism as a Western style for dominating, restructuring, and having authority over the Orient. . . . My contention is that without examining Orientalism as a discourse we cannot possibly understand the enormously systematic discipline by which European culture was able to manage—and even produce—the Orient politically, sociologically, ideologically, scientifically, and imaginatively during the post-Enlightenment period. (1979, 3)

Since its publication, *Orientalism* has sparked a number of creative studies and inquiries about representations of the Third World in various contexts, although few have dealt explicitly with the question of development. Nevertheless, the general questions some of these works raised serve as markers for the analysis of development as a regime of representation. In his excellent book *The Invention of Africa*, the African philosopher V. Y. Mudimbe, for example, states his objective thus: "To study the theme of the foundations of discourse about Africa . . . [how] African worlds have been established as realities for knowledge" (1988, xi) in Western discourse. His concern, moreover, goes beyond "the 'invention' of Africanism as a scientific discipline" (9), particularly in anthropology and philosophy, in order to investigate the "amplification" by African scholars of the work of critical European thinkers, particularly Foucault and Lévi-Strauss. Although Mudimbe finds that even in the most Afrocentric perspectives the Western epistemological order continues to be both context and referent, he nevertheless finds some works in which critical European insights are being carried even further than those works themselves anticipated. What is at stake for these latter works, Mudimbe explains, is a critical reinterpretation of African history as it has been seen from Africa's (epistemological, historical, and geographical) exteriority, indeed, a weakening of the very notion of Africa. This, for Mudimbe, implies a radical break in African anthropology, history, and ideology.

Critical work of this kind, Mudimbe believes, may open the way for "the process of refounding and reassuming an interrupted historicity within representations" (183), in other words, the process by which Africans can have greater autonomy over how they are represented and how they can construct their own social and cultural models in ways not so mediated by a Western episteme and historicity—albeit in an increasingly transnational context. This notion can be extended to the Third World as a whole, for what is at stake is the process by which, in the history of the modern West, non-European areas have been systematically organized into, and transformed according to, European constructs. Representations of Asia, Africa, and Latin America as Third World and underdeveloped are the heirs of an illustrious genealogy of Western conceptions about those parts of the world.[3]

Timothy Mitchell unveils another important mechanism at work in European representations of other societies. Like Mudimbe, Mitchell's goal is to explore "the peculiar methods of order and truth that characterise the modern West" (1988, ix) and their impact on nineteenth-century Egypt. The setting up of the world as a picture, in the model of the world exhibitions of the last century, Mitchell suggests, is at the core of these methods and their political expediency. For the modern (European) subject, this entailed that s/he would experience life as if s/he were set apart from the physical world, as if s/he were a visitor at an exhibition. The observer inevitably "enframed" external reality in order to make sense of it; this enframing took place according to European categories. What emerged was a regime of objectivism in which Europeans were subjected to a double demand: to be detached and objective, and yet to immerse themselves in local life.

This experience as participant observer was made possible by a curious trick, that of eliminating from the picture the presence of the European observer (see also Clifford 1988, 145); in more concrete terms, observing the (colonial) world as object "from a position that is invisible and set apart" (Mitchell 1988, 28). The West had come to live "as though the world were divided in this way into two: into a realm of mere representations and a realm of the 'real'; into exhibitions and an external reality; into an order of mere models, descriptions or copies, and an order of the original" (32). This regime of order and truth is a quintessential aspect of modernity and has been deepened by economics and development. It is reflected in an objectivist and empiricist stand that dictates that the Third World and its peoples exist "out there," to be known through theories and intervened upon from the outside.

The consequences of this feature of modernity have been enormous. Chandra Mohanty, for example, refers to the same feature when raising the questions of who produces knowledge about Third World women and from what spaces; she discovered that women in the Third World are represented in most feminist literature on development as having "needs" and "problems" but few choices and no freedom to act. What emerges from such modes of analysis is the image of an average Third World woman, constructed through the use of statistics and certain categories:

> This average third world woman leads an essentially truncated life based on her feminine gender (read: sexually constrained) and her being "third world" (read: ignorant, poor, uneducated, tradition-bound, domestic, family-oriented, victimized, etc.). This, I suggest, is in contrast to the (implicit) self-representation of Western women as educated, as modern, as having control over their own bodies and sexualities, and the freedom to make their own decisions. (1991b, 56)

These representations implicitly assume Western standards as the benchmark against which to measure the situation of Third World women. The result, Mohanty believes, is a paternalistic attitude on the part of Western women toward their Third World counterparts and, more generally, the perpetuation of the hegemonic idea of the West's superiority. Within this discursive regime, works about Third World women develop a certain coherence of effects that reinforces that hegemony. "It is in this process of discursive homogenization and systematization of the oppression of women in the third world," Mohanty concludes, "that power is exercised in much of recent Western feminist discourse, and this power needs to be defined and named" (54).[4]

Needless to say, Mohanty's critique applies with greater pertinence to mainstream development literature, in which there exists a veritable underdeveloped subjectivity endowed with features such as powerlessness, passivity, poverty, and ignorance, usually dark and lacking in historical agency, as if waiting for the (white) Western hand to help subjects along and not infrequently hungry, illiterate, needy, and oppressed by its own stubbornness, lack of initiative, and traditions. This image also universalizes and homogenizes Third World cultures in an ahistorical fashion. Only from a certain Western perspective does this description make sense; that it exists at all is more a sign of power over the Third World than a truth about it. It is important to highlight for now that the deployment of this discourse in a world system in which the West has a certain dominance over the Third World has profound political, economic, and cultural effects that have to be explored.

The production of discourse under conditions of unequal power is what Mohanty and others refer to as "the colonialist move." This move entails specific constructions of the colonial / Third World subject in/through discourse in ways that allow the exercise of power over it. Colonial discourse, although "the most theoretically underdeveloped form of discourse," according to Homi Bhabha, is "crucial to the binding of a range of differences and discriminations that inform the discursive and political practices of racial and cultural hierarchization" (1990, 72). Bhabha's definition of colonial discourse, although complex, is illuminating:

> [Colonial discourse] is an apparatus that turns on the recognition and disavowal of racial/cultural/historical differences. Its predominant strategic function is the creation of a space for a "subject peoples" through the production of knowledges in terms of which surveillance is exercised and a complex form of pleasure/unpleasure is incited. . . . The objective of colonial discourse is to construe the colonized as a population of degenerate types on the basis of racial origin, in order to justify conquest and to establish systems of administration and instruction. . . . I am referring to a form of governmentality that in marking out a "subject nation," appropriates, directs and dominates its various spheres of activity. (1990, 75)

Although some of the terms of this definition might be more applicable to the colonial context strictly speaking, the development discourse is governed by the same principles; it has created an extremely efficient apparatus for producing knowledge about, and the exercise of power over, the Third World. This apparatus came into existence roughly in the period 1945 to 1955 and has not since ceased to produce new arrangements of knowledge and power, new practices, theories, strategies, and so on. In sum, it has successfully deployed a regime of government over the Third World, a "space for 'subject peoples'" that ensures certain control over it.

This space is also a geopolitical space, a series of imaginative geographies, to use Said's (1979) term. The development discourse inevitably contained a geopolitical imagination that has shaped the meaning of development for more than four decades. For some, this will to spatial power is one of the most essential features of development (Slater 1993). It is implicit in expressions such as First and Third World, North and South, center and periphery. The social production of space implicit in these terms is bound with the production of differences, subjectivities, and social orders. Despite the correctives introduced to this geopolitics—the decentering of the world, the demise of the Second World, the emergence of a network of world cities, the globalization of cultural production, and so on—they continue to function imaginatively in powerful ways. There is a relation among history, geography, and modernity that resists disintegration as far as the Third World is concerned, despite the important changes that have given rise to postmodern geographies (Soja 1989).

To sum up, I propose to speak of development as a historically singular experience, the creation of a domain of thought and action, by analyzing the characteristics and interrelations of the three axes that define it: the forms of knowledge that refer to it and through which it comes into being and is elaborated into objects, concepts, theories, and the like; the system of power that regulates its practice; and the forms of subjectivity fostered by this discourse, those through which people come to recognize themselves as developed or underdeveloped. The ensemble of forms found along these axes constitutes development as a discursive formation, giving rise to an efficient apparatus that systematically relates forms of knowledge and techniques of power.[5]

The analysis will thus be couched in terms of regimes of discourse and representation. Regimes of representation can be analyzed as places of encounter where identities are constructed and also where violence is originated, symbolized, and managed. This useful hypothesis, developed by a Colombian scholar to explain nineteenth-century violence in her country, building particularly on the works of Bakhtin, Foucault, and Girard, conceives of regimes of representation as places of encounter of languages of the past and languages of the present (such as the languages of "civilization" and "barbarism" in postindependence Latin America), internal and external languages, and languages of self and other (Rojas de Ferro 1994). A similar encounter of regimes of representation took place in the late 1940s with the emergence of development, also accompanied by specific forms of modernized violence.[6]

The notion of regimes of representation is a final theoretical and methodological principle for examining the mechanisms for, and consequences of, the construction of the Third World in/through representation. Charting regimes of representation of the Third World brought about by the development discourse represents an attempt to draw the "cartographies" (Deleuze 1988) or maps of the configurations of knowledge and power that define the post–World War II period. These are also cartographies of struggle, as Mohanty (1991a) adds. Although they are geared toward an understanding of the conceptual maps that are used to locate and chart Third World people's experience, they also reveal— even if indirectly at times—the categories with which people have to struggle. *Encountering Development* provides a general map for orienting oneself in the discourses and practices that account for today's dominant forms of sociocultural and economic production of the Third World. . . .

This, one might say, is a study of developmentalism as a discursive field. Unlike Said's study of Orientalism, however, I pay closer attention to the deployment of the discourse through practices. I want to show that this discourse results in concrete practices of thinking and acting through which the Third World is produced. The example I chose for this closer investigation is the implementation of rural development, health, and nutrition programs in Latin America in the 1970s and 1980s. Another difference in relation to *Orientalism* originates in Homi Bhabha's caution that "there is always, in Said, the suggestion that colonial power is possessed entirely by the colonizer, given its intentionality and unidirectionality" (1990, 77). This is a danger I seek to avoid by considering the variety of forms with which Third World people resist development interventions and how they struggle to create alternative ways of being and doing.

Like Mudimbe's study of Africanism, I also want to unveil the foundations of an order of knowledge and a discourse about the Third World as underdeveloped. I want to map, so to say, the invention of development. Instead of focusing on anthropology and philosophy, however, I contextualize the era of development within the overall space of modernity, particularly modern economic practices. From this perspective, development can be seen as a chapter of what can be called an anthropology of modernity, that is, a general investigation of Western modernity as a culturally and historically specific phenomenon. If it is true that there is an "anthropological structure" (Foucault 1975, 198) that sustains the modern order and its human sciences, it must be investigated to what extent this structure has also given rise to the regime of development, perhaps as a specific mutation of modernity. A general direction for this anthropology of modernity has already been suggested, in the sense of rendering "exotic" the West's cultural products in order to see them for what they are: "We need to anthropologize the West: show how exotic its constitution of reality has been; emphasize those domains most taken for granted as universal (this includes epistemology and economics); make them seem as historically peculiar as possible; show how their claims to truth are linked to social practices and have hence become effective forces in the social world" (Rabinow 1986, 241).

The anthropology of modernity would rely on ethnographic approaches that look at social forms as produced by historical practices combining knowledge and power; it would seek to examine how truth claims are related to practices and symbols that produce and regulate social life. As we will see, the production of the Third World through the articulation of knowledge and power is essential to the development discourse. This does not preclude the fact that from many Third World spaces, even the most reasonable among the West's social and cultural practices might look quite peculiar, even strange. Nevertheless, even today most people in the West (and many parts of the Third World) have great difficulty thinking about Third World situations and people in terms other than those provided by the development discourse. These terms—such as overpopulation, the permanent threat of famine, poverty, illiteracy, and the like—operate as the most common signifiers, already stereotyped and burdened with development signifieds. Media images of the Third World are the clearest example of developmentalist representations. These images just do not seem to go away. This is why it is necessary to examine development in relation to the modern experiences of knowing, seeing, counting, economizing, and the like.

## Deconstructing Development

The discursive analysis of development started in the late 1980s and will most likely continue into the 1990s, coupled with attempts at articulating alternative regimes of representation and practice. Few works, however, have undertaken the deconstruction of the development discourse.[7] James Ferguson's recent book on development in Lesotho (1990) is a sophisticated example of the deconstructionist approach. Ferguson provides an in-depth analysis of rural development programs implemented in the country under World Bank sponsorship. Further entrenchment of the state, the restructuring of rural social relations, the deepening of Western modernizing influences, and the depoliticization of problems are among the most important effects of the deployment of rural development in Lesotho, despite the apparent failure of the programs in terms of their stated objectives. It is at the level of these effects, Ferguson concludes, that the productivity of the apparatus has to be assessed.

Another deconstructionist approach (Sachs 1992) analyzes the central constructs or key words of the development discourse, such as market, planning, population, environment, production, equality, participation, needs, poverty, and the like. After briefly tracing the origin of each concept in European civilization, each chapter examines the uses and transformation of the concept in the development discourse from the 1950s to the present. The intent of the book is to expose the arbitrary character of the concepts, their cultural and historical specificity, and the dangers that their use represents in the context of the Third World.[8] A related, group project is conceived in terms of a "systems of knowledge" approach. Cultures, this group believes, are characterized not only by rules and values but also by ways of knowing. Development has relied exclusively on one knowledge system, namely, the modern Western one. The dominance of this knowledge system has dictated the marginalization and disqualification of non-Western knowledge systems. In these latter knowledge systems, the authors conclude, researchers and activists might find alternative rationalities to guide social action away from economistic and reductionistic ways of thinking.[9]

In the 1970s, women were discovered to have been "bypassed" by development interventions. This "discovery" resulted in the growth during the late 1970s and 1980s of a whole new field, women in development (WID), which has been analyzed by several feminist researchers as a regime representation, most notably Adele Mueller (1986, 1987a, 1991) and Chandra Mohanty. At the core of these works is an insightful analysis of the practices of dominant development institutions in creating and managing their client populations. Similar analyses of particular development subfields—such as economics and the environment, for example—are a needed contribution to the understanding of the function of development as a discourse and will continue to appear.[10]

A group of Swedish anthropologists focus their work on how the concepts of development and modernity are used, interpreted, questioned, and reproduced in various social contexts in different parts of the world. An entire constellation of usages, modes of operation, and effects associated with these terms, which are profoundly local, is beginning to surface. Whether in a Papua New Guinean village or in a small town of

Kenya or Ethiopia, local versions of development and modernity are formulated according to complex processes that include traditional cultural practices, histories of colonialism, and contemporary location within the global economy of goods and symbols (Dahl and Rabo 1992). These much-needed local ethnographies of development and modernity are also being pioneered by Pigg (1992) in her work on the introduction of health practices in Nepal.

Finally, it is important to mention a few works that focus on the role of conventional disciplines within the development discourse. Irene Gendzier (1985) examines the role political science played in the conformation of theories of modernization, particularly in the 1950s, and its relation to issues of the moment such as national security and economic imperatives. Also within political science, Kathryn Sikkink (1991) has more recently taken on the emergence of developmentalism in Brazil and Argentina in the 1950s and 1960s. Her chief interest is the role of ideas in the adoption, implementation, and consolidation of developmentalism as an economic development mode.[11] The Chilean Pedro Morandé (1984) analyzes how the adoption and dominance of North American sociology in the 1950s and 1960s in Latin America set the stage for a purely functional conception of development, conceived of as the transformation of "traditional" into a "modern" society and devoid of any cultural considerations. Kate Manzo (1991) makes a somewhat similar case in her analysis of the shortcomings of modernist approaches to development, such as dependency theory, and in her call for paying attention to "countermodernist" alternatives that are grounded in the practices of Third World grassroots actors. The call for a return of culture in the critical analysis of development, particularly local cultures, is also central to this book.

The goal here is to contribute to the liberation of the discursive field so that the task of imagining alternatives can be commenced (or perceived by researchers in a new light) in those spaces where the production of scholarly and expert knowledge for development purposes continues to take place.

## Anthropology and the Development Encounter

In the introduction to his well-known collection on anthropology's relation to colonialism, *Anthropology and the Colonial Encounter* (1973), Talal Asad raised the question of whether there was not still "a strange reluctance on the part of most professional anthropologists to consider seriously the power structure within which their discipline has taken shape" (5), namely, the whole problematic of colonialism and neocolonialism, their political economy and institutions. Does not development today, as colonialism did in a former epoch, make possible "the kind of human intimacy on which anthropological fieldwork is based, but insure[s] that intimacy should be one-sided and provisional" (17), even if the contemporary subjects move and talk back? In addition, if during the colonial period "the general drift of anthropological understanding did not constitute a basic challenge to the unequal world represented by the colonial system" (18), is this not also the case with the development system? In sum, can we not speak with equal pertinence of "anthropology and the development encounter"?

It is generally true that anthropology as a whole has not dealt explicitly with the fact that it takes place within the post–World War II encounter between rich and poor nations established by the development discourse. Although a number of anthropologists have opposed development interventions, particularly on behalf of indigenous people,[12] large numbers of anthropologists have been involved with development organizations such as the World Bank and the United States Agency for International Development (U.S. AID). This problematic involvement was particularly noticeable in the decade 1975–1985 and has been analyzed elsewhere (Escobar 1991). As Stacy Leigh Pigg (1992) rightly points out, anthropologists have been for the most part either inside development, as applied anthropologists, or outside development, as the champions of the authentically indigenous and "the native's point of view." Thus they overlook the ways in which development operates as an arena of cultural contestation and identity construction. A small number of anthropologists, however, have studied forms and processes of resistance to development interventions (Taussig 1980; Fals Borda 1984; Scott 1985; Ong 1987; see also Comaroff 1985 and Comaroff and Comaroff 1991 for resistance in the colonial context).

The absence of anthropologists from discussions of development as a regime of representation is regrettable because, if it is true that many aspects of colonialism have been superseded, representations of the Third World through development are no less pervasive and effective than their colonial counterparts. Perhaps even more so. It is also disturbing, as Said has pointed out, that in recent anthropological literature "there is an almost total absence of any reference to American imperial intervention as a factor affecting the theoretical discussion" (1989, 214; see also Friedman 1987; Ulin 1991). This imperial intervention takes place at many levels— economic, military, political, and cultural—which are woven together by development representations. Also disturbing, as Said proceeds to argue, is the lack of attention on the part of Western scholars to the sizable and impassioned critical literature by Third World intellectuals on colonialism, history, tradition, and domination—and, one might add, development.

The number of Third World voices calling for a dismantling of the entire discourse of development is fast increasing.

The deep changes experienced in anthropology during the 1980s opened the way for examining how anthropology is bound up with "Western ways of creating the world," as Strathern (1988, 4) advises, and potentially with other possible ways of representing the interests of Third World peoples. This critical examination of anthropology's practices led to the realization that "no one can write about others any longer as if they were discrete objects or texts." A new task thus insinuated itself: that of coming up with "more subtle, concrete ways of writing and reading . . . new conceptions of culture as interactive and historical" (Clifford 1986, 25). Innovation in anthropological writing within this context was seen as "moving [ethnography] toward an unprecedentedly acute political and historical sensibility that is transforming the way cultural diversity is portrayed" (Marcus and Fischer 1986, 16).

This reimagining of anthropology, launched in the mid-1980s, has become the object of various critiques, qualifications, and extensions from within its own ranks and by feminists, political economists, Third World scholars, Third World feminists, and anti-postmodernists. Some of these critiques are more or less pointed and constructive than others, and it is not necessary to analyze them in this introduction.[13] To this extent, "the experimental moment" of the 1980s has been very fruitful and relatively rich in applications. The process of reimagining anthropology, however, is clearly still under way and will have to be deepened, perhaps by taking the debates to other arenas and in other directions. Anthropology, it is now argued, has to "reenter" the real world, after the moment of textualist critique. To do this, it has to rehistoricize its own practice and acknowledge that this practice is shaped by many forces that are well beyond the control of the ethnographer. Moreover, it must be willing to subject its most cherished notions, such as ethnography, culture, and science, to a more radical scrutiny (Fox 1991).

Strathern's call that this questioning be advanced in the context of Western social science practices and their "endorsement of certain interests in the description of social life" is of fundamental importance. At the core of this recentering of the debates within the disciplines are the limits that exist to the Western project of deconstruction and self-critique. It is becoming increasingly evident, at least for those who are struggling for different ways of having a voice, that the process of deconstructing and dismantling has to be accompanied by that of constructing new ways of seeing and acting. Needless to say, this aspect is crucial in discussions about development, because people's survival is at stake. As Mohanty (1991a) insists, both projects—deconstruction and reconstruction—have to be carried out simultaneously. This simultaneous project could focus strategically on the collective action of social movements: they struggle not only for goods and services but also for the very definition of life, economy, nature, and society. They are, in short, cultural struggles.

As Bhabha wants us to acknowledge, deconstruction and other types of critiques do not lead automatically to "an unproblematic reading of other cultural and discursive systems." They might be necessary to combat ethnocentrism, "but they cannot, of themselves, unreconstructed, represent that otherness" (Bhabha 1990, 75). Moreover, there is the tendency in these critiques to discuss otherness principally in terms of the limits of Western logocentricity, thus denying that cultural otherness is "implicated in specific historical and discursive conditions, requiring constructions in different practices of reading" (Bhabha 1990, 73). There is a similar insistence in Latin America that the proposals of postmodernism, to be fruitful there, have to make clear their commitment to justice and to the construction of alternative social orders.[14] These Third World correctives indicate the need for alternative questions and strategies for the construction of anticolonialist discourses (and the reconstruction of Third World societies in/through representations that can develop into alternative practices). Calling into question the limitations of the West's self-critique, as currently practiced in much of contemporary theory, they make it possible to visualize the "discursive insurrection" by Third World people proposed by Mudimbe in relation to the "sovereignty of the very European thought from which we wish to disentangle ourselves" (quoted in Diawara 1990, 79).

The needed liberation of anthropology from the space mapped by the development encounter (and, more generally, modernity), to be achieved through a close examination of the ways in which it has been implicated in it, is an important step in the direction of more autonomous regimes of representation; this is so to the extent that it might motivate anthropologists and others to delve into the strategies people in the Third World pursue to resignify and transform their reality through their collective political practice. This challenge may provide paths toward the radicalization of the discipline's reimagining started with enthusiasm during the 1980s.

## Notes

1. For an interesting contemporary analysis of this document, see Frankel (1953, 82–110).

2. Some trends in the 1960s and 1970s were critical of development, although, as will become clear shortly, they were unable to articulate a rejection of the

discourse that struck at its roots. Among these, it is important to mention Paulo Freire's "pedagogy of the oppressed" (Freire 1970); the birth of Liberation Theology at the Latin American Bishops' Conference held in Medellín in 1964; and the critiques of "intellectual colonialism" (Fals Borda 1970) and economic dependency (Cardoso and Faletto 1979) of the late 1960s and early 1970s. The most perceptive cultural critique of development was by Illich (1969). All of these critiques were important for the discursive approach of the 1980s and 1990s analyzed in this chapter.

3. "According to the same learned white man [Ivan Illich], the concept that is currently named 'development' has gone through six stages of metamorphosis since late antiquity. The perception of the outsider as the one who needs help has taken on the successive forms of the barbarian, the pagan, the infidel, the wild man, the 'native,' and the underdeveloped" (Trinh 1989, 54). See Hirschman (1981, 24) for a similar idea and set of terms. It should be pointed out, however, that the term *underdeveloped*—linked from a certain vantage point to equality and the prospects of liberation through development—can be seen in part as a response to more openly racist conceptions of "the primitive" and "the savage." In many contexts, however, the new term failed to correct the negative connotations implied by the earlier qualifiers: The "myth of the lazy native" (Alatas 1977) is still alive today in many quarters.

4. Mohanty's work can be situated within a growing critique by feminists, especially Third World feminists, of ethnocentrism in feminist scholarship and the feminist movement. See also Mani (1989); Trinh (1989); Spelman (1988); and hooks (1990).

5. The study of a discourse along these axes is proposed by Foucault (1986, 4). The forms of subjectivity that development produced are not explored in this chapter in a significant manner. An illustrious group of thinkers, including Frantz Fanon (1967, 1968), Albert Memmi (1967), Ashis Nandy (1983), and Homi Bhabha (1990), have produced increasingly enlightening accounts of the creation of subjectivity and consciousness under colonialism and postcolonialism.

6. On the violence of representation, see also de Lauretis (1987).

7. Article-length analyses of development as discourse include Escobar (1984, 1988); Mueller (1987b); Dubois (1991); Parajuli (1991); and St-Hilaire (1993).

8. The group responsible for this "dictionary of toxic words" in the development discourse includes Ivan Illich, Wolfgang Sachs, Barbara Duden, Ashis Nandy, Vandana Shiva, Majid Mumma, Gustavo Esteva, and myself among others.

9. This group, convened under the sponsorship of the United Nations World Institute for Development Economics Research (WIDER) and headed by Stephen Marglin and Frédérique Apffel-Marglin, has been meeting for several years and includes some of the people mentioned in the previous note. One edited volume has already been published as a result of the project (Apffel-Marglin and Marglin 1990), and a second one (Apffel-Marglin and Marglin 1994) is in press.

10. A collection by Jonathan Crush (Queens University, Canada) on discourses of development is in the process of being compiled; it includes analy-

ses of "languages of development" (Crush 1994). Discourse analyses of development fields is the subject of the project "Development and Social Science Knowledge," sponsored by the Social Science Research Council (SSRC) and coordinated by Frederick Cooper (University of Michigan) and Randall Packard (Tufts University). This project began in the spring of 1994 and will probably continue for several years.

11. Sikkink rightly differentiates her institutional-interpretive method from "discourse and power" approaches, although her characterization of the latter reflects only the initial formulation of the discursive approach. I feel that both methods—the history of ideas and the study of discursive formations—are not incompatible. Although the former method pays attention to the internal dynamics of the social generation of ideas in ways that the latter sometimes overlooks (thus giving the impression that development models are just "imposed on" the Third World, not produced from the inside as well), the history of ideas tends to ignore the systematic effects of discourse production, which in important ways shapes what counts as ideas in the first place. For a differentiation between the history of ideas and the history of discourses, see Foucault (1972, 135–98; 1991b).

12. This is the case with the organization Cultural Survival, for example, and its advocacy anthropology (Maybury-Lewis 1985). Its work, however, recycles some problematic views of the anthropologist speaking on behalf of "the natives" (Escobar 1991). See also Price (1989) for an example of anthropologists opposing a World Bank project in defense of indigenous peoples.

13. See, for instance, Ulin (1991); Sutton (1991); hooks (1990); Said (1989); Trinh (1989); Mascia-Lees, Sharpe, and Cohen (1989); Gordon (1988, 1991); and Friedman (1987).

14. Discussions on modernity and postmodernity in Latin America are becoming a central focus of research and political action. See Calderón, ed. (1988); Quijano (1988, 1990); Lechner (1988); García Canclini (1990); Sarlo (1991); and Yúdice, Franco, and Flores, eds. (1992). For a review of some of these works, see Montaldo (1991).

## References

Alatas, Syed Hussein. 1977. *The Myth of the Lazy Native*. London: Frank Cass.
Apffel-Marglin, Frédérique, and Stephen Marglin, eds. 1990. *Dominating Knowledge: Development, Culture, and Resistance*. Oxford: Clarendon Press.
———. 1994. *Decolonizing Knowledge: From Development to Dialogue*. Oxford: Clarendon Press.
Asad, Talal. 1973. "Introduction." In *Anthropology and the Colonial Encounter*, edited by Talal Asad, 9–20. Atlantic Highlands, N.J.: Humanities Press.
Bhabha, Homi. 1990. "The Other Question: Difference, Discrimination, and the Discourse of Colonialism." In *Out There: Marginalization and Contemporary Cultures*, edited by Russell Ferguson, et al., 71–89. New York: New Museum of Contemporary Art; and Cambridge: MIT Press.

Calderón, Fernando, ed. 1988. *Imágenes Desconocidas: La Modemidad en la Éncrucijada Postmoderna*. Buenos Aires: CLACSO.

Cardoso, Fernando Henrique and Enzo Faletto. 1979. *Dependency and Development in Latin America*. Berkeley: University of California Press.

Clifford, James. 1986. "Introduction: Partial Truths." In *Writing Culture: The Poetics and Politics of Ethnography*, edited by James Clifford and George Marcus, 1–27. Berkeley: University of California Press.

———. 1988. *The Predicament of Culture*. Cambridge: Harvard University Press.

Comaroff, Jean. 1985. *Body of Power, Spirit of Resistance*. Chicago: University of Chicago Press.

Comaroff, Jean, and John Comaroff. 1991. *Of Revelation and Revolution*. Chicago: University of Chicago Press.

Crush, Jonathan, ed. 1994. *Discourses of Development*. New York: Routledge.

Dahl, G., and A. Rabo, eds. 1992. *Kam-Ap or Take-Off: Local Notions of Development*. Stockholm: Stockholm Studies in Social Anthropology.

De Lauretis, Teresa. 1987. *Technologies of Gender*. Bloomington: Indiana University Press.

Deleuze, Gilles. 1988. *Foucault*. Minneapolis: University of Minnesota Press.

Deleuze, Gilles, and Félix Guattari. 1987. *A Thousand Plateaus*. Minneapolis: University of Minnesota Press.

Diawara, Manthia. 1990. "Reading Africa through Foucault: V. Y. Mudimbe's Reaffirmation of the Subject." *October* 55:79–104.

Dubois, Marc. 1991. "The Governance of the Third World: A Foucauldian Perspective of Power Relations in Development." *Alternatives* 16 (1): 1–30.

Escobar, Arturo. 1984. "Discourse and Power in Development: Michel Foucault and the Relevance of His Work to the Third World." *Alternatives* 10 (3): 377–400.

———. 1988. "Power and Visibility: Development and the Invention and Management of the Third World." *Cultural Anthropology* 3 (4): 428–43.

———. 1991. "Anthropology and the Development Encounter: The Making and Marketing of Development Anthropology." *American Ethnolog*ist 18 (4): 16–40.

Fals Borda, Orlando. 1984. *Resistencia en San Jorge*. Bogotá: Carlos Valencia Editores.

Fanon, Frantz. 1967. *Black Skin, White Masks*. New York: Grove Press.

———. 1968. *The Wretched of the Earth*. New York: Grove Press.

Ferguson, James. 1990. *The Anti-Politics Machine: "Development," Depoliticization, and Bureaucratic Power in Lesotho*. Cambridge: Cambridge University Press.

Foucault, Michel. 1972. *The Archaeology of Knowledge*. New York: Harper Colophon Books.

———. 1975. *The Birth of the Clinic*. New York: Vintage Books.

———. 1986. *The Use of Pleasure*. New York: Pantheon Books.

———. 1991b. "Politics and the Study of Discourse." In *The Foucault Effect*, edited by Graham Burchell, Colin Gordon, and Peter Miller, 53–72. Chicago: University of Chicago Press.

Fox, Richard, ed. 1991. *Recapturing Anthropology: Working in the Present.* Santa Fe, N.M.: School of American Research.

Frankel, Herbert. 1953. *The Economic Impact on Underdeveloped Societies.* Cambridge: Harvard University Press.

Freire, Paulo. 1970. *Pedagogy of the Oppressed.* New York: Herder and Herder.

Friedman, Jonathan. 1987. "Beyond Otherness or: The Spectacularization of Anthropology." *Telos* 71:161–70.

García Canclini, Néstor. 1990. *Culturas Híbridas: Estrategias pars Entrar y Salir de la Modernidad.* México, D.F: Grijalbo.

Gendzier, Irene. 1985. *Managing Political Change: Social Scientists and the Third World.* Boulder: Westview Press.

Gordon, Deborah. 1988. "Writing Culture, Writing Feminism: The Poetics and Politics of Experimental Ethnography." *Inscriptions* 3/4:7–26.

———. 1991. "Engendering Ethnography." Ph.D. diss. Board of Studies in History of Consciousness, University of California, Santa Cruz.

Hirschman, Albert. 1981. *Essays in Trespassing: Economics to Politics and Beyond.* Cambridge: Cambridge University Press.

hooks, bell. 1990. *Yearning: Race, Gender, and Cultural Politics.* Boston: South End Press.

Illich, Ivan. 1969. *Celebration of Awareness.* New York: Pantheon Books.

Lechner, Norbert. 1988. *Los Patios Interiores de la Democracia. Subjetividad y Política.* Santiago: FLACSO.

Mani, Lata. 1989. "Multiple Mediations: Feminist Scholarship in the Age of Multinational Reception." *Inscriptions* 5:1–24.

Manzo, Kate. 1991. "Modernist Discourse and the Crisis of Development Theory." *Studies in Comparative International Development* 26 (2): 3–36.

Marcus, George, and Michael Fischer. 1986. *Anthropology as Cultural Critique.* Chicago: University of Chicago Press.

Mascia-Lees, Frances, F. P. Sharpe, and C. Ballerino Cohen. 1989. "The Postmodernist Turn in Anthropology: Cautions from a Feminist Perspective." *Signs* 15 (1): 7–33.

Maybury-Lewis, David. 1985. "A Special Sort of Pleading: Anthropology at the Service of Ethnic Groups." In *Advocacy and Anthropology: First Encounters,* edited by Robert Paine, 131–48. St. John's, New Foundland: Memorial University of New Foundland.

Memmi, Albert. 1967. *The Colonizer and the Colonized.* Boston: Beacon Press.

Mitchell, Timothy. 1988. *Colonising Egypt.* Cambridge: Cambridge University Press.

Mohanty, Chandra. 1991a. "Cartographies of Struggle: Third World Women and the Politics of Feminism." In *Third World Women and the Politics of Feminism,* edited by Chandra Mohanty, Ann Russo, and Lourdes Torres, 1–47. Bloomington: Indiana University Press.

———. 1991b. "Under Western Eyes: Feminist Scholarship and Colonial Discourses." In *Third World Women and the Politics of Feminism*, edited by Chandra Mohanty, Ann Russo, and Lourdes Torres, 51–80. Bloomington: Indiana University Press.

Montaldo, Graciela. 1991. "Estrategias del Fin de Siglo." *Nueva Sociedad*, no. 116: 75–87.

Morandé, Pedro. 1984. *Cultura y Modernización en América Latina*. Santiago: Universidad Católica.

Mudimbe, V. Y. 1988. *The Invention of Africa*. Bloomington: Indiana University Press.

Mueller, Adele. 1986. "The Bureaucratization of Feminist Knowledge: The Case of Women in Development." *Resources for Feminist Research* 15 (1): 36–38.

———. 1987a. "Peasants and Professionals: The Social Organization of Women in Development Knowledge." Ph.D. diss. Ontario Institute for Studies in Education.

———. 1987b. "Power and Naming in the Development Institution: The 'Discovery' of 'Women in Peru.'" Paper presented at the Fourteenth Annual Third World Conference, Chicago.

———. 1991. "In and Against Development: Feminists Confront Development on Its Own Ground." Photocopy.

Nandy, Ashis. 1983. *The Intimate Enemy: Loss and Recovery of Self under Colonialism*. Delhi: Oxford University Press.

———. 1998. "Shamans, Savages, and the Wilderness: On the Audibility of Dissent and the Future of Civilizations." *Alternatives* 14 (3): 263–78.

Ong, Aihwa. 1987. *Spirits of Resistance and Capitalist Discipline*. Albany: SUNY Press.

Parajuli, Pramod. 1991. "Power and Knowledge in Development Discourse." *International Social Science Journal* 127:173–90.

Pigg, Stacy Leigh. 1992. "Constructing Social Categories through Place: Social Representations and Development in Nepal." *Comparative Studies in Society and History* 34 (3): 491–513.

Price, David. 1989. *Before the Bulldozer*. Washington, D.C.: Cabin John Press.

Quijano, Aníbal. 1988. *Modernidad, Identidad y Utopía en América Latina*. Lima: Sociedad y Política Ediciones.

———. 1990. "Estélica de la Utopía." *David y Goliath* 57:34–38.

Rabinow, Paul. 1986. "Representations Are Social Facts: Modernity and Post-Modernity in Anthropology." In *Writing Culture: The Poetics and Politics of Ethnography*, edited by James Clifford and George E. Marcus, 234–61. Berkeley: University of California Press.

Rojas de Ferro, María Cristina. 1994. "A Political Economy of Violence." Ph.D. diss., Carleton University, Ottawa.

Sachs, Wolfgang. 1992. "Environment." In *The Development Dictionary*, edited by Wolfgang Sachs, 26–37.

Said, Edward. 1979. *Orientalism*. New York: Vintage Books.

———. 1989. "Representing the Colonized: Anthropology's Interlocutors." *Critical Inquiry* 15:205–25.

Sarlo, Beatriz. 1991. "Un Debate Sobre la Cultura." *Nueva Sociedad* 116:88–93.

Scott, James. 1985. *Weapons of the Weak: Everyday Forms of Peasant Resistance.* New Haven: Yale University Press.

Sikkink, Kathryn. 1991. *Ideas and Institutions: Developmentalism in Brazil and Argentina.* Ithaca: Cornell University Press.

Slater, David. 1993. "The Geopolitical Imagination and the Enframing of Development Theory." Photocopy.

Soja, Edward. 1989. *Postmodern Geographies.* London: Verso.

Spelman, Elizabeth. 1988. *Inessential Woman: Problems of Exclusion in Feminist Thought.* Boston: Beacon Press.

St-Hilaire, Colette. 1993. "Canadian Aid, Women and Development." *The Ecologist* 23 (2): 57–63.

Strahm, Rudolf. 1986. *Por Qué Somos Tan Pobres?* México, D.F.: Secretaría de Educación Pública.

Strathern, Marilyn. 1988. *The Gender of the Gift.* Berkeley: University of California Press.

Sutton, David. 1991. "Is Anybody Out There? Anthropology and the Question of Audience." *Critique of Anthropology* 11 (1): 91–104.

Taussig, Michael. 1980. *The Devil and Commodity Fetishism in South America.* Chapel Hill: University of North Carolina Press.

Trinh T. Minh-ha. 1989. *Woman, Native, Other.* Bloomington: Indiana University Press.

Truman, Harry. [1949] 1964. Public Papers of the Presidents of the United States: Harry S. Truman. Washington, D.C.: U.S. Government Printing Office.

Ulin, Robert. 1991. "Critical Anthropology Twenty Years Later: Modernism and Post-modernism in Anthropology." *Critique of Anthropology* 11 (1): 63–89.

United Nations, Department of Social and Economic Affairs. 1951. Measures for the Economic Development of Underdeveloped Countries. New York: United Nations.

Urla, Jacqueline. 1993. "Cultural Politics in the Age of Statistics: Numbers, Nations, and the Making of Basque Identities." *American Ethnologist* 20 (4): 818–43.

Yúdice, George, Jean Franco, and Juan Flores, eds. 1992. *On Edge: The Crisis of Contemporary Latin American Culture.* Minneapolis: University of Minnesota Press.

CATHERINE V. SCOTT

# 15. Tradition and Gender in
# Modernization Theory

THERE DOES NOT SEEM to be much more to write about modernization
theory of the 1950s and 1960s. Numerous critics have taken early modernization theorists such as Rostow (1960), Parsons (1960), and Inkeles
(1969) to task for their ethnocentrism, naive optimism, and "failure to
recognize the political implications of economic dependency upon the
West" (Randall and Theobald 1985, 33). Other critics pointed to modernization theory's reliance upon evolutionary and linear notions of social and political change and its reductionism and oversimplification of
the development process (e.g., Portes 1976; Tipps 1976).[1] However, upon
closer inspection it is evident that modernization theory was mainly
criticized for its empirical content, lack of predictive ability, definitional
shortcomings, and Western bias. Virtually no questions were asked about
the way in which challenges to modernization were framed, and the extent to which the dichotomies of traditional and modern depended upon
conceptions of gender, gender differences, and the devaluation of "the
feminine."

Embedded within constructs of traditional society are ideas about
women, family, and community that function as points of contrast for
modernization theorists' idealization of a rational, forward-looking,
male-dominated public sphere. Conceptions of linear time also play an
important role for modernization theorists, with tradition and the feminine viewed as part of the past. As Inkeles and Smith (1974, 3–4) put it,
"Mounting evidence suggests that it is impossible for a state to move into
the twentieth century if its people continue to live in an earlier era." For
development theorists seeking to construct the antinomy of tradition and
modernity, it is important to distance one from the other and stress the
importance of autonomy and separation of men from the household and
the feminine traits associated with it.

There are three major themes evident in the work of theorists as diverse
as Alex Inkeles and W. W. Rostow. The first is an unconscious and pervasive psychological preoccupation with separation and differentiation
from the household. This distancing is accomplished by the presentation
of tradition as a bundle of characteristics that also have historically been
used to subordinate women and denigrate the social relations associated
with females, especially mothers. It is interesting to note that some early
critics of modernization theory argued that it undertheorized tradition

and presented it as a static and "residual concept" (Randall and Theobald 1985, 35). This chapter will argue that the powerful imagery and the descriptions of idealized modernity provided by early modernization theorists were laden with such significant demarcations of constructed gender differences that explicit explorations of tradition were unnecessary.

A second theme evident in early modernization theory is the reliance on the public/private distinction in discussions of modernity and tradition. Modernity, rationality, technological progress, and good government are achieved in a public realm inhabited by autonomous men. With the exception of the Comparative Politics Committee of the Social Science Research Council (ssrc), which displayed some ambivalence toward tradition and called for more exploration of the content of traditional societies, early modernization theorists viewed tradition, and the values associated with tradition and women, as absolutely incompatible with modern institutions.

Finally, early modernization theorists rely, implicitly or explicitly, upon evolutionary models of social and political change, which provide an important lens for viewing their ideas about development, modernization, and gender. In their reliance upon an evolutionary model, they inevitably portray development as a struggle for dominance over nature, and implicitly, over women. Moreover, in using an evolutionary model, they portray development as the ever-widening ability of men to create and transform their environment. Within this linear framework of evolutionary social and political change, women are "left behind," confined to the household and denied citizenship. Women's continued subordination in fact defines male citizenship.

## Sexism and Modernization Theory

The argument here is that modernization theorists brought deeply held masculinist and dualistic views of the world of tradition and modernity that relied upon configurations of the public and private spheres, the household, and evolutionary progress. It is important and useful also to note that this literature consistently purported to present a universal model of the modernization process that was, in fact, partial and based on an (often idealized) version of masculine modernity. Women are either invisible, treated paternalistically, or used as a litmus test for determining the degree of "backwardness" of a particular Third World country. A startling example of invisibility is the project that interviewed six thousand men in Argentina, Chile, East Pakistan (Bangladesh), India, Israel, and Nigeria in order to examine the effects of factory life on modern attitudes (Inkeles 1969; Inkeles and Smith 1974). They report that budget limitations and the concentration of men in industrial jobs explain the gender of the sample (Inkeles and Smith 1974, 311). But, surely, would not women be included in the cultivator and nonindustrial worker category, two other categories of respondents interviewed in each of the countries? The authors never explain why only men were included in these categories as well. They also make the interesting assertion, "We are firmly convinced that the overwhelming majority of the psychosocial indicators we used to identify the modern man would also discriminate effectively among women" (Inkeles, Smith, et al. 1983, 123). This directly contradicts their reporting on the low correlations concerning modern attitudes about political life and attitudes about the family.

As an example of striking paternalism, Daniel Lerner (1958, 29) took Zilla K. along as an interviewer when he returned to the village of Balgat, Turkey, in 1954 (he had been there four years before). This is his description of her hiring:

> I had "ordered" her through a colleague, at Ankara University, "by the numbers": thirtyish, semi-trained, alert, compliant with instructions, not sexy enough to impede our relations with the men of Balgat but chic enough to provoke the women. A glance and a word showed that Zilla filled the requisition.

Rostow (1960, 91) speculated about what lies beyond the state of high mass-consumption reached by societies such as the United States and worried about the onset of pervasive boredom—for men. Women, on the other hand, "will not recognize the reality of the problem" because of their involvement in childrearing: "The problem of boredom is a man's problem, at least until the children have grown up."

The comparison of the liberated and independent woman of the West with the tradition-bound woman of the Third World also informs many accounts of the psychosocial requisites of modernity. When women are discussed by the modernization theorists in any specific way they are presented in remarkably flat terms, and often uniformly oppressed by men and family structures. Lerner (1958, 199) notes that "traditional women are content to accept the role and status assigned them," as the "stolid guardians of custom and routine." Women who represent modern values in Middle Eastern societies such as Lebanon yearn for the greater educational and career opportunities available to women in the West. The Western media provides a constant reminder to Middle Eastern women of their restricted opportunities. In a puzzling analogy, Lerner (1958, 204) notes that "as the American housewife uses soap operas to fill her day and satisfy her needs, so this young Lebanese woman finds gratification through borrowed experiences." While implicitly acknowledging that viewing soap operas might represent frustration and denied opportunities for middle-class U.S. women, Lerner never explicitly challenges the media's juxtaposition of the "enlightened and independent woman" of the West with the backward and traditional woman of the Middle East. McClelland (1976, 399–400) makes a similar contrast:

> A crucial way to break with tradition and introduce new norms is via the emancipation of women. . . . The most general explanation lies in the fact that women are the most conservative members of a culture. They are less subject to influences outside the home than the men and yet they are the ones who rear the next generation and give it the traditional values of the culture.

Inkeles and Smith, et al. describe "most of the traditional societies and communities of the world" as "if not strictly patriarchal, at least vigorously male dominated" (Inkeles and Smith, et al. 1983, 26). While traditional man is reluctant to accept women's freedom, modern man is willing to "allow women to take advantage of opportunities outside the confines of the household" (Inkeles and Smith 1974, 77, 291). In a later work they predicted that "the liberating forces of modernization would act on men's attitudes and incline them to accord to women status and rights more nearly equal to those enjoyed by men" (Inkeles, Smith, et al. 1983, 42). Such contrasts not only serve to establish a Western sense of difference and superiority (and complacency about women's rights in the West); they also mark women, in Mohanty's (1991, 56) terms, as "third world (read: ignorant, poor, uneducated, tradition-bound, domestic, family-oriented, victimized, etc.)." As the most "backward" group in society, women serve as an implicit contrast between Western modernity and non-Western tradition.

## "Becoming Modern": The Syndrome of Modern Male Citizenship

Randall and Theobald (1985, 15) place early modernization theories into one of two categories: psychocultural or structural-functional. Psychocultural approaches examine the attitudinal prerequisites of modernity, while structural-functional approaches focus on the institutional changes needed for modernity. Inkeles (1969), Inkeles and Smith (1974), and Inkeles, Smith, et al. (1983) adopt a psychocultural approach to modernization. In the study of six thousand men in the six countries listed above, Inkeles and Smith locate a syndrome of participant citizenship, "attitudes and capacities" necessary to realize "nation-building and institution-building" in the Third World (1974, 3).

294 *Catherine V. Scott*

TABLE 1  Eighteenth- and Nineteenth-Century Dichotomies

| Traditional | Modern |
| --- | --- |
| Nature | Culture |
| Woman | Man |
| Physical | Mental |
| Mothering | Thinking |
| Feeling and superstition | Abstract knowledge and thought |
| Country | City |
| Darkness | Light |
| Nature | Science and civilization |

*Source:* Jordanova 1980, 44.

Inkeles and Smith (1974, 19–24) argue that twelve traits define modern man (*sic*). In addition, they argue that modernity is also characterized by a host of other orientations toward religion, the family, and social stratification (1974, 25). Their analytic and topical characteristics of the modern man are summarized in table 1. Feminist critics of the Western philosophical tradition have noted the persistent denigration of the feminine within that tradition. Lloyd (1984, 2–3), for example, notes that in the triumph of reason over darkness, the early Greeks used symbolic associations of the female as what needed "to be shed in developing culturally prized rationality." Rooney (1991, 91) and Jordanova (1980) have noted the images of battle or struggle that are common in discussions of reason and unreason. Jordanova's (1980, 44) presentation of the dichotomies that emerged in the biomedical sciences in the eighteenth and nineteenth centuries showed similarities with the contrasts between traditional and modern man presented by Inkeles and Smith (see table 1).

Jordanova (1980, 44) suggests that the oppositions contain an important gender dimension and connotations of battle: the struggle between the forces of tradition and modernity was also a struggle between the sexes, with the increasing assertion of masculinity over "irrational, backward-looking women" applauded as inevitable. Furthermore, she shows how science and medicine used sexual metaphors that portrayed nature as a woman to be penetrated, unclothed, and unveiled by masculine science (Jordanova 1980, 45).

Inkeles and Smith replicate these Enlightenment dichotomies in their comparison of traditional and modern men (see table 2). In the larger study, they present case studies from East Pakistan (Bangladesh) of a traditional man and a modern one (Inkeles and Smith 1974, 73–80).

TABLE 2 Traditional Man and Modern Man

| Traditional | Modern |
|---|---|
| Not receptive to new ideas | Open to new experience |
| Rooted in tradition | Change orientation |
| Only interested in things that touch him immediately | Interested in outside world |
| Denial of different opinions | Acknowledges different opinions |
| Uninterested in new information | Eager to seek out new information |
| Oriented toward the past | Punctual: oriented toward the present |
| Concerned with the short term | Values planning |
| Distrustful of people beyond the family | Calculability; trust in people to meet obligations |
| Suspicious of technology | Values technical skills |
| High value placed on religion and the sacred | High value placed on formal education and science |
| Traditional patron-client relations prevail | Respect for the dignity of others; belief that rewards should be distributed according to rules |
| Particularistic | Universalistic |
| Fatalistic | Optimistic |

*Source:* Inkeles and Smith 1974, 19–34.

Ahmadullah, the traditional man, "was relatively passive, even fatalistic, and very much dependent on outside forces, above all on the intervention of God." He said he could do nothing in the face of an unjust law, and he preferred living in the "closed and unchanging world" of the village. Nuril, on the other hand, had lived in the city for ten years, approved of women acquiring more education, was open to meeting new people and having new experiences, and believed that "the outcome of things depended very much on himself, and [that] others bore responsibility for their individual actions." As Inkeles noted in his earlier study (1969, 1122–23), the modern man possesses an orientation toward politics that recognizes the necessity and desirability of a "rational structure of rules and regulation."

Juxtaposed with the village, family, and kinship structures stands the factory, a "school in rationality" (Inkeles 1969, 1140). The factory is an exemplar of efficiency, innovation, planning, punctuality, rules and formal procedures, and objective standards for assessing skills and output (Inkeles and Smith 1974, 158–163). City life, they argued, also has a powerful indirect effect on creating modern attitudes because cities have greater concentrations of schools, factories, and mass media (Inkeles and Smith 1974, 228).

In addition to an uncritical perspective on the nature of factory work in both the First and Third Worlds, the description by Inkeles and Smith of the benefits of factory work relies upon a liberal framework of contractual obligation and individualism that reflects a masculinist standpoint and preoccupation with autonomy. Hirschmann (1989, 1237) argues that this is especially evident in symbolic language that reflects desires for dominance and nonreciprocal recognition. In describing modern man's experiences as a "shift from the more traditional settings of village, farm, and tribe to city residence, industrial employment, and national citizenship" (Inkeles and Smith 1974, 156), psychocultural theorists of modernization juxtapose community, family, and kinship with the modern, and it is women who stand at the center of the traditional community. The factory serves as the emblem of scientific progress and technological prowess that promises to shatter any resistance to rationalized relationships in the public realm. This liberal and masculinist conception of freedom entails nonrecognition of the female and the relationships she represents. Freedom requires not only moving beyond the household: subordination of the household becomes the means of achieving freedom (Hirschmann 1989, 1235). Women were not only excluded from the samples because they worked in factories, but because they resided in the very location that undermines the institutions that "train men in active citizenship" (Inkeles 1969, 1141).

Daniel Lerner's *The Passing of Traditional Society* (1958) is another representative of the psychocultural approach. Lerner presents the parable of modern Turkey through the story of the Grocer and the Chief, two men interviewed in the village of Balgat, near Ankara, in 1950 and 1954. The Chief "was a man of few words on many subjects," who "audits his life placidly, makes no comparisons, thanks God." The Grocer, on the other hand, perceived his story as "a drama of self versus village," a man whose "psychic antennae were endlessly seeking the new future here and now" (Lerner 1958, 22, 24).

Lerner's contrasts between traditional and modern society (1958, 44) echo Enlightenment thinkers and Inkeles and Smith: "village versus town, land versus cash, illiteracy versus enlightenment, resignation versus ambition, piety versus excitement." In modern societies, personal mobility is a "first-order value," and a modern society "has to encourage rationality, for the calculus of choice shapes individual behavior and conditions its reward" (Lerner 1958, 48). Empathy is the mechanism that accompanies the transformation of traditional man, i.e., "the capacity to see oneself in the other fellow's situation" (Lerner 1958, 50). Empathy takes place through both projection ("assigning the object certain preferred attributes of the self") and introjection ("attributing to the self certain desirable attributes of the object") (Lerner 1958, 49). Identification with others is a key component of modern man's personality.

Chodorow has noted the importance of negative identification, differentiation, and nonrecognition in human development, and these themes recur in Lerner's definition of modern man's development and "maturation." Differentiation is defined relationally, and because men have "conflictual core gender identity problems," it is important to maintain a rigid boundary between the masculine and feminine: "Boys and men come to deny the feminine identification within themselves and those feelings they experience as feminine; feelings of dependence, relational needs, emotions generally" (1989, 109–10). The development of masculine identity as outlined by object relations theory resonates in Lerner's (1958, 410) definition of modern man's solitary struggle against forces represented by the village, "the passive, destitute, illiterate and altogether 'submerged' mass which looms so large in its [the Middle East's] sociological landscape."

McClelland's (1976, 107) chief goal was to determine the extent to which a "culture or nation has adapted more or less rapidly to modern civilization, with its stress on technology, the specialization of labor, and the factory system." McClelland and his colleagues developed a measure of "*n* achievement" (shorthand for need achievement) through content analysis of achievement-related stories written by male college students, folk tales from various cultures, and children's stories. He explicitly links high *n* achievement with boys who had mothers who encouraged independence yet at the same time provided warmth and affection. Reporting on earlier findings that attempted to demonstrate a link between socialization and the propensity for high achievement, McClelland (1976, 46) summarized: "The mothers of the sons with high *n*-achievement have set higher standards for their sons; they expect self-mastery at an early age." Thus he not only touches on themes within object relations theory, he literally claims that characteristics of mothering (along with other factors) are influential in determining whether a society develops. McClelland (1976, 404–95) also warns about father-dominance in producing low achievement, because "the boy is more likely to get his conception of the male role from his relationship with the father rather than his mother and therefore, to conceive of himself as a dependent, obedient sort of person if his father is strong and dominating" (McClelland 1976, 353). It is in his relationship with the mother that the boy obtains a sense of independence and autonomy, but only from mothers who are "controlled and moderate in warmth and affection" (McClelland 1976, 405).

From these observations, McClelland hypothesizes about how to bring about development. First, "other-directedness" is essential (McClelland 1976, 192). The "authority of tradition" must be replaced and men must learn to pay attention to newspapers, local political parties, and the radio, a "new voice of authority." Development, in other words, requires a shift in allegiance from the private to the public realm. Second, *n* achievement needs to be increased, and McClelland speculates about the prospects for decreasing father-dominance, protestant conversion, and a reorganization of fantasy life (McClelland 1976, 406–18). Finally, McClelland suggests that existing *n* achievement resources could be used more efficiently to encourage "young men with high *n* achievement to turn their talents to business or productive enterprise" (McClelland 1976, 418).

Rostow's (1960) *Stages of Economic Growth* introduces both the concept of evolutionary stages of societal development and attitudinal prerequisites as crucial for understanding political development. He conceptualizes the evolutionary path of development as composed of five stages: tradition, societies poised to "take-off," the "take-off" into modernity itself, the drive to maturity, and the age of high mass-consumption. Traditional societies are characterized by Rostow (1960, 4) as "pre-Newtonian" because they are located on the other side of "that watershed in history when men came widely to believe that the external world was subject to a few knowable laws and was capable of productive manipulation." The "frame of mind" conducive to modern science was nonexistent in these pre-Newtonian societies, which possessed a "long-term fatalism" and a "ceiling on the productivity of their economic techniques" (Rostow 1960, 5). During the time before take-off, "limited bursts" of entrepreneurial activity and "enclaves of modernity" emerge, spurred by "enterprising men" who are willing to "take risks in pursuit of profit or modernization" (Rostow 1960, 6–7). Rostow presents us with the image of energetic men emerging from rural backwardness and leaving the bonds of tradition to transform and manipulate the forces of nature:

> Man need not regard his physical environment as virtually a given factor by nature and providence, but as an ordered world which, if rationally understood, can be manipulated in ways which yield productive change and, in one dimension at least, progress. (1960, 19)

Rostow contrasts the world of family, mother, and household with the modern world of markets, technology, and science. In fact, traditional societies become eligible for take-off when "men come to be valued in society not with their connection with clan or class . . . but for their individual ability to perform certain specific, increasingly specialized functions" (Rostow 1960, 19). This requires attitudinal changes toward science, propensities to calculate and take risks, and a willingness to work (Rostow 1960, 20). Rostow appeals to male heroic leadership in his analysis of the key take-off from tradition to modernity. He juxtaposes this new elite with "the old land-based elite" which is mired in agrarian practices and worldviews that do not regard "modernization as a possible task" (Rostow 1960, 26).

## Equality, Capacity, and Differentiation: Structural Explanations of Modernity

Structural-functional frameworks for explaining modernization and development share the dualistic and gendered framework employed by psychocultural approaches. Parsons's pattern-variables of universal social roles represent the most famous structural-functional approach (Parsons 1960). Parsons makes distinctions between affectively neutral and affective actions, universal and particular orientations, specific and diffuse obligations, self-oriented rather than collectively oriented behavior, and achievement and ascriptive criteria for recognizing performance, with modernity characterizing the first category in each paired concept. In addition to opposing tradition and modernity in this way, Parsons also unconsciously "feminizes" traditional society in that the terms he uses to define "traditional" Third World societies have also been used to juxtapose male rationality and inherent superiority with "lower order" female passions and instincts.

One of Parsons's most interesting distinctions is between universal and particular modes of categorizing social objects. While modern man is able to use a set of norms and standards that apply to all objects in a particular class, traditional man treats them in terms of their standing in some particular relationship to him (Bluh 1982, 88–90). As Devereux (1961, 41) explains it,

> Whether someone is a good doctor, a competent secretary, or a beautiful woman are presumably matters to be determined on universalistic grounds. But while certain modes of behavior might be evoked toward beautiful women or deserving children in general, where one's own wife or child is involved, one is committed in many special ways, regardless of beauty or desert.

Parsons (1960, 119) identifies the major obstacles to economic development as a "combination of 'traditionalism' and a strong pressure to reproduce the existing pattern of economic organization wherever opportunity exists for its expansion." Developed societies, on the other hand, have legal systems that embody principles of universalism and specificity, contracts and property, and occupational roles that free individuals "from ties and imperatives which would interfere seriously with economic production" (Parsons 1960, 147).

While Parsons is obviously enumerating the values and institutions compatible with capitalist development, he is also denigrating characteristics of traditional societies that have historically been associated with women and devalued in the Western tradition. The individual's quest for modernity is a battle against the village, family, and "tribe." These communities are not only averse to economic productivity and the capitalist work ethic; they represent, symbolically, female attributes and the relationship and structures historically associated with them.

Parsons (1964, 356) also relied explicitly upon an evolutionary model of development. He argued that certain "organizational complexes" were necessary for societies to emerge from "primitiveness." In order to evolve along the scale of development, societies require stratification, necessary to emerge from the "seamless web" of relationships that characterize societies governed by strong kinship and family ties (Parsons 1964, 342). Stratification is necessary in order to create avenues of upward mobility for leaders who wish to marshal the resources for development (Parsons 1964, 34). Extending this functionalist and evolutionary framework even further, Parsons argues that legitimation is necessary in order to justify hierarchy; bureaucracy is necessary in order to develop enhanced capacity for change; money and markets are necessary as "the great mediator of the instrumental use of goods and services" (Parsons 1964, 349).

Parsons makes a number of key assumptions in this presentation of evolutionary development. He assumes that power must be concentrated in hierarchies of stratification in order for political development to occur. Kinship and family ties, on the other hand, are clearly "rigid" and incapable of generating power to effect change. Second, Parsons assumes that development will be a divisive process, as one group will inevitably seek power and try to achieve dominance over another. He never doubts that struggles and drives for dominance will increase the "long-run adaptive capacity" of societies evolving toward the development ideal (Parsons 1964, 340). He assumes that despite "severe dislocations" resulting from such struggles, the societies undergoing evolutionary development are experiencing an inevitable and "natural" process. Why is stratification necessary in order to marshal resources? Why is a concentration of power necessary for productive change to take place? While Parsons produces a legitimation of capitalism and markets, he also implicitly presents male/culture triumphant over female/nature. The evolutionary universals considered fundamental for understanding development are also justifications for male dominance and denigration of traits historically associated with women. Liberal ideology and evolutionary functionalism join together to legitimate masculinist individualism and domination and exclusionary practices with regard to the household.

The SSRC Committee on Comparative Politics also worked with a structural-functional framework. They located a "development syndrome" as increasing equality, capacity, and differentiation, the evolution of which produces inevitable strains and tensions in traditional and transitional societies. They defined the obstacles to the achievement of development as "crises" of identity, legitimacy, participation, penetration, and distribution, faced by every society in "building both state and nation" (Pye 1971a, viii). Coleman (1971, 73) pointed out that the SSRC participants chose to view development as an evolutionary process—as an

> open-ended increase in the capacity of political man to initiate and institutionalize new structures and supporting cultures, to cope with or resolve problems, to absorb and adapt to continuous change, and to strive purposively and creatively for the attainment of new goals.

The SSRC committee characterized development and modernization as conflictual processes laden with the potential for setbacks. While a "complete reversion to a traditional pre-modern system" had not yet occurred, there had been "abortive developments" in many Third World countries (Coleman 1971, 84). In fact, Coleman (1971, 100) argued that "arrested development" may begin to affect an increasing number of Third World countries as they became stalled at a certain level of differentiation, equality, and capacity.

Abortive or arrested development occurs as a result of conflicts in the resolution of the five "crises" listed above. For example, fragmentation challenges state capacity for integration; demands for equality challenge the state's ability to solve the participation crisis; challenges to state capacity create legitimacy and identity crises, and so forth. As Pye (1971b, 106) notes, these crises, if unresolved, create difficulties in achieving a "higher level of performance" on the part of Third World leaders. Pye's (1971b, 110) discussion of the identity crisis is especially worth noting. He argues that identity crises take place over four different issues: territory, class, ethnicity, and social change. In a revealing discussion of the legacy of colonialism, Pye (1971b, 122) discusses the ambivalence these leaders feel toward colonial relationships of dependence:

> All humans must experience complete and protracted dependency, for it is the very mark of human-ness. . . . With development people have ambivalent feelings toward their sentiments related to dependency, and the search for individual identity inevitably involves an assertion for independence in which any continuing cravings for dependence must be veiled.

These themes, which resemble theories of male psychological development, recur throughout the SSRC committee's discussion of modernization and development. Anxiety, crises, capacity, performance, individual
identity, and overcoming the challenges posed by tradition represent the
"syndrome" of development in both becoming a mature male adult and a
mature, national society.

Furthermore, modernization is conceived as a masculine triumph, the
result of enhanced capacity, improved performance, and effective penetration. LaPalombara (1971, 206) conceives of the crisis of penetration as a
"test" of the "organizational, technological, and/or diplomatic capabilities
of an existing governing elite." It refers to "whether [leaders] can get what
they want from people over whom they seek to exercise power" (LaPalombara 1971, 209). Gaining compliance, creating new organizations, and
breaking down old loyalties are all aspects of the penetration challenge.

Reliance on an evolutionary model of modernization is also evident
in the SSRC committee's work (Coleman 1971, 75). Images of struggle
are deeply embedded in evolutionary models of society that emerged
in the nineteenth century. Gross and Averill (1983, 81) note that the importance of competition in evolutionary theory was used by men in the
twentieth century to impose order on the perceived chaos associated with
reproduction:

> Evolution and natural selection, as products of nineteenth century thought,
> coincide with other reflections of men's anxiety about women, most plainly
> displayed in their preoccupation with her reproductive ability: her uncon-
> trolled sexuality, her ("pathological") reproductive physiology, even her
> (hysterical) psychology. The nineteenth century medicalization of women's
> reproductive capacities . . . parallels the emphasis on domination and com-
> petition in nature as the main restraints over unbridled chaos in the orderly
> evolution of species.

Structural-functional theories of evolutionary development can also be
viewed as expressing masculine concern and preoccupation with the
fragmentation, particularism, and even chaos of traditional society. The
modernization process is laden with conflict, a "ceaseless straining and
tugging between the development processes and the requirement that
the political system maintain itself" (Pye 1971b, 101). Assumptions that
"abortive" or "arrested" development are inevitable and demand specialization, hierarchy, and enhanced capacity in order for progress to take
place, constitute a masculinist version of modernity and development.

The view of development as an evolutionary struggle to achieve greater
capacity to dominate nature coincides with the liberal underpinnings of

modernization theory, which also relies upon assumptions of scarcity. In Macpherson's phrase, liberalism makes a "maximizing claim" (1973, 4) in that it provides a framework for realizing individual desires. Modernization theory shares this view of humans as "essentially a bundle of appetites demanding satisfaction." Macpherson (1973, 18–19) argues that liberal theory pits unlimited desire against scarcity, with unlimited desire conceived of as rational and morally acceptable rather than deplored as greed: "the chief purpose of man is an endless battle against scarcity" (Macpherson 1973, 18). Evolutionary paradigms of development also assume that "scarcity is inevitable and in turn demands competition, which is expressed in dominance relationships that make for evolutionary 'progress'" (Gross and Averill 1983, 82).

## Modernization Revisionists

The initial criticisms of modernization theory came from theorists who continued to characterize modernity and tradition in dualistic terms, but who insisted on the recognition of the continuing salience of caste, ethnicity, and other "particularistic" characteristics of traditional societies. Rudolph and Rudolph, for example, pointed to the resilience of caste organizations in Indian politics, adhered to the meaning of tradition and modernity held by earlier theorists:

> Modernity assumes that local ties and parochial perspectives give way to universal commitments and cosmopolitan attitudes; the truths of utility, calculation, and science take precedence over those of the emotions, the sacred, and the non-rational . . . that mastery rather than fatalism orient their attitude toward the material and human environment. (1967, 3–4)

In addition, theories about patrimonialism and the "soft state" reinforced the appreciation of traditional institutions and represented a departure from early modernization theory because they challenged the assumption that Western institutions could be duplicated in post-colonial societies. Samuel Huntington also challenged the naive optimism of modernization theorists.

## Creating a (Gendered) Political Order

Samuel Huntington's *Political Order in Changing Societies* (1968) attempted to temper the optimism of modernization theorists by pointing to the potential problems that could occur during the modernization process. *Political Order* opens with a Manichaean vision of international politics (1968, 1). On one side of the divide are modern polities such as the United States, Great Britain, and the Soviet Union, with their "effective bureaucracies, well-organized political parties . . . and reasonably effective procedures for regulating succession and controlling political conflict." On the other side there are the governments of Asia, Africa, and Latin America, where "increasing ethnic and class conflict, recurring rioting and mob violence, frequent military coup d'état . . . the loss of authority by legislatures and courts" constitute the norms of public life. Modernization theory, Huntington argues, could not anticipate this outcome because it assumed that economic development and social mobilization would lead to political development. Huntington (1968, 41) argued that the very process of modernization threatened the political order. Thus the task of developing countries should be to create political institutions that "derive their interests not from the extent to which they represent the interests of the people or any other group but the extent to which they have distinct interests of their own apart from other groups" (1968, 27). Like modernization theorists, Huntington argues that political institutions must attain autonomy, but he presents the achievement of autonomy as a struggle that must be waged against the confusion, chaos, and alienation that accompany modernization (1968, 30–31). Order must be created: "Men may, of course, have order without liberty, but they cannot have liberty without order" (1968, 7–8).

Huntington takes a grim view of the world of politics. In defending his calls for order in societies with weak institutions, he writes that "politics is a Hobbesian world of unrelenting competition of social forces—between man and man, family and family, clan and clan, region and region, class and class—a competition unmediated by more comprehensive political organizations" (1968, 24). He implies that modernizing societies lack the kind of rationality required to create political institutions to regulate conflict and that a government ordered along the lines of the military is most likely to create viable political institutions:

> Unity, esprit, morale, and discipline are needed in governments as well as regiments. . . . The problems of creating coherent political organizations are more difficult but not fundamentally different from those involved in the creation of coherent military organizations. . . . Discipline and development go hand in hand. (1968, 33–34)

Huntington's dualistic brand of thinking, his endorsement of authoritarian government, and his praise for the coherence and discipline of military modes of governance make his work one of the most remarkable and striking in development theory and perhaps suggest a reason for the wide appeal and readership of the book. His formulation of the challenges of development also has a distinctively masculinist tenor. For Huntington, the ends of politics should be toward creating political institutions that, once in place, are capable of regulating political conflicts that arise (1968, 9, 11). In proposing his well-run polity with coherent, adaptable, autonomous, and complex political institutions, he explicitly contrasts the world of "natural" communities, "the isolated clan, family, or tribe," with political institutions and organizations that are consciously created through "political action" and "political labor" (1968, 10–11). "Natural" communities are simple, spontaneous, and less diverse; political communities are man-made political structures that do not depend upon the obligations and relationships that exist in "natural" communities. They are based on purely instrumental and calculating imperatives. Furthermore, Huntington's emphasis upon teamwork, command, and discipline, and his analogy between creating coherent political organizations and coherent military structures reflect a distinctly heroic, atomistic, and individualistic approach to governance. While earlier modernization theorists conceived of tradition and the socially constructed feminine values associated with it as something to be transcended, Huntington views it as a dangerous force to be tamed and disciplined.

Huntington is a transitional figure in development theory. He made his case for a politics of order by challenging the naïve belief in evolutionary progress exhibited by many earlier modernization theorists of the early 1960s. Dependency theorists reacted against Huntington's call for order and his neglect of international politics and economics in affecting political stability in the Third World.

## Conclusion

Early modernization theory has been discredited for being unscientific and sexist in its focus on male heads of households. While these are valid criticisms, it is also important to examine the implicit assumptions, concerns, unstated preoccupations, and avoidances of these early theorists. Despite its official demise, early modernization theory's conceptual foundations continue to have pervasive power. This power helps explain the difficulty encountered by those who seek to challenge its masculinist worldview.

There is no need to speculate about the early upbringing of these male theorists to recognize the links they make between traditional society, on the one hand, and the denial of power and autonomy for the household and women, on the other. As Hirschmann (1989, 1232) notes, if self-conceptions are gender-related, accompanying worldviews will also differ by gender. And as Flax (1983, 246) contends, "patriarchy by definition imputes political, moral, and social *meanings* to sexual differentiation."

Object relations theory provides a powerful means for understanding the themes of differentiation, autonomy, identity, and suppression of the household's characteristics that play so heavily in modernization theory. For these theorists, modernization requires self-propelled men to leave the household, abandon tradition, and assume their rightful place among other rational men. Women and the household are conceived of as part of the past that contains the dangerous worldview that nature is unalterable and that man is powerless in his efforts to control it. Only the SSRC committee displayed a hint of ambivalence about the compatibility of traditional and modern institutions. Coleman (171, 86–89) argued that, at times, traditional institutions can facilitate modernization, but only to the extent that traditional society has some characteristics that resemble the modern one! This is rarely the case, according to modernization theory, as traditional societies are composed of "highly pluralistic traditions that reflect very different, frequently conflicting or incongruent cultural patterns" (Coleman 1971, 87). Societies are modern to the extent that they overcome tradition, but abortive or arrested development are constant possibilities. Being "stuck" on the evolutionary scale toward national maturity reflects anxiety over the ability of state leaders to resolve their ambivalence toward dependence. Modernization is the triumph of penetration, identity, and legitimation, and the subordination of tradition, nature, and the "feminine."

Theories of modernization also replicate the public/private split that has occupied such a prominent place in Western political thought. It is a complicated tradition that at times has treated the private sphere and females as inferior and derivative and at other times complementary to the "male paradigm of excellence" (Lloyd 1984, 75). Modernization theory does not present tradition and the household as a different type of rationality that possesses its own form of excellence. Inkeles and Smith and others judge all traditional societies against the idealized standard of (male) rationality, universal norms, and achievement criteria. Tradition always fails to meet these standards and in fact its persistence threatens the public life of male citizenship. The dichotomous comparisons of traditional and modern man made by Inkeles and Smith are almost a caricature of Enlightenment dichotomies that portrayed tradition as embodying ignorant peasant women tenaciously clinging to kin and family in the face of the benevolent progress offered by technology and science.

Finally, it is important to note that the themes of struggle and mastery over idealized and feminized tradition dovetail with a liberal conception of society, functionalism, evolutionary metaphors, and the possibility of human engineering in development. Haraway (1978, 26) defines human engineering as "the project of design and management of human material for efficient, rational functioning in a scientifically organized society." Parsons's audacious attempt to pinpoint the necessary societal functions for the emergence from "primitiveness," the "seamless web" of kinship and clan, is a classic example of the wholesale adoption of a Darwinian framework of social change to human societies. Images of struggle and surging toward the achievement of a "break through" (Parsons 1964, 357) from the morass of tradition and the chaos of the household permeate Parsons's theory and modernization theory generally. The only way to achieve "adaptive capacity" is to deny tradition. Evolution as a cultural product serves to justify industrial capitalism and the subordination of women. It serves a similar function in theories of modernization.

There has been a resurgence of interest in the study of tradition in development theory, for a number of reasons. In general, assertion of ethnic and "primordial" sentiments and the role of culture and values in the East Asian "economic miracles" has led to an appreciation of the continuing importance of tradition as both obstacle and facilitator of development (Banuazizi 1987, 284). In African studies, the failure of Afro-Marxist regimes and the reign of structural adjustment programs from the World Bank and International Monetary Fund (IMF) have led to a "profound period of revisionism" (Shaw 1991, 193). Africa's continued marginal position in the international division of labor has led to a search for explanations that ostensibly go beyond theories of modernization and dependence. However, a significant body of this theory constitutes an attempt to revive modernization theory's gendered dichotomies. My chapter "From Modernization Theory to the 'Soft State' in Africa" in *Gender and Development* will show how theories of the African "soft state" seek to demonstrate the extent to which the characteristics of traditional/feminine society have become predominant in state practices. For theorists of the "soft state," the task of modernization is not only that of ridding society of its tradition, softness, and "femininity," but also that of similarly changing the state: the modern state is the hardened, masculine state.

## Note

1. The role of the Cold War in shaping U.S. development theory and practice is explored in different ways by Packenham (1973) and Gendzier (1985).

## References

Banuazizi, Ali. 1987. "Social-Psychological Approaches to Development." In Myron Weiner and Samuel P. Huntington, eds., *Understanding Political Development*. Boston: Little, Brown.

Bluh, B. J. 1982. *Parson's General Theory of Social Action*. Granada Hills, California: NBS.

Chodorow, Nancy. 1989. *Feminism and Psychoanalytic Theory*. New Haven: Yale University Press.

Coleman, James C. 1971. "The Development Syndrome: Differentiation-Equality-Capacity." In Leonard Binder, et al., eds., *Crises and Sequences of Political Development*. Princeton: Princeton University Press.

Devereux, Edward C., Jr. 1961. "Parson's Sociological Theory." In Max E. Black, ed., *The Social Theories of Talcott Parsons*. Englewood Cliffs, New Jersey: Prentice-Hall.

Flax, Jane. 1983. "Political Philosophy and the Patriarchal Unconscious: A Psychoanalytic Perspective on Epistemology and Metaphysics." In Sandra Harding and Merrill Hintikka, eds., *Discovering Reality: Perspectives on Epistemology, Metaphysics, Methodology, and Philosophy of Science*. Dordrecht, Holland: D. Reidel.

Gendzier, Irene. 1985. *Managing Political Change. Social Scientists and the Third World*. Boulder: Westview Press.

Gross, Michael, and Mary Beth Averill. 1983. "Evolution and Patriarchal Myths of Scarcity and Competition." In Sandra Harding and Merrill B. Hintikka, eds., *Discovering Reality: Feminist Perspectives on Epistemology, Metaphysics, and Philosophy of Science*. Dordrecht, Holland: D. Reidel.

Haraway, Donna. 1978. "Animal Sociology and a Natural Economy of the Body Politic, Part 1: A Political Physiology of Dominance." *Signs* 4: 21–36.

Hirschmann, Nancy. 1989. "Freedom, Recognition, and Obligation: A Feminist Approach to Political Theory." *American Political Science Review* 83: 122–1245.

Huntington, Samuel P. 1968. *Political Order in Changing Societies*. New Haven: Yale University Press.

Inkeles, Alex. 1969. "Participant Citizenship in Six Developing Countries." *American Political Science Review* 63: 1120–1141.

Inkeles, Alex, and David H. Smith. 1974. *Becoming Modern: Individual Change in Six Developing Countries*. Cambridge: Harvard University Press.

Inkeles, Alex, David H. Smith, et al. 1983. *Exploring Individual Modernity*. New York: Columbia University Press.

Jordanova, L. J. 1980. "Natural Facts: A Historical Perspective on Science and Sexuality." In Carol P. MacCormack and Marilyn Strathern, eds., *Nature, Culture, and Gender*. London: Cambridge University Press.

LaPalombara, Joseph. 1971. "Penetration: A Crisis of Governmental Capacity." In Leonard Binder, et al., eds., *Crises and Sequences of Political Development*. Princeton: Princeton University Press.

Lerner, Daniel. 1958. *The Passing of Traditional Society: Modernizing the Middle East*. New York: Free Press.

Lloyd, Genevieve. 1984. *The Man of Reason: "Male" and "Female" in Western Philosophy*. Minneapolis: University of Minnesota Press.

Macpherson, C. B. 1973. *Democratic Theory: Essays in Retrieval*. Oxford: Clarendon Press.

Mamdani, Mahmood. 1985. "A Great Leap Backward: A Review of Goran Hyden's *No Shortcuts to Progress*." *Ufamahu* 14: 178–195.

McClelland, David C. 1976. *The Achieving Society* reprint. New York: Irvington.

Mohanty, Chandra Talpade. 1991b. "Under Western Eyes: Feminist Scholarship and Colonial Discourses." In Mohanty, Ann Russo, and Lourdes Torres, eds., *Third World Women and the Politics of Feminism*. Bloomington, Indiana: Indiana University Press.

Packenham, Robert A. 1973. *Liberal America and the Third World*. Princeton: Princeton University Press.

Parsons, Talcott. 1960. *Structure and Process in Modern Societies*. New York: Free Press.

———. 1964. "Evolutionary Universals in Society." *American Journal of Sociology* 29: 339–58.

Portes, Alejandro. 1976. "On the Sociology of National Development: Theories and Issues." *American Journal of Sociology* 82: 55–85.

Pye, Lucian. 1971a. "Foreword." In Leonard Binder, et al., *Crises and Sequences of Political Development*. Princeton: Princeton University Press.

———. 1971b. "Identity and the Political Culture." In Leonard Binder, et al., *Crises and Sequences of Political Development*. Princeton: Princeton University Press.

Randall, Vicky, and Robin Theobald. 1985. *Political Change and Underdevelopment: A Critical Introduction to Third World Politics*. Durham: Duke University Press.

Rooney, Phyllis. 1991. "Gendered Reason: Sex Metaphors and Conceptions of Reason." *Hypatia* 6: 77–104.

Rostow, W. W. 1960. *The Stages of Economic Growth: A Non-Communist Manifesto*. Cambridge, England: Cambridge University Press.

Rudolph, Lloyd I., and Susan Hoeber Rudolph. 1967. *The Modernity of Tradition: Political Development in India*. Chicago: University of Chicago Press.

Shaw, Timothy. 1991. "Reformism, Revisionism, and Radicalism in African Political Economy During the 1990s." *Journal of Modern African Studies* 29: 191–212.

Tipps, Dean. 1976. "Modernization Theory and the Comparative Study of Societies." In Cyril E. Black, ed., *Comparative Modernization*. New York: Free Press.

# 16. Security and Survival

## *Why Do Poor People Have Many Children?*

ON THE SURFACE, fears of a population explosion are borne out by basic demographic statistics. In the twentieth century the world has experienced an unprecedented increase in population. In 1900 global population was 1.7 billion, in 1950 it reached 2.5 billion, and today roughly 5.7 billion people inhabit the earth. Three quarters of them live in the so-called Third World. The United Nations predicts that world population will reach 6 billion by the end of the century and will eventually stabilize at about 11.6 billion between 2150 and 2200, though such long-term demographic projections are notoriously imprecise.

Initially, this rapid increase in population was due in part to some very positive factors: Advances in medicine, public health measures, and better nutrition meant that more people lived longer. However, in other cases, notably in Africa, it may have been a response to colonialism, as indigenous communities sought to reconstitute themselves after suffering high death rates from slavery, diseases introduced from Europe, and oppressive labor conditions. In many countries colonialism also disrupted traditional methods of birth spacing.[1]

In most industrialized countries, the decline in mortality rates was eventually offset by declines in birth rates, so that population growth began to stabilize in what is called the "demographic transition." Most industrialized countries have now reached the "replacement level"[2] of fertility, and in some the population is actually declining.

Today birth rates are also falling in virtually every area of the Third World. In fact, the *rate* of world population growth has been slowing since the mid-1960s. Population growth rates are highest in sub-Saharan Africa—about 2.9 percent in 1994—but are considerably less than that in Asia (1.9 percent) and Latin America (2.0 percent). It is also important to remember that despite higher rates of growth, Africa contains a relatively small share of the world's population—India will have more births in 1994 than all fifty sub-Saharan nations combined.

The United Nations estimates that by 2045 most countries will have reached replacement-level fertility. The reason population growth still seems to be "exploding" is that a large proportion of the present population is composed of men and women of childbearing age. Half of the world's people are under the age of twenty-five. Barring major catastrophes, an inevitable demographic momentum is built into our present numbers, but this should be a subject of rational planning, not public paranoia. The truth is that the population "explosion" is gradually fizzling out.

Nevertheless, there is still considerable discrepancy between birth rates in the industrialized world and birth rates in many Third World countries, particularly in sub-Saharan Africa. Conventional wisdom has it that Third World people continue to have so many children because they are ignorant and irrational—they exercise no control over their sexuality, "breeding like rabbits." This "superiority complex" of many Westerners as well as some Third World elites is one of the main obstacles in the way of meaningful discussion of the population problem. It assumes that everyone lives in the same basic social environment and faces the same set of reproductive choices. Nothing is further from the truth.

In many Third World societies, having a large family is an eminently rational strategy of survival. Children's labor is a vital part of the family economy in many peasant communities of Asia, Africa, and Latin America. Children help in the fields, tend animals, fetch water and wood, and care for their younger brothers and sisters, freeing their parents for other tasks. Quite early in life, children's labor makes them an asset rather than a drain on family income. In Bangladesh, for example, boys produce more than they consume by the age of ten to thirteen, and by the age of fifteen their total production has exceeded their cumulative lifetime consumption. Girls likewise perform a number of valuable economic tasks, which include helping their mothers with cooking and the post-harvest processing of crops.[3]

In urban settings children often earn income as servants, messenger boys, etc., or else stay home to care for younger children while their parents work. Among the Yoruba community in Nigeria, demographer John Caldwell found that even urban professional families benefit from many children through "sibling assistance chains." As one child completes education and takes a job, he or she helps younger brothers and sisters move up the educational and employment ladder, and the connections and the influence of the family spread.[4]

In recent years, however, urbanization has been associated with fertility decline in a number of countries for both positive and negative reasons. For those with ample resources, living in an urban area can mean greater access to education, health and family planning services, and the kind of information and media that promote a smaller family norm. Since the debt crisis and economic recession of the 1980s, however, the quality of life of the urban poor has deteriorated in many countries. High unemployment or work in insecure, low-wage occupations mean that poor people simply do not have enough financial resources to support a large family. Brazil has experienced such a distress-related fertility decline. The government's failure to institute agrarian reforms in rural areas forced people to flee the poverty of the countryside, only to face the harsh realities of urban slums.[5]

Security is another crucial reason to have many children. In many Third World societies, the vast majority of the population has no access to insurance schemes, pension plans, or government social security. It is children who care for their parents in their old age; without them one's future is endangered. The help of grown children can also be crucial in surviving the periodic crises—illness, drought, floods, food shortages, land disputes, political upheavals—which, unfortunately, punctuate village life in most parts of the world.[6]

By contrast, parents in industrialized countries and their affluent counterparts among Third World urban elites have much less need to rely on children either for labor or old-age security. The economics of family size changes as income goes up, until children become a financial burden instead of an asset. When children are in school, for example, they no longer serve as a source of labor. Instead parents must pay for their education, as well as for their other needs, which cost far more in a high-consumption society than in a peasant village. And there is often no guarantee that parents' investment will buy the future loyalty of a grown child. As economist Nancy Folbre notes, "The 'gift' of education, unlike a bequest, cannot be made contingent upon conformity to certain expectations. Once given, it can hardly be revoked."[7]

In industrialized societies personal savings, pension plans, and government programs replace children as the basic forms of social security. These social changes fundamentally alter the value of children, making it far more rational from an economic standpoint to limit family size. Folbre also argues that as the value of children decreases, male heads of households are more willing to allow their wives to work outside the home, since the contribution of their wages to the family economy now exceeds the value of their household work.[8] This further spurs fertility decline.

Son preference can be another important motive for having large families. The subordination of women means that economically and socially daughters are not valued as highly as sons in many cultures, particularly in South Asia, China, and parts of the Middle East. Not only does daughters' domestic work have less prestige, but daughters typically provide fewer years of productive labor to their parents, because in many societies, they marry and leave home to live with their in-laws shortly after puberty.

Son preference, combined with high infant and child mortality rates, means that parents must have many children just to ensure that one or two sons survive.[9] A computer simulation found that in the 1960s an Indian couple had to bear an average of 6.3 children to be confident of having one son who would survive to adulthood.[10] Son preference can also lead to skewed sex ratios, so that there are more males than females in a given population. Although at birth boys outnumber girls by a ratio of about 105 to 100, this discrepancy soon disappears, all else being equal, because biologically girls have better survival rates beginning in the first months after birth. In China, much of South Asia, and parts of North Africa, however, discrimination against girls means that there continues to be fewer girls than boys, and women than men, in the population. Discrimination takes many forms, from more typical forms of unbenign neglect, such as giving girls less food and health care, to female infanticide and sex-selective abortion. According to one estimate, more than 100 million women are "missing" throughout the world as the result of such discrimination. The situation is particularly serious in North India and China.[11]

High infant and child mortality rates are major underlying causes of high birth rates. Each year in developing countries more than 12 million children die before reaching their fifth birthday. The average infant mortality rate is more than 71 deaths per 1,000 live births in developing countries as a whole, and over 100 in sub-Saharan Africa. By comparison, it is only 14 in industrialized countries.[12] In recent decades there has been some progress in reducing infant and child mortality, but not nearly enough.

High infant mortality means that parents cannot be sure their children will survive to contribute to the family economy and to take care of them in their old age. The poor are thus caught in a death trap: They have to keep producing children in order that some will survive. Most countries that have achieved a low birth rate have done so only after infant and child mortality has declined.

High infant mortality is primarily caused by poor nutrition, both of the mother and of the child. In situations of chronic scarcity, women often eat last and least, with a profound impact on infant health. Inadequately nourished mothers typically give birth to underweight babies, and low birth weight has been identified as the "greatest single hazard for infants," increasing their vulnerability to developmental problems and their risk of death from common childhood illnesses.[13] Severely undernourished women also give lower-quality breast milk; for a woman to breast-feed successfully without damaging her health, she must increase her calorie and nutrient intake by up to 25 percent, an impossibility for many poor women.[14]

Breast-feeding, in fact, has relevance to the population issue on several different but interconnected levels. In many countries, increases in infant mortality have been linked to the switch from breast-feeding to bottle feeding. Infant formula lacks the antibodies in breast milk that help to protect babies from disease. Poor women, moreover, often cannot afford a steady supply of formula and dilute it with too much water.

Proper sterilization of bottles, nipples, and drinking water is also problematic in poor households. As a result, Third World infants breast-fed for less than six months are *five* times more likely to die in the second six months of life than those who have been breast-fed longer. Overall the mortality rate for bottle-fed infants in the Third World is roughly double that for breast-fed infants.[15]

Intensive sales campaigns by multinational corporations, such as the Swiss-based Nestlé Company and American Home Products, bear a large share of the responsibility for the shift from breast- to bottle-feeding. Advertisements with pictures of plump, smiling white babies and the penetration of local health establishments by company representatives have helped convince women that they will have healthier children if they switch to formula. An international campaign against these formula "pushers" finally led the World Health Organization (WHO) in 1980 to establish a voluntary code of conduct, setting standards for the advertising and marketing of formula. Every country signed on, except for the United States.

Although the code has brought improvements in terms of mass advertising, the companies still widely promote formula through health establishments. A study of breast-feeding in four Third World cities found that formula marketers had close commercial ties with physicians, pharmacists, and midwives, and that as a result, women who used Western-type health and maternity services tended to introduce formula earlier.[16] The WHO code is clearly not strong enough.

The decline in breast-feeding is also due to the fact that more women are now employed outside the home in occupations and workplaces that actively discourage the practice or do not offer support. Extended maternity leave with pay, on-site daycare, and flexible scheduling would greatly facilitate breast-feeding, but unfortunately, in most countries, these are not high social priorities despite their obvious benefits to women and children.

In addition to giving protection against infection, another key advantage of breast-feeding is that it happens to be one of the world's most effective natural contraceptives. It frequently causes lactational amenorrhea, the suppression of ovulation and menstruation by the release of the hormone prolactin. Each month of breast-feeding adds up to three weeks to the interval between births, and women who breast-feed often, whenever the baby wants, delay the return to fertility even longer.[17]

Hence, a decline in breast-feeding without a corresponding use of effective contraception means that pregnancies are more closely spaced, and the close spacing of births is itself a major cause of infant mortality. The relationship can also go the other way: A child's death means a woman stops breast-feeding and resumes fertility sooner, with a higher risk of the next child's death. This vicious biological circle may be one of the key reasons high infant mortality and high fertility go hand in hand.[18]

The last (but not least) cause of high birth rates is the subordination of women. Male dominance in the family, patriarchal social mores, the systematic exclusion of women from the development process, and the absence of decent birth control services combine to force many women into having more children than they want. The social environment, in effect, leaves them little or no reproductive choice.

Behind the demographic statistics, then, lies a reality unfamiliar to many middle-class people, who do not have to worry from day to day about who will help in the fields, who will take care of them when they are old and sick, or how many children they need to have in order to ensure that a few survive to adulthood. High birth rates are often a distress signal that people's survival is endangered. Yet the proponents of population control put the argument the other way around, insisting that people are endangering their own survival—and the survival of future generations—by having so many children. This is the basis of the Malthusian philosophy that has defined the dimensions of the population problem for so long.

## Notes

1. See, for example, Marc H. Dawson, "Health, Nutrition, and Population in Central Kenya, 1890–1945," in Dennis D. Cordell and Joel W. Gregory, eds., *African Population and Capitalism: Historical Perspectives* (Boulder: Westview Press, 1987). Basic demographic statistics taken from U.S. Bureau of the Census, Report WP/94, World Population Profile: 1994 (Washington, D.C.: Government Printing Office, 1994) and William K. Stevens, "Feeding a Booming Population Without Destroying the Planet," *New York Times*, 5 April 1994.

2. (Crude) birth rate is the number of births per thousand people in a given year.

(Crude) mortality or death rate is the number of deaths per thousand people in a given year.

Replacement-level fertility is the level of fertility at which women on the average are having only enough daughters to "replace" themselves in a given population.

Population growth rate is the rate at which a population is growing or declining in a given year from natural increase and net migration, computed as a percentage of the base population.

3. Mead T. Cain, "The Economic Activities of Children in a Village in Bangladesh," *Population and Development Review*, vol. 3, no. 3 (September 1977).

4. John C. Caldwell, *Theory of Fertility Decline* (London: Academic Press, 1982), 69.

5. Thais Corall, "Brazil: A Failed Success Story," paper presented to the Conference on Multilateral Population Assistance, Oslo, Norway, 25 May 1994.

6. See Mead T. Cain, "Risk and Insurance: Perspectives on Fertility and Agrarian Change in India and Bangladesh," *Population and Development Review*, vol. 7, no. 3 (September 1981) and, by the same author, "Fertility as an Adjustment to Risk," Population Council, Center for Policy Studies Working Papers, No. 100 (New York: October 1983). Also see Caldwell, *Theory of Fertility Decline*.

7. Nancy Folbre, "Of Patriarchy Born: The Political Economy of Fertility Decisions," *Feminist Studies*, vol. 9, no. 2 (Summer 1983), 274. For more on the costs of children—and the political economy of who bears those costs, see Folbre, *Who Pays for the Kids: Gender and the Structures of Constraint* (London and New York: Routledge, 1994).

Also see Caldwell, *Theory of Fertility Decline*, for more on how education changes the value of children. Japan provides an interesting example of how the role of children as a source of security changes. In 1950, at the beginning of Japan's industrial boom, a survey showed that over half the population expected to be supported by children in their old age. By 1961, after a decade of rapid growth, this figure had already declined to 27 percent, and the birth rate had also fallen dramatically. Japan example from William W. Murdoch, *The Poverty of Nations: The Political Economy of Hunger and Population* (Baltimore: Johns Hopkins University Press, 1980), 29.

8. Folbre, "Of Patriarchy Born."

9. Infant mortality rate is the number of deaths of infants under one year of age per thousand live births in a given year.

Child mortality rate is the number of deaths of children aged one to four per thousand children in this age group in a given year.

10. Indian example from Frances Moore Lappé and Joseph Collins, *Food First: Beyond the Myth of Scarcity* (New York: Ballantine Books, 1979), 32.

11. Amartya Sen, "More Than 100 Million Women Are Missing," *New York Review of Books*, 20 December 1990.

12. United Nations Development Program (UNDP), *Human Development Report 1993* (New York: Oxford University Press, 1993), Tables 3 and 4, 140–43.

13. José Villar and José M. Belizan, "Women's Poor Health in Developing Countries: A Vicious Circle," in Patricia Blair, ed., *Health Needs of the World's Poor Women* (Washington, D.C.: Equity Policy Center, 1981).

14. Isabel Nieves, "Changing Infant Feeding Practices: A Woman-Centered View," in Blair, ed., *Health Needs*.

15. Ann Wigglesworth, "Space to Live," background article, *The State of World Population 1983* Press File, prepared by the New Internationalist Publications Cooperative for the UNFPA (Oxford: 1983); Lappé and Collins, *Food First*, 337.

16. Beverly Winikoff, *The Infant Feeding Study: Summary*, report submitted to AID by the Population Council (New York: Population Council, n.d.).

17. Maggie Jones, "The Biggest Contraceptive in the World," *New Internationalist*, no. 110 (April 1982).

18. See Wigglesworth, "Space to Live"; James P. Grant, *State of the World's Children 1982–83* (New York: UNICEF, 1983); and Kathleen Newland, *Infant Mortality and the Health of Societies*, Worldwatch Paper No. 47 (Washington, D.C.: Worldwatch Institute, December 1981).

# 17. Call for a New Approach

THE COMMITTEE ON WOMEN, Population, and the Environment is an alliance of women activists, community organizers, health practitioners, and scholars of diverse races, cultures, and countries of origin working for women's empowerment and reproductive freedom, and against poverty, inequality, racism, and environmental degradation. Issued in 1992, their statement "Women, Population and the Environment: Call for a New Approach" continues to gather individual and organizational endorsements from around the world.

## Call for a New Approach

We are troubled by recent statements and analyses that single out population size and growth as a primary cause of global environmental degradation.

We believe the major causes of global environmental degradation are:

—Economic systems that exploit and misuse nature and people in the drive for short-term and short-sighted gains and profits.

—The rapid urbanization and poverty resulting from migration from rural areas and from inadequate planning and resource allocation in towns and cities.

—The displacement of small farmers and indigenous peoples by agribusiness, timber, mining, and energy corporations, often with encouragement and assistance from international financial institutions, and with the complicity of national governments.

—The disproportionate consumption patterns of the affluent the world over. Currently, the industrialized nations, with 22 percent of the world's population, consume 70 percent of the world's resources. Within the United States, deepening economic inequalities mean that the poor are consuming less, and the rich more.

—Technologies designed to exploit but not to restore natural resources.

—Warmaking and arms production which divest resources from human needs, poison the natural environment and perpetuate the militarization of culture, encouraging violence against women.

Environmental degradation derives thus from complex, interrelated causes. Demographic variables can have an impact on the environment, but reducing population growth will not solve the above problems. In many countries, population growth rates have declined yet environmental conditions continue to deteriorate.

Moreover, blaming global environmental degradation on population growth helps to lay the groundwork for the re-emergence and intensification of top-down, demographically driven population policies and programs which are deeply disrespectful of women, particularly women of color and their children.

In Southern countries, as well as in the United States and other Northern countries, family planning programs have often been the main vehicles for dissemination of modern contraceptive technologies. However, because so many of their activities have been oriented toward population control rather than women's reproductive health needs, they have too often involved sterilization abuse; denied women full information on contraceptive risks and side effects; neglected proper medical screening, follow-up care, and informed consent; and ignored the need for safe abortion and barrier and male methods of contraception. Population programs have frequently fostered a climate where coercion is permissible and racism acceptable.

Demographic data from around the globe affirm that improvements in women's social, economic, and health status, and in general living standards, are often keys to declines in population growth rates. We call on the world to recognize women's basic right to control their own bodies and to have access to the power, resources, and reproductive health services to ensure that they can do so.

National governments, international agencies, and other social institutions must take seriously their obligation to provide the essential prerequisites for women's development and freedom. These include:

1. Resources such as fair and equitable wages, land rights, appropriate technology, education, and access to credit.
2. An end to structural adjustment programs, imposed by the IMF, the World Bank, and repressive governments, which sacrifice human dignity and basic needs for food, health, and education to debt repayment and "free market," male-dominated models of unsustainable development.
3. Full participation in the decisions which affect our own lives, our families, our communities, and our environment, and incorporation of women's knowledge systems and expertise to enrich these decisions.
4. Affordable, culturally appropriate, and comprehensive health care and health education for women of all ages and their families.
5. Access to safe, voluntary contraception and abortion as part of broader reproductive health services which also provide pre- and post-natal care,

infertility services, and prevention and treatment of sexually transmitted diseases including HIV and AIDS.

6. Family support services that include child-care, parental leave and elder care.

7. Reproductive health services and social programs that sensitize men to their parental responsibilities and to the need to stop gender inequalities and violence against women and children.

8. Speedy ratification and enforcement of the U.N. Convention on the Elimination of All Forms of Discrimination Against Women as well as other UN conventions on human rights.

People who want to see improvements in the relationship between the human population and natural environment should work for the full range of women's rights; global demilitarization; redistribution of resources and wealth between and within nations; reduction of consumption rates of polluting products and processes and of non-renewable resources; reduction of chemical dependency in agriculture; and environmentally responsible technology. They should support local, national, and international initiatives for democracy, social justice, and human rights.

JENNY REARDON

# 18. The Human Genome Diversity Project

*What Went Wrong?*

BY ALL ACCOUNTS, no one expected it.

In the summer of 1991, leading population geneticists and evolutionary biologists from the United States proposed a project to sample and archive the world's human genetic diversity (Cavalli-Sforza et al. 1991).[1] The proposed survey, they argued, promised "enormous leaps" in our understanding of "who we are as a species and how we came to be" (ibid., 491; Human Genome Diversity Project 1992, 1). To realize these promised advances in knowledge, proponents urged the scientific community to act swiftly. Social changes that facilitated the mixing of populations, they warned, threatened the identity of groups of greatest importance for understanding human evolutionary history—"isolated indigenous populations" (Cavalli-Sforza et al. 1991). To unravel the mysteries of human origins and migrations, these valuable gene pools would need to be sampled before they "vanished" (ibid.). The resulting time pressure, and the tens of millions of dollars it would take to conduct the survey, posed substantial challenges. Proponents recognized these constraints. It crossed nobody's mind that the project might one day be accused of inventing a new form of colonialism.

Initially, the proposal captured the imaginations of leaders in the human genomics community worldwide. The Human Genome Organization (HUGO), an international body responsible for coordinating activities within the Human Genome Project, formed a committee to investigate how to carry the initiative forward. The National Science Foundation (NSF), the National Human Genome Research Center (NHGRC), the National Institute of General Medical Sciences (NIGMS) and the Department of Energy (DOE) provided funds for three planning workshops. With this support in place, by the end of 1992 organizers had every reason to believe that what had become known as the Human Genome Diversity Project would begin operation by 1994.[2]

Their expectations, however, were disappointed. Far from winning support, in a series of events that many organizers have found inexplicable and even bizarre, the Diversity Project became the target of vociferous outrage and opposition shortly after the initiative's second planning workshop in October 1992.[3] In May 1993 some physical anthropologists accused the initiative of using twenty-first-century technology to propagate the concepts of nineteenth-century racist biology (Lewin 1993). In

June of that year, indigenous leaders from fourteen United Nations member states drafted a declaration calling for an immediate halt to the initiative. In July the Third World Network charged the Project with violating the human rights of indigenous peoples by turning them into objects of scientific research and "material for patenting" (Native-L 1993a). And in December the World Congress of Indigenous Peoples dubbed the initiative the "Vampire Project," a project more interested in collecting the blood of indigenous peoples than in their well-being (Indigenous Peoples Council on Biocolonialism 1998). By 1998 over a hundred groups advocating for the rights of tribes in the United States and indigenous groups worldwide had signed declarations condemning the Project (ibid.).

In the aftermath of these events, the puzzle for many scientists, ethicists, and government officials who seek to study human genetic differences is how this seemingly beneficent and well-intentioned initiative came to be so stigmatized.[4] The Project's leaders included some of biology's most respected, socially conscious scientists—scientists who had devoted significant energy over many decades to fighting racism and promoting human rights. Mary-Claire King, a medical and population geneticist, used genetic techniques to assist the Abuelas de Plaza de Mayo (Grandmothers of the May Plaza) in their effort to identify grandchildren kidnapped during Argentina's Dirty War. Luca Cavalli-Sforza, a human population geneticist, debated William Shockley, a Stanford physicist who called for the sterilization of women from "inferior races," during the race and IQ debates of the 1970s. Robert Cook-Deegan, a physician and geneticist, worked for Physicians for Human Rights. These were not self-seeking researchers who sought to extract the blood of indigenous peoples for the sake of financial and political gain. They were scientists who sincerely hoped to create a project that would deepen the stores of human knowledge while fighting racism and countering Eurocentrism (Bowcock and Cavalli-Sforza 1991, Cavalli-Sforza et al. 1994). It would be historically inaccurate, and morally insensitive, to understand the Diversity Project as an extension of older racist practices by labeling the initiative the product of white scientists wielding the power of science to objectify and exploit marginalized groups. The story of the Project is more complicated. It raises questions that cannot be resolved so easily.

I argue in *Race to the Finish* that, far from being a straightforward story about the powerful exploiting the powerless, the Diversity Project debates raise fundamental questions about how to understand the very constitution of power and its relationship to science in an age when scientific claims about the human—in particular, its genomes—increasingly influence decisions about how humans should regulate and conduct their lives. Dominant analytic frameworks in the social sciences assume power distorts science and the work of scientists—leading them, for example, to produce racist ideologies. Science, in reverse, is the antidote to power; it produces truth that counters ideologies. In the case of the Diversity Project, however, this understanding of the oppositions between science and power, or truth and ideology, proved inadequate. Claims that the Project would lead to the end of racism by producing reliable scientific knowledge were just as unconvincing as some of the critics' claims that the Project would propagate racism and colonialism by exploiting the genes of indigenous peoples.

In order to understand the Diversity Project debates, a different understanding of science and its relation to power is needed. In place of a framework that casts science and power as already-formed entities that oppose each other, the simultaneous emergence of novel forms of knowing, and of governing the human, evident in this initiative, challenges us to find conceptual tools that will draw into view the ways in which knowledge and power form together. The Diversity Project raised fundamental questions about how to characterize human genetic diversity for the purpose of understanding human evolution and history. Yet, these questions about how to order and classify an aspect of nature to advance human understanding proved inseparable from an allied set of questions about how to organize human differences for the purposes of creating credible and legitimate systems of governance. The conceptions of science and power upon which many Project organizers relied did not bring these entanglements into sharp focus. Thus, organizers were continually caught off guard when questions about power—for example, questions about how to make authoritative claims about human diversity—turned out to be embedded in what they viewed as merely a scientific, humanistic, and anti-racist endeavor to understand the history and evolution of the human species.

Although the Diversity Project has ceased to move forward in its original form, the contentious questions it raised endure in their importance. Human genetic-variation research now tops the agendas of both private and public research institutions. In October 2002, the National Human Genome Research Institute (NHGRI) announced the launching of a $100 million public-private effort to map human genetic variation, the International Haplotype Map (HapMap) Project (http://genome.gov/10005336). (Countless other initiatives speckle the life sciences landscape as researchers interested not just in human evolution, but also in medicine and public health, seek to understand the human—its diseases, health, and potential—using the new powerful tools of the genomic revolution. Far from transcending the problems raised by the Diversity Project, these current efforts have only generated similar troubling questions (Couzin 2002). By returning to the Diversity Project debates, *Race to the Finish* seeks to bring into view and clarify the underlying contestations over the nature of knowledge, power, and expertise that were at stake in this effort to catalog human genetic diversity, and that continue to create discomfort today.

## Race, Expertise, and Power

At the center of these contestations is a broader struggle over the meaning of race in science, medicine, and the modern state. As the last millennium ended, efforts to use racial categories in biomedical research and public health generated fundamental questions. Is race an obsolete concept that should be left behind by the operations of liberal democratic societies and scientific and medical research, insofar as this is possible, as some cultural and scientific critics have argued (Appiah 1990; Freeman 1998; Gilroy 2000; Wilson et al., 2001)? Or should race be understood as a positive category that designates both cultural and national belonging, and contributes to public health efforts to reduce the burden of disease (Du Bois 1961 [1903]; Cruse 1968; NIH 1994; Risch 2002)? Can past misconduct and inequities in government, including the provision of health services, be overcome by transcending the concept of race, or, conversely, must this concept be actively employed to overcome those racial structures that continue to oppress?

Diversity Project organizers found themselves in a peculiarly ambiguous and paradoxical position with respect to these questions. On the one hand, they claimed that the Diversity Project would help "combat the scourge of racism" by demonstrating that "there is no absolute 'purity'" and no "documented biological superiority of any race, however defined" (Cavalli-Sforza et al. 1994, 1, 10). Rather than promote racial division, Project organizers promised the initiative would demonstrate "humanity's diversity and its deep and underlying unity" (Cavalli-Sforza et al. 1994, 1). As one step toward these goals, one of the Project's main scientific leaders, the human population geneticist Luca Cavalli-Sforza, advocated abandoning the category of 'race' in favor of the categories 'group' and 'population' (Cavalli-Sforza et al. 1994, 11).

Yet, as we will see, at the same time that proposers of the Project disavowed the use of the category of race, they found themselves being accused by many of reinscribing old racial categories, and even of being racist. Some physical anthropologists argued that the Project employed categories that carried forward notions of racial purity (Lewin 1993; Marks 1995). Numerous indigenous rights groups charged Diversity Project organizers with continuing a long tradition of the West's use of racial science to justify its exploitation of the powerless (Mead 1996; Indigenous

Peoples Council on Biocolonialism 1998). In an ironic turn, in the face of these critiques, some Project organizers began to explicitly employ racial categories. Representing what appeared to be a turnaround from the earlier disavowal of race, some leaders of the initiative now argued that the Project would include the genomes of African Americans and other "major ethnic groups," and in this way would serve as an "affirmative action" response to the Human Genome Project (Weiss 1993).

As the chapters in *Race to the Finish* illustrate, underlying these debates about race were fundamental questions about how knowledge of the human should be produced in a genomic age, and who possesses the expertise needed to participate in this pursuit. How could human beings come to know their own species—its history and evolution—in an age when novel technologies enabled a purportedly molecular vision of human existence? What role, if any, could studies of human genetic differences play? How should such studies be designed? Which of the human sciences, if any, could provide the organizing concepts and methods? Human population genetics? Physical anthropology? Cultural anthropology? All of them?

These questions about the constitution of the right kind of knowledge were connected to questions about the nature of power. As Michel Foucault demonstrated through his studies of madness, the clinic, and the prison, the human sciences play central roles in constituting techniques and procedures for directing human behavior in the modern epoch (Foucault 1973, 1975, 1976). This modern age witnessed the entanglement of rules that govern what can count as knowledge with rules that determine which human lives can be lived. The result was the emergence of a new kind of power, what Foucault named *biopower* (Foucault 1976).

This close relationship between the rules that structure power and knowledge (a relationship Foucault highlighted through use of the contraction *power-knowledge*) was tellingly revealed in the attempts to plan the Diversity Project. Not surprisingly, a scientific endeavor that promised "enormous leaps in our grasps of human origins, evolution, prehistory, and potential" provoked questions about the role the initiative might play in producing human subjects who could be governed in novel, and potentially oppressive, ways (Cavalli-Sforza et al. 1991, 491). Would Diversity Project organizers' seemingly benevolent efforts to include groups in the design and the regulation of the sampling initiative facilitate their inclusion in an ethical manner, or would these efforts only threaten the sovereignty rights of tribes in the United States and marginalize other minority groups? Would the sampling and potential patenting of indigenous DNA bring benefits to indigenous groups through royalty sharing and medical advances, or would it merely enable their further objectification and exploitation? And who, if anyone, could speak for tribes and major ethnic groups in the United States, and indigenous groups across the globe, on these novel issues raised by a global survey of human genetic diversity? As these questions would make clear, at stake in the struggles over the Diversity Project were not only the validation of a research project, but also the resolution of some of the central social debates in an age defined by the emergence of genomics, globalization, deepening tensions between the (global) North and South, and a renewed struggle over the status and definition of race in the United States.

## Co-Production: A Framework for Studying Emergence

To address these connections more directly, *Race to the Finish* tells the story of the Project from within an analytic framework in science and technology studies that seeks to understand how scientific knowledge and social order are produced simultaneously—or, in a word, co-produced (Jasanoff et al. 2004).[5] By linking critical studies of scientific knowledge and analyses of political institutions and social structures, I along with other scholars working within this analytic framework am attempting to clarify the ways in which bringing technoscientific phenomena or objects into being (objects, for instance, like 'human genetic diversity') requires the simultaneous production of scientific ideas and practices *and* other social practices—such as norms of ethical research and credible systems of governance—that support them.

The resulting fine-grained portrayals of the mutual constitution of natural and political order have done much to undo grand narratives that have either celebrated science as an ideal polity (Merton 1942), or condemned it for reinscribing hegemonic and oppressive political orders (Habermas 1975). In the place of totalizing broad-stroked accounts, we have gained richer, more useful pictures of science and technology (s&t) that neither simply sanctify nor condemn, but rather bring to light the locally contingent and often ambivalent roles the creation of s&t plays in the ongoing human struggle to produce societal arrangements in which humane and meaningful lives can be lived.

On a terrain prone to polarization, the co-production idiom is a particularly valuable resource for countering analyses that too easily celebrate enlightened, objective studies of human genetic diversity, or too readily dismiss them as ideological and racist. It provides a framework that enables us to see that all efforts to organize such studies necessarily entail the production of both conceptual order and societal interests, and that the two domains are inextricably linked—indeed, inseparable. Such critical vision requires that the analyst give causal primacy to neither 'society' nor 'science,' but rather engage in a "symmetrical probing of the constitutive elements" of both (Jasanoff et al. 2004). The results are accounts that resist both technological and social determinism, and easy pronouncements of right and wrong (Rabinow 1999).

The co-production framework is also appropriate for this study because of its utility for studying emergent phenomena. It is at the point of emergence, when actors are deciding how to recognize, name, investigate, and interpret new objects, that one can most easily view the ways in which scientific ideas and practices and societal arrangements come into being together (Jasanoff et al. 2004; Daston 2000; Latour 1993). This is especially the case when the epistemological and normative implications of the emerging object are contested, and when the effort to establish its legitimacy and meaning spans multiple cultural contexts (Jasanoff et al. 2004). All these conditions hold for 'human genetic diversity', making it an especially rewarding site for analysis.

Scholarship guided by this analytic framework runs against dominant ideas about science in the academy and in society that conventionally have been linked to the Enlightenment. Enlightenment thinkers such as Voltaire, Rousseau, and Thomas Jefferson believed that through science and reason "men" could discover universal truths about Nature. It was on the basis of these truths, they argued, that man could recover from the blinding effects of dogmatic beliefs and uninterrogated traditions and achieve the enlightened stance required to build good and just governments. Later, Marxists would hail "scientific" thinking about political economy and the class struggle as a critical tool for piercing the veil of ideology.[6] In other words, science and reason, they believed, could be used to work against the corrupting influences of power divorced from truth.

Organizers of the Diversity Project shared this Enlightenment vision. They believed that their Project would generate scientific data that could oppose racist ideologies in society. Specifically, the first proposers of the Diversity Project joined critical race theorists, American historians, and cultural and biological anthropologists in their belief that population genetics demonstrated the biological meaninglessness of socially meaningful racial categories (Cavalli-Sforza 1989, Cavali-Sforza et al. 1994). Human population genetics, they accordingly argued, was a powerful antidote to the ideologies propagated by the use of these unscientific categories. Data generated by this discipline would lead to the final demise of the category of race in science and its replacement by the scientifically rigorous category population (Cavalli-Sforza et al. 1994).

This analysis of race as an obsolete concept in the life sciences illustrates what might be called a debunking critique of ideology. This form of critique seeks to reveal hidden connections between discourse and social power.[7] The goal is to problematize the truth claims of a given discourse by demonstrating that these claims are the product of dominant, often reprehensible, social interests—and thus constitute ideology.[8] As noted above, this mode of critique—one that relies upon a clear distinction between power and knowledge—did not prove effective in the case of the Diversity Project. It provided explanations that no one found satisfactory.

Rather than dwelling on the opposition of truth on the one hand and power on the other, co-productionist work demonstrates that scientific knowledge and political order come into being together. Thus, government officials, policy makers, and academics cannot simply turn to scientific research (such as human genetic diversity research) for independent and objective answers to social problems, such as racism; nor can these social actors govern without the aid of science (or systematic knowledge-seeking), for scientific knowledge and ethical and political decisions about human diversity can only be made together. Human genetic diversity simply cannot become an object of study absent social and moral choices about what we want to know and who we want to become. Moral and social choices about the directions of human inquiry are not possible without cognitive frameworks that frame the human and its possible variations.

The story of the Diversity Project, I argue, can be more effectively told in this idiom of simultaneous emergence. This language allows one to step back from uncritical pronouncements about knowledge or assertions about power in order to ask a prior set of questions about how each is implicated in the formation of the other. To draw these processes of formation into focus, in this book I ask specifically about how categories used to classify human diversity in nature and those used to order relevant aspects of social practice shape, entail, and refer to each other: How do societal arrangements affect the kinds of categories that scientists can use to characterize human diversity? How do these categories in turn "loop back" to produce new societal arrangements (Hacking 1999)? What deliberative engagements between scientists, policy makers, research subjects, and citizens enable a socially meaningful biological category like population or race to work? As *Race to the Finish* demonstrates, formulating and answering these questions renders visible the micro-processes by which social life and cognitive understandings gain form and meaning together. The result is a rich, empirically grounded account of the entangled processes through which knowledge and social order sustain each other in contemporary societies.

Such an account promises to bring into view the tensions and paradoxes surrounding race raised by the Diversity Project. Just as co-productionist studies provide a way out of dualistic modes of understanding science (either it produces truth, or it is distorted to produce ideology), so too they provide a way out of the dichotomy that is prevalent in current analyses of race in the academy. Most conventional accounts portray race as either a valid category of research that can help produce legitimate knowledge (Dobzhanksy 1937 [1951]; UNESCO 1952; Risch et al. 2002; Burchard et al. 2003), or as a social construct (i.e., ideology) propagated by the powerful (Gates 1986; Fields 1990; Cooper et al. 2003). By adopting a co-productionist approach, this study demonstrates that race defies simple categorization as either the reflection of scientific truth or social ideology. As the chapters in *Race to the Finish* demonstrate, a socially important category like race is likely to generate scientific attention. What makes this ordering tool of interest to scientists is precisely what makes it of interest to the law, criminal justice system, and institutions of higher learning— through centuries of use it has become a tool that draws into focus differences between humans deemed meaningful. At the same time, use of a socially meaningful category like race can never be refined so that it acts only to elucidate natural reality. As much as biologists have tried over the last several decades to constrict race to apolitical scientific purposes, the use of race is never neutral.[9] It is always tied to questions with political and social salience. Some of these questions—about, for example, resource allocation—are relatively benign. Others, such as questions about the creation of new forms of racism, are of great political significance.

Rather than a category that respects any demarcation between the scientific and social realm, race has traveled vigorously and often across the boundaries of science and society, reality and ideology, throughout the twentieth century. In the process, it has been stabilized and destabilized, made and remade. Analyzing the Diversity Project debates using the co-productionist framework allows us to draw into focus the simultaneously scientific and social dimensions of contemporary attempts to define and construct this category, and promises to create new insights into the dilemmas of categorizing human difference that confront us in the genomic age.

At the same time, analyzing race in this manner promises new understandings of the processes by which natural and social order come into being together. In many important senses, struggles over the meaning of race and racism are at once contestations over who has the expertise needed to represent the truth about human diversity in nature, and contestations over who has the authority to create just representations of human diversity in society. This is particularly true in the United States where debates over how to define race and racism are perennially at the center of battles over the nature of truth and justice. Especially in the post–World War II era, to be deemed a racist was to risk losing one's status as a speaker of truth *and* one's authority as a credible voice in society

(Southern 1971; Baldwin 1998). As we will see, at stake in the contestations over what constituted racial categories, and the role—if any—these categories should play in ordering the Diversity Project were not just answers to fundamental questions about what will count as the truth about human nature and its diversity, but also answers to fundamental questions about who will count as legitimate members of particular societies. Who will be represented? Whose voices will be deemed authoritative or at least worth hearing? What structural effects will result? In short, focusing on the debates about race and racism sparked by the Diversity Project enables us to bring into view the processes by which knowledge and social order form together. It is only by bringing these processes into view, and making them available for critical debate and understanding, that we will begin to make sense of and meaningfully address debates sparked by the Diversity Project.

Excavating the History of Race and Science

At the same time that *Race to the Finish* is an interpretive project concerned with how objects (e.g., 'human genetic diversity') gain both scientific and broader social meanings, it is also necessarily a historical one. Objects do not come into being de novo. Rather, they are the products of long historical processes that embed past contestations and settlements (Daston 2000). Bringing these historical processes into view is vital to the task of understanding why some phenomena gain the material and intellectual support needed to persist, while others fade away.

This is especially true in the case of 'human genetic diversity.' As we will see, the efforts of Project organizers to render this phenomenon an object of sustained scientific inquiry raised historically entrenched questions that were in the deepest sense simultaneously scientific and social: What kinds of human diversities matter? Cultural or biological? The diversities of individuals or groups? If biological, how should the biological realm be defined? If groups, how should these groups be defined, by whom and for what purposes? In the interwar and post–World War II era, questions about the role that ideas and practices of race should play in resolving these questions have been at the center of debate. Is race, many have asked, one such grouping category that can be used to order human genetic diversity? Or is this term too confused and/or dangerous to be employed by scientists or any other actors in society?

Discussion and debate of these questions as they connected to the Diversity Project ostensibly began in the context of efforts to organize the Human Genome Project. In October 1990 the United States National Institutes of Health and Department of Energy authorized $3 billion to be spent over fifteen years to sequence a single human genome.[10] One of the main assumptions underlying this endeavor was that all human genomes were enough alike to create such a record. The initial proponents of the Human Genome Diversity Project challenged this assumption. As the human population geneticist and founding father of the Project, Luca Cavalli-Sforza, explained: "Each group has its differences and each person has differences. If we don't understand that diversity, we're missing a lot that's important" (Cavalli-Sforza quoted in Rensberger 1993). One consequence, he argued, would be the continuation of the "Eurocentric" bias of studies of human genetic diversity, studies that to date had "been made on Caucasoid samples for obvious reasons of expediency" (Bowcock and Cavalli-Sforza 1991, 491).

Yet the question about the value of studying diversity far predated the genomics era. During much of the first half of the twentieth century, it lay at the heart of a debate between experimental and population geneticists about how to study human evolution. Classical experimental geneticists assumed for the purposes of their research that all genomes were essentially the same. Based on this assumption of essential similarity, this group of scientists sought to discover the purportedly universal basic mechanisms that regulated and controlled all life. In contrast, prominent evolutionary biologists interested in genetics—most notably, the population geneticist Theodosius Dobzhansky—began their work and research from the assumption that all individuals are unique. For these scientists, diversity did not constitute deviation; rather, it was a normal and critical part of the natural world that provided the biological material upon which natural selection acted. Far from being a secondary concern, genetic diversity proved central to understanding evolution (Provine and Mayr 1982).

Within the group of scientists who agreed that genetic diversity would be a meaningful object of study, another debate emerged about how this diversity should be ordered. As we will see in the first part of *Race to the Finish*, at the center of the debate was a struggle over whether and how to use race for the purposes of studying and interpreting human biological differences. This struggle first arose during the interwar period, sparked by the rise of eugenics and Nazi science. Following World War II, the use of the category of race in biology (and throughout the sciences) would be the subject of review by newly emerging international institutions. Most notably the United Nations Educational, Scientific and Cultural Organization (UNESCO) drafted a Statement on Race in 1950 that would be followed by a Second Statement published in 1952 (UNESCO 1952).

Most historians of biological beliefs about race hold that these UNESCO Statements on Race mark the beginning of the decline of race as a meaningful biological category, and its emergence as a sociological category. On their account, by mid-century scientists had lifted the veil of ideology that previously shrouded biological studies of human diversity. The legitimate science of population genetics eclipsed the ideological science of race biology; population replaced race as the category that biologists believed most usefully organized their analyses of human diversity (Stepan 1982; Barkan 1992).

Yet despite these histories, many, including the well-regarded founding father of population genetics, Dobzhansky, would continue to find race useful (Provine et al. 1981). "Races," he argued in his classic *Genetics and the Origin of Species*, "may be defined as Mendelian populations of a species which differ in the frequencies of one or more genetic variants, gene alleles, or chromosomal structures." He in turn defined a "Mendelian population" as "a reproductive community of individuals which share in a common gene pool" (Dobzhansky 1951 [1937], 138, 15). Dobzhansky argued that anthropologists' failure to recognize that races were Mendelian populations led to their "endless disagreement" and confusion about the meaning of this term (ibid., 140). Debates on this score would continue between and among population geneticists and physical anthropologists in the coming decades.

In many ways, these debates among and between geneticists and anthropologists can be seen as precursors to the Diversity Project debates. Although they were not as drawn out or as formal as the later debates, they did involve not only the carving out of a conceptual order, but also the creation of a social space that could support research on the formation of human races. This effort took place at a time when segregation and race-based lynchings in the American South, eugenics movements, and, above all, World War II had sparked fundamental critiques of purportedly scientific studies of race. In this new environment, it would no longer be permissible to conduct research that might be associated with Nazi race science. Even the term *race* had started to become taboo. However, far from leading to the end of the history of race in science, as many historians on anthropology and biology have claimed, this new environment fostered efforts by geneticists and physical anthropologists to carve out an expert space in which their studies of human diversity and race formation could continue.

To date, however, scholars have largely overlooked these continuing debates. By their account, as noted above, uses of race in science ended at mid-century once progressive scientists revealed the concept's ideological underbelly. This rendering of the history, however, is too simple. Uses of racial categories in science did not come to an end following World War II. To the contrary, scientific debates about race proved just as persistent and contentious as did the parallel social debates. To understand the controversies surrounding the Diversity Project, these debates must be unearthed from the archival records.

My goal in conducting this historical analysis, however, is not merely to excavate scientists' continuing debates about the proper definition and uses of race following World War II, debates that shaped the terrain upon which Diversity Project organizers would attempt to build their initiative. Rather, in keeping with the co-production idiom, my aim is to reveal the inextricable links between these debates about race, and broader social debates about the role this category should play in classifying human differences in society. As in the sciences, in post–World War II societies more broadly, debates about how to understand the human encountered tensions between discourses of sameness (most notably, universal humanism), and discourses of difference (race, ethnicity, and nationhood). Discourses of sameness and unity, such as universal humanism, gained strength. Many viewed these discourses as antidotes to the logics of differentiation that led to the slaughter of millions of innocent lives during World War II. Instead of a world divided into superior and inferior individuals and groups, universal humanism imagined the "united family of man" (Haraway 1989, 198). This doctrine became embodied in institutions such as the new international organization, the United Nations, and official documents, such as the Universal Declaration of Human Rights (United Nations 1949 [1948]).

Simultaneously, discourses of difference persisted. Claims about racial and ethnic difference, for example, continued to play primary roles in the reconstruction and maintenance of national and global political orders. Reminiscent of debates about diversity in the natural order, however, efforts to define race for the purposes of creating social order raised many questions. How should race be determined? For what purposes? How, if at all, does race differ from ethnicity? In the wake of the Civil Rights movement and the adoption of "affirmative action" policies in the United States, these struggles over the determination of racial and ethnic differences proved to be as much about access to social goods as they were about the denial of these social goods to certain groups. Some discourses of race and ethnicity helped to articulate policies designed to open up education and job opportunities for minorities, while others continued to undergird discriminatory practices (Executive Order 10925, 1961).

These changes were connected to a broader set of political transformations that accompanied the emergence of identity politics in the 1960s. Social and political theorists have celebrated the rise of social movements constructed around claims about identity. This strategic use of identity, they have argued, represented a departure from old models of identity formation in which the state classified and identified people (Melucci 1989). Race in this new era emerged as a powerful resource used by citizens to build civic identities that could be mobilized to make demands against the state.

In order to make sense of the Diversity Project debates, the initiative must be understood in the context of these broader social struggles and transformations. Far from transcending the dilemmas and tensions generated by the liberatory and discriminatory effects of recognizing human differences, today studies of human genetic variation play central roles in negotiating them.[11] Some identify genomics with the liberatory trends of identity politics. In recent years, for example, scholars of biomedicine and genomics have observed that novel genomic and biomedical techniques and ideas enable new identity formations (Epstein 1996; Rabinow 1999; Kaufmann 2001).[12] Some argue this biologization of identity is fundamentally different from older reductionist and eugenic models (Rabinow 1999, 9).[13] For example, the anthropologist of science Paul Rabinow believes that the rigid, oppressive nineteenth-century categories (such as race) have been replaced by categories that are "inherently manipulable and reformable," and offer new possibilities for life (Rabinow 1999, 1996). Many have heralded the category of population brought into being by population genetics as one such category with liberatory potential (Haraway 1989; Stepan 1982).

Some also believe that defining racial and ethnic groups in genetic terms is a necessary component of progressive affirmative action policies. Specifically, in recent years scientists and public health officials have argued that genetic research on particular ethnic populations will help bring more minorities into science, as well as make it possible for the biomedical sciences to address the medical needs of minority communities (DNA Learning Center 1992; Nickens 1993; Pollack 2003). Some also hope genetics might offer a less contestable method for determining that an individual is "Native American," thus creating new possibilities for gaining access to the resources that come from tribal membership in the United States (Yona 2000).

But others view the rise of human genomics as less liberating. Since the early days of planning the Human Genome Project, many have expressed concern that this new burgeoning field of research could extend discriminatory practices into new realms—such as the provision of health and life insurance—and provide these practices with new, more powerful and impenetrable (i.e., scientific) sources of legitimacy (Billings 1992; Natowicz 1992; Juengst 1995). Others fear that genetic diversity research will lead to a new eugenics (Guerrero 1998). Still others fear that race-based discrimination based on the old biology might be replaced by discrimination based on the new group categories of population genetics—for example, characterizing 'demes' might lead to "demic discrimination" (Juengst 1995).[14]

Huge investments of venture capital into genomics have also spawned new fears about the commodification of life and the transformation of biology into a tool of global capitalism (Rural Advancement Foundation International 1993). Using the terms "biocolonialism" and "bioprospecting," scholars and activists alike have observed links between exploitative capitalist practices and the emergence of biological (in particular, genetic) diversity as a site of informational and commercial value (Mead 1996; Whitt 1998). For example, they point to the extraction of the non-Western world's plant resources as a critical element of colonial expansion since the sixteenth century (Kloppenburg 1988; Shiva 1997). In the contemporary era, many argue that pharmaceutical companies, in conjunction with government-sanctioned research, are extending these practices from the domain of plants to the domain of humans (Hayden 1998; Harry 1995; Rural Advancement Foundation International 1993).

Finally, claims about the distinctness of Native peoples have long been at the center of debates about the extent and legitimacy of their claims to sovereignty rights (Deloria 1985).[15] Some worry that genetic research will not serve to legitimate tribal membership claims, but rather will be used to undermine these claims by highlighting the similarities between Native peoples and other inhabitants of the Americas (Indigenous Peoples Council on Biocolonialism 2000).

As Diversity Project organizers only too slowly began to recognize, the Project was entangled in these broader questions about the role genomics might play in constructing identity and novel forms of governance. There were three moves that organizers made to respond to critics and stabilize the Project: (1) diversify the experts involved in the planning of the Project; (2) adapt existing tools in Western biomedical ethics to fit the "group" contexts in which the Project would be operating; (3) include "major ethnic groups" in the United States and indigenous groups worldwide in the design and conduct of research. Yet each move proved unable to assuage critics as they respectively failed to engage with questions about how the Project's effort to order human diversity in nature for the purposes of scientific research would be inextricably tied to questions about how to classify human differences in society. Instead, these frameworks presumed the prior existence of groups in nature and society. Further, they presupposed the existence of the expertise needed to discern and represent these groups. Finally, they assumed that geneticists, anthropologists, and health workers working with populations possessed this expertise. Thus, they did not draw into sharp enough focus fundamental questions about the existence of groups in nature and society, the role of race in ordering and defining these groups, and the locus of expertise for answering these questions.

In short, these frameworks enabled too much to escape Project organizers' critical attention. Despite their best intentions, in responding to their critics, organizers often inadvertently acted to exclude too many people from the debates—both scientists who held different views, as well as people who had not yet been deemed experts in the official sense, but whose lives were being affected by the genomic revolution, and whose knowledge might have provided important insights into what it meant to interpret and define human diversity using the tools of scientific (genetic) experts. Consequently the debate remained too narrow. Focused primarily on the practical task of organizing a DNA sampling initiative, proponents rarely asked the more fundamental and, I would argue, more compelling and relevant questions generated by the genomic revolution, of which their Project would be a central part: What kind of human is brought into being via genomic analysis? What are this human's possible variations? Who can speak for humanness in this genomic age? Who will decide what kind of lives can be lived and not lived? . . .

I seek to take a step back and clarify the fundamental issues that underlie contemporary debates about genomic studies of human diversity. My goal is to bring into critical view the heretofore unconscious processes through which human genomic research reconfigures both nature and society. My hope is to generate understandings that can lead to more reflective human futures.

## Notes

In this book I use single quotation marks to indicate categories (like 'race,' 'population,' and 'group') and objects of study (like 'human genetic diversity'). To keep the use of quotation marks to a minimum, I often write out "the concept of" or "the category of" instead of using quotation marks. When introducing new concepts and categories I might use both methods (for example, in this case, "the categories of 'group' and 'population'"). Double quotation marks are reserved for direct quotes and in place of "so-called."

1. This group of scientists included Luca Cavalli-Sforza (Stanford population geneticist), Allan C. Wilson (UC Berkeley biochemist and molecular evolutionary biologist), Mary-Claire King (then a UC Berkeley medical and population geneticist), Charles Cantor (principal scientist of the Department of Energy's Human Genome Project), and Robert Cook-Deegan (medical geneticist and

policy adviser for James Watson, then director of the Human Genome Project). Their call for the survey appeared in the journal *Genomics*.

2. The term "Human Genome Diversity Project" is highly contested and filled with logics and politics that could and should be explored. In this book, I use the term to refer to the definitions provided by the mostly American organizers of the Project who tried to generate U.S. support for a global survey of human genetic diversity. The Diversity Project has since been multiply defined in regions across the globe. Those who resist the Project provide some of these definitions. Others are provided by the Project's regional committees, which now exist in North America, Europe, Africa, South America, and China. Some would argue that the Diversity Project is moving forward in some of these other regions of the globe (Greely 2001).

3. Additionally, the Diversity Project has been the subject of inconclusive national and international reviews. In October 1995, the United Nations Educational Scientific and Cultural Organization's International Bioethics Committee reviewed the Human Genome Diversity Project. *Nature* magazine reported an unfavorable review. (See Butler 1995). In interviews with the author, Diversity Project leaders contested this interpretation. In October 1997, the National Research Council released its long anticipated review of the Diversity Project, *Evaluating Human Genome Diversity*. Again, an inconclusive interpretation of the review resulted. *Science* reported support for the Project (Pennisi 1997). *Nature* reported a negative review. (See "Diversity Project" 1997).

4. In the mid- and late 1990s, ethicists, policy makers, and scientists involved in efforts to create projects to study human genetic variation at the National Institutes of Health attempted to distance their studies from the Diversity Project. In the last few years, however, those attempting to accommodate these new projects (such as the Environmental Genome Project) have increasingly recognized that to move forward, their projects must address similar problems as those raised by the Diversity Project (Sharp 2002, Interview S).

5. For examples of works that exemplify this framework, see Shapin and Schaffer 1985, Ezrahi 1990, Jasanoff, et al. 1995, Wynne 1996, Hilgartner 2000, Daston 2000, Jasanoff 2004.

6. On most accounts the term "ideology" itself did not emerge until 1796 when de Tracy used it to talk about a science of ideas (Williams 1983, 154). Since its emergence, it has been defined in a variety of different ways.

7. "Discourse" here is defined as a set of rules and practices that differentiate objects, *and* gain material institutional support. I also hold "discourse" to define the field of possibilities—what can and cannot be contemplated. This definition is Foucauldian in inspiration. For a full explanation of this meaning of the term as I use it, see Foucault 1972, 79–105.

8. Many have characterized this view of ideology as "functional false consciousness." For an overview, see Sturr 1998, 2–10.

9. The same would be true of any category asked to take the place of "race." To gain institutional support, any category used to classify human beings for the

purposes of scientific research would have to be socially meaningful. If socially meaningful, then the category will not only act to discern truth, it will have ordering effects in society.

10. For a detailed account of the emergence of the Human Genome Project, see Cook-Deegan 1994.

11. One prominent site where these different visions of what biologization (in particular, geneticization) will mean for humanity (its identity and potential) stands in great public tension today is Iceland. There, a private company, deCODE genetics, proposed to create a national database that would contain medical, genealogical, and genetic details about all the citizens of this small island nation. This proposal has stirred the passions and ire of many. Celebrated scientists such as Richard Lewontin as well as ethicists and legal scholars in the United States have joined the consumer group Mannvernd in arguing that the database would violate human rights (Coghian 1998). Others argue that the critique is unwarranted. In particular, Rabinow has argued that the moralizing tenor of critics has prevented a critical assessment of what is going on in Iceland. Critics have, he argues, fallen into the trap of humanism described by Michel Foucault: providing solutions to problems that have not been adequately posed (Rabinow 1999, 15).

12. See also the special issue of the *Social Studies of Science* (Fall 1998) on "Contested Identities."

13. Rabinow has coined the term "biosociality" to refer to this new locus of identity (Rabinow 1996).

14. Just as those argued that definitions of race based on culture would not necessarily end racism, so too these critics argue that the new purportedly liberatory categories of population genetics could just as easily be used for racist ends (Smith 1994).

15. For key legal decisions see *Cherokee Nation v. Georgia*, 1831; *Worcester v. Georgia*, 1832.

## References

Appiah, Anthony. 1990. "Racism." In *Anatomy of Racism*. Ed. by David Theo Goldberg (Minneapolis: University of Minnesota Press): 3–16.

Baldwin, Kate. 1998. "Black Like Who? Cross-Testing the 'Real' Lines of John Howard Griffin's *Black Like Me*." *Cultural Critique*: 103–43.

Barkan, Elazar. 1992. *The Retreat of Scientific Racism* (Cambridge: Cambridge University Press).

Billings, Paul R. 1992. "Discrimination as a Consequence of Genetic Testing." *American Journal of Human Genetics* 50: 476–82.

Bowcock, Anne, and Luca Cavalli-Sforza. 1991. "The Study of Variation in the Human Genome." *Genomics* 11 (Summer): 491–98.

Burchard, E., et al. 2003. "The Importance of Race and Ethnic Background in Biomedical Research." *NEJM* 348(12): 1170–75.

Butler, Declan. 1995. "Genetic Diversity Panel Fails to Impress International Ethics Panel." *Nature* 377 (5 October): 373.

Cavalli-Sforza, L. Luca. 1989. Letter to Robert Cook-Deegan (September 22).

Cavalli-Sforza, L. Luca, Paolo Menozzi, and Alberto Piazzi. 1994. *The History and Geography of Human Genes* (Princeton, NJ: Princeton University Press).

Cavalli-Sforza, L. Luca, Allan C. Wilson, Charles R. Cantor, Robert M. Cook-Deegan and Mary-Claire King. 1991. "Call for a Worldwide Survey of Human Genetic Diversity: A Vanishing Opportunity for the Human Genome Project." *Genomics* 11 (Summer): 490–91.

Coghian, Andy. 1998. "Selling the Family Secrets." *New Scientist* (5 December): 20–21.

Cook-Deegan, Robert. 1994. *The Gene Wars* (New York: W. W. Norton & Company).

Cooper, Richard S., Jay S. Kaufman, and Ryk Ward. 2003. "Race and Genomics." *NEJM* 348(12): 1166–70.

Couzin, Jennifer. 2002. "New Mapping Project Splits the Community." *Science* 296 (24 May): 1391–93.

Cruse, Harold. 1968. *The Crisis of the Negro Intellectual* (New York: William Morrow).

Daston, Lorraine, ed. 2000. *Biographies of Scientific Objects* (Chicago: University of Chicago Press).

Deloria, Vine Jr. 1985. *American Indian Policy in the Twentieth Century* (Norman: University of Oklahoma Press).

"Diversity Project 'Does Not Merit Federal Funding.'" 1997. *Nature* 389 (23 October): 774.

DNA Learning Center. 1992. "Extending Our Expertise to Minority and Disadvantaged Settings." *DNA Learning Center Annual Report* (Cold Spring Harbor): 376–78.

Dobzhansky, Theodosius. 1951 (1937). *Genetics and the Origin of Species*, 3rd ed. (New York: Columbia University Press).

Du Bois, W. E. B. 1961 (1903). *The Souls of Black Folk* (Greenwich, CT: Fawcett).

Epstein, Steven. 1996. *Impure Science* (Berkeley: University of California Press).

Ezrahi, Yaron. 1990. *The Descent of Icarus* (Cambridge: Harvard University Press).

Fields, Barbara. 1990. "Slavery, Race and Ideology in the United States of America." *New Left Review* 181 (May/June): 95–188.

Foucault, Michel. 1966. *The Order of Things: An Archaeology of the Human Sciences* (New York: Vintage).

———. 1972. *The Archaeology of Knowledge* (New York: Pantheon Books).

———. 1973. *The Birth of the Clinic: An Archaeology of Medical Perception* (New York: Vintage Books).

———. 1975. *Discipline and Punish: The Birth of the Prison* (London: Allen Lane).

———. 1976. *The History of Sexuality, Volume I: An Introduction* (London: Allen Lane).

Freeman, Harold P. 1998. "The Meaning of Race in Science—Considerations for Cancer Research." *Cancer* 82(1): 219–25.

Gates, Henry Louis Jr., ed. 1986. *"Race," Writing, and Difference* (Chicago: University of Chicago Press).

Gilroy, Paul. 2000. *Beyond Camps: Race, Identity and Nationalism at the End of the Colour Line* (London, England: The Penguin Press).

Greely, Henry T. 2001. "Human Genome Diversity: What about the Other Human Genome Project?" *Genetics* (March): 222–27.

Guerrero, M. A. J. 1998. "Eugenics Coding and American Racisms: Focus on the Human Genome Diversity Project as 'The Great Spiritual Ripoff.'" Paper presented to the First North American Conference on Genetics and Native Peoples, Polson, MT (11–12 October).

Habermas, J. 1975. *Legitimation Crisis* (Boston: Beacon Press).

Hacking, Ian. 1999. *The Social Construction of What?* (Cambridge, MA: Harvard University Press).

Haraway, Donna. 1989. *Primate Visions* (New York: Routledge).

Harry, Debra. 1995. "Patenting of Life and Its Implications for Indigenous Peoples." *Information about Intellectual Property Rights* 7: 1–2.

Hayden, Corrine P. 1998. "A Biodiversity Sampler for the Millennium." In *Reproducing Reproduction: Kinship, Power, and Technological Innovation*. Ed. by Sarah Franklin and Helen Ragone (Philadelphia: University of Pennsylvania Press): 173–206.

Hilgartner, Stephen. 2000. *Science on Stage: Expert Advice as Public Drama* (Stanford: Stanford University Press).

Human Genome Diversity Project. 1992. *Human Genome Diversity Workshop 1* (Stanford, California: Stanford).

Indigenous Peoples Council on Biocolonialism. 1998. "Resolution by Indigenous Peoples." http://www.ipcb.org/resolutions/index.htm (accessed 5 March 2002).

———. 2000. *Indigenous Research Protection Act*. http://www.ipcb.org/pub/irpaintro.html (accessed 4 March 4 2002).

Jasanoff, J., G. E. Markle, J. C. Peterson, and T. Pinch, eds. 2004. "Ordering Knowledge, Ordering Society." In *States of Knowledge: The Co-Production of Science and Social Order*. Ed. by Sheila Jasanoff (London: Routledge).

Juengst, Eric. 1995. "Lineage, Land Tenure, and Demic Discrimination: The Social Risks of the Human Genome Diversity Project." Paper presented at A Public Symposium on Genetics and the Human Genome Project: Where Scientific Cultures and Public Cultures Meet, Stanford University (3–4 November).

Kaufmann, Alain. 2001. "Mapping at the Genethon Laboratory: The French Muscular Distrophy Association and the Politics of the Gene." A paper presented to The Mapping Cultures of Twentieth-Century Genetics, Max-Planck Institute, Berlin, Germany (1–4 March).

Kloppenburg, Jack. 1988. *First the Seed: The Political Economy of Plant Biotechnology* (Cambridge, UK: Cambridge University Press).

Latour, Bruno. 1993. *We Have Never Been Modern* (Cambridge: Harvard University Press).

Lewin, Roger. 1993. "Genes from a Disappearing World." *New Scientist* (May 29): 25–29.

Marks, Jonathan. 1995. "The Human Genome Diversity Project: Good for if Not Good as Anthropology?" *Anthropology Newsletter* (April): 72.

Mead, Aroha Te Pareake. 1996. "Genealogy, Sacredness, and the Commodities Market." *Cultural Survival Quarterly* (Summer): 464–50.

Melucci, Alberto. 1989. *Nomads of the Present: Social Movements and Individual Needs in Contemporary Society*. Ed. by John Keane and Paul Mier (Philadelphia: Temple University Press).

Merton, Robert. 1942. *The Sociology of Science* (Chicago: University of Chicago Press).

National Institutes of Health. 1994. NIH Guidelines on the Inclusion of Women and Minorities as Subjects in Clinical Research. *Federal Register* 59 (Part VIII): 14508–13.

Native-L. 1993a. "Call to Stop Human Genome Project." http://nativenet .uthscsa.edu/archive/nl/9307/0046.html (accessed 6 October 1999).

———. 1993b. "Human Genome Diversity Project." http://nativenet.uthscsa .edu/archive/nl/9307/0046.html (accessed 6 October 1999).

Natowicz, M. R. 1992. "Genetic Discrimination and Law." *American Journal of Human Genetics* 50: 465–75.

Nickens, H. 1993. "Minority Health Research Issues." *Science, Technology, & Human Values* 18: 506–10.

Pennisi, Elizabeth. 1997. "NRC OKs Long-Delayed Survey of Human Genome Diversity." *Science* 278 (24 October).

Pollack, Andrew. 2003. "Large DNA File to Help Track Illness in Blacks." *New York Times* (27 May): A1.

Provine, William, R. C. Lewontin, John A. Moore, and Bruce Wallace. 1981. *Dobzhansky's Genetics of Natural Populations* (New York: Columbia University Press).

Provine, William, and Ernst Mayr, eds. 1982. *The Evolutionary Synthesis: Perspectives on the Unification of Biology* (Cambridge: Harvard University Press).

Rabinow, Paul. 1996. "Fragmentation and Redemption in Late Modernity." In *Essays on Anthropology of Reason* (Princeton: Princeton University Press).

———. 1999. *French DNA: Trouble in Purgatory* (Chicago: University of Chicago Press).

Rensberger, Boyce. 1993. "Tracking the Parade of Mankind Via Clues in the Genetic Code." *Washington Post* (22 February): A3.

Risch, Neil, et al. 2002. "Categorization of Humans in Biomedical Research: Genes, Race and Disease." *Genome Biology* 3:2007.1–2007.12.

Rural Advancement Foundation International. 1993. "Patents, Indigenous Peoples, and Human Genetic Diversity." *RAFI Communique* (May).

Shapin, Steven, and Simon Schaffer. 1985. *Leviathan and the Air-Pump: Hobbes, Boyle, and the Experimental Life* (Princeton: Princeton University Press).

Sharp, Richard. 2002. "How Informed Can Informed Consent Be?" Talk presented at the Human Genetics, Environment, and Communities of Color: Ethical and Social Implications Conference, West Harlem, New York (4 February).

Shiva, Vandana. 1997. *Biopiracy: The Plunder of Nature and Knowledge* (Toronto: Between the Lines).

Smith, Anna Marie. 1994. *New Rights Discourse on Race and Sexuality: Britain, 1968–1990* (Cambridge: Cambridge University Press).

Southern, David. 1971. *An American Dilemma Revisited: Myrdal's Study through a Quarter Century* (Ann Arbor, Michigan: University Microfilms).

Stepan, Nancy Leys. 1982. *The Idea of Race in Science* (Hamden, CT: Archon Books).

Sturr, Christopher J. 1998. *Ideology, Discursive Norms, and Rationality*. Ph.D. diss., Cornell University, Ithaca, New York.

UNESCO. 1952. *The Race Concept: Results of an Inquiry* (Paris: UNESCO).

United Nations. 1949 (1948). *Universal Declaration of Human Rights* (Washington, DC: U.S. Department of State).

Weiss, Kenneth M. 1993. "Letter Dated April 30, 1993, to Senator Akaka." In Committee on Governmental Affairs, *Human Genome Diversity Project: Hearing before the Committee on Governmental Affairs,* United States Senate, 103rd Congress 1st Session, 26 April: 43–44.

Whitt, Laurie Anne. 1998. "Biocolonialism and the Commodification of Knowledge." Paper presented to The First North American Conference on Genetics and Native Peoples, Polson, MT (October 11–12).

Williams, Raymond. 1983. *Keywords: A Vocabulary of Culture and Society* (New York: Oxford University Press).

Wilson, James, et al. 2001. "Population Genetic Structure of Variable Drug Response." *Nature Genetics* 29(3): 265–69.

Wynne, Brian. 1996. "Misunderstood Misunderstandings: Social Identities and Public Uptake of Science." In *Misunderstanding Science? The Public Reconstruction of Science and Technology.* Ed. by A. Irwin (Cambridge: Cambridge University Press).

Yona, Nokwisa. 2000. "DNA Testing, Vermont H. 809, and the First Nations," *Native American* Village. http://www.google.com/search?q=cache:205.25375.10/native/special/ht809200.html+Nokwisa+Yona%2BDNA&hl=en. (accessed 21 February 2001).

CORI HAYDEN

# 19. Bioprospecting's
# Representational Dilemma

THIS IS ABOUT THE POLITICALLY and epistemologically charged endeavor of turning medicinal plants and "traditional knowledge" into pharmaceutical products. In Mexico, where I conducted ethnographic research
on one such endeavor in the late 1990s, the project of eliciting pharmaceuticals from traditional or popular remedies has belonged as much to
vaunted national(ist) scientific traditions (Lozoya 1984) as to processes
we might diagnose as an extractive colonialism or the "needs" of transnational capital, foreign drug companies, and researchers from abroad. Here,
as elsewhere, efforts to tease out the "efficacy" of traditional knowledge
have not been framed as a particularly symmetrical project. Plant- and
ethnobotanically-guided drug discovery—whether conducted by, for example, the researchers in Mexico's National Medical Institute in the late
nineteenth century, by the U.S. National Cancer Institute in the 1950s, or
by the San Francisco–based bioprospecting company Shaman Pharmaceuticals in the 1990s—has been an effort to render "traditional" or popular medicine actionable in terms established by the exigencies of industrial
drug discovery, biochemistry, and intellectual property. As such, it has focused on the form of the isolated, bioactive chemical compound.

Certainly, many researchers I know in Mexico and in the U.S. take an
equitable view if these wrenching efforts at transformation or corrobora-
tion "fail," often pointing to the narrowness or inadequacy of biochemical models rather than simply concluding that traditional remedies or
particular plants simply "don't work" in the ways people say they do. But
the ideological charge of the overall project does not thereby evaporate:
plant- and ethnobotanically-guided drug discovery is, by definition, an
effort that relies on biomedical models and "strong" patent provisions as
its ultimate source of legitimation and value. The uneven epistemological
weight that is built-in to this "corroborative" project significantly com-
plicates the well-intentioned efforts of many activist ethnobotanists and
chemists to "prove" the veracity of traditional knowledge by demonstrating that it *really* works—in pharmacological terms (Adams 2002).

Such questions about belief, knowledge, and efficacy are well-worn in
anthropology and in the sociology of knowledge. But, in recent years—
more specifically, since the early 1990s—this already heavily-freighted
political–technical project has become differently charged. The 1992 UN

Convention on Biological Diversity (CBD) ushered in some significant shifts in the relationship between indigenous knowledge and pharmaceutical research and development, particularly when such articulations cross north–south lines. Giving institutional form—albeit uneasily—to many years of activist and policy efforts, the CBD included a novel and hotly contested benefit-sharing mandate, in which corporations are required to ensure some form of "equitable returns" to source countries and source communities, should they desire continued access to Southern biogenetic resources and cultural knowledge. In the wake of the CBD, plants and "traditional knowledge," formerly considered part of the global commons and hence ostensibly (though this is of course a point of serious contest) "free for the taking," have become different sorts of resources. Once unencumbered, they now come with potential claimants attached. The ethic and (soft) mandate of benefit-sharing indexes, in part, what we might call a certain Lockean activism; that is, an idea that plants come with the innovative labor of "local" or "indigenous" people *already embedded*. These people, long excluded from the rewards granted to patent-holders on new drugs, should thus have the right to stake a claim should these plants prove to be the key to, for example, the ever elusive "cure for cancer" (Kloppenburg 1988; ISE 1988).

As we might imagine, in this context, the already overdetermined question of the pharmacological efficacy of traditional remedies has taken on new technico-political life. Bringing traditional knowledge (and its academic proxy, ethnobotanical knowledge) into the process of drug discovery is not simply meant as a short-cut to new pharmaceutical products. It is now also meant to bring dividends—in the form of compensation or other types of material reward—to the people who provided this knowledge. The "valorizations" at stake for resource providers (a dreadful term) are not simply symbolic or epistemological, but potentially *material* as well.

In the years leading up to and then following the 1992 CBD, one of the key mechanisms for setting (the promise of) this enhanced chain of value productions in motion has been the benefit-sharing or bioprospecting contract—a formalized arrangement in which corporate access to plants, microbes, and "knowledge" now include provisions for giving something back. In agreements implemented in Mexico and beyond, these benefit-sharing provisions have taken many forms: promises of royalty-sharing, community development funds or projects, up-front access fees, technology transfer, infrastructure-building. They have also set up a number of different kinds of actors as potential benefit-recipients, including indigenous or local "communities," scientists in the source country, nongovernmental organizations, groups of traditional healers, and national biodiversity institutes.

There is much in these arrangements that would seem ripe for an analysis framed by something called postcolonial science studies. As other contributions to this issue highlight, any attempt to think about a postcolonial science studies will necessarily run into the problem of definitions—we face significant questions about what will count as either postcolonial or science studies, not to mention what the combination itself might mean. In this chapter, I want to try and think about one of many potential versions of just such a combined analytic project through a critical engagement with bioprospecting practice. Broadly stated, I am interested in the work—simultaneously biochemical and political—that "local" or "ethnobotanical" knowledge is expected or asked to do as it travels into drug discovery circuits, and, ostensibly, back out again in the form of benefits-to-be-shared. In a formulation that I draw as much from science studies as from postcolonial studies, I pose this question as a matter of the *representational capacities of knowledge*—or, differently stated, as a question about the 'interests' that local knowledge is expected to bear, represent, or animate in these agreements. It is an abstract formulation, but as we shall see below it comes alive in the conduct of one particular U.S.–Mexico benefit-sharing collaboration, in which the promise of benefits to local resource providers depends in large measure on whether or not the plants and knowledge they provide actually succeed in becoming a patented drug. It is this dual notion of efficacy that interests me here, in which the bioactive potential of "local knowledge" holds the promise of activating benefit-sharing claims for "local people." Taking at face value, for now, the terms set by one prospecting agreement, this article poses a simple ethnographic question: what *is* the contribution of "local" knowledge to the process of drug development? How, in turn, does the relationship between ethnobotanical leads and pharmaceutical measures of value bear on the claims and entitlements that prospecting participants might expect in the future?

## On "Representation"

The idea that people have an interest in their knowledge is not of course limited to the terrain of contemporary bioresource appropriation. Concern with the representational work that knowledge does is, arguably, what brings certain strands of postcolonial studies and science and technology studies into conversation with each other—and what seems to drive them apart, as well. To start, we might consider one of the key arguments of a specific strand of science studies. Actor–network theory (ANT), associated with the work of Bruno Latour, Michel Callon, Annemarie Mol, John Law, and others, holds as one of its primary suppositions the argument that scientific knowledge does not simply represent (in the sense of depict) "nature," but it also represents (in the political sense) dense and diffuse webs of "social interests" (Latour 1993, 27; Callon 1986; Callon and Law 1982). That is, in order for facts and artifacts to become authoritative, they must mobilize, or gather in their name, a robust network of interested parties. The interests in question may belong to scientists, microbes, scallops, or any one of a number of potential "allies," human or non-human.

This argument that "representation" always carries a double meaning—as depiction (a picture of) and as proxy (an act of speaking for)—has a long lineage in social theory.[1] Echoing a distinctly Marxian formulation while deliberately casting aside Marx's central preoccupation with class, Latour and other actor–network theorists insist that speaking of (nature) is always also a mode of speaking for (interests). In *We Have Never Been Modern* (1993), Latour argues that the task of science studies should be to reunite these two meanings and domains of representation, by working to identify the (social/political) interests of the "actants" who come to be wrapped up in, and thus represented by, technoscientific facts and artifacts.

Postcolonial studies scholars, particularly those who have developed critical conversations between poststructuralism and Marxist subaltern studies, take this question of speaking-of and speaking-for somewhere else altogether.[2] Arguing that intellectual knowledge production (especially that which seeks to "restore a voice to the silenced") is always complicit in the very power relations it hopes to critique, writers such as Gayatri Spivak and Dipesh Chakrabarty insist on a differently inflected understanding of the doubleness of representation (Spivak 1988; Chakrabarty 1999). Here, the question is framed with reference to the historiography of the colonized or the subaltern: the problem with "representation"—speaking *for* the subaltern by speaking *of* her (restoring her voice to the historical record)—is fundamentally, its impossibility.[3]

A group of Latin American scholars working in conversation with the South Asian subaltern studies group has eloquently taken up this point. Alberto Moreiras, for example, argues that the distinctive task of a subaltern political and analytical stance is in fact to be "counter-representational"—to insist on the impossibility of representation. "Subalternism is on the side of negation," he argues: it is a very different stance than a populism, for example, in which scholars might actually claim to "represent the people" (Moreiras 1997, quoted in Beverley 1999; see also Moreiras 2001; Rabasa 2001).[4] [This is not to say, of course, that questions about agency and representation have not been rehearsed in STS, in characteristically idiosyncratic ways: recalling Callon (1986), we might ask, can the scallop speak?][5] In short, a subaltern studies approach entails highlighting the epistemological and political violences that are constitutive of efforts to "speak of" and (thus) to "speak for" the marginalized or excluded.

We might say, then, that actor–network theory shades toward the positivities if not the positivist dimensions of knowledge production as an act of representation, while much postcolonial studies in South Asia and Latin America (specifically, the work of subaltern studies scholars) asks us to attend to its negativities—its violences and elisions. As such, ANT and the kind of postcolonial/subaltern studies developed by Spivak, Moreiras, and others seem implicated in two fundamentally distinct, if not opposing, ways of thinking about the political/social/epistemological work that knowledge does. "It" (knowledge) includes, and shores up; it elides, and does violence. Where, and how, might these divergent theoretical perspectives take us with regard to bioprospecting?

One route would certainly be to address historiographic and epistemological questions about the effacements enacted through the "translation" of traditional knowledge into pharmacological value.[6] What, for example, is lost and gained (and by whom) when complex indigenous ontologies of illness and therapeutics are rendered in the very different language of the single chemical compound and its patentable effects (Adams 2002; Lozoya and Zolla 1984)? Many potential kinds of assimilation and effacement are at stake in these "corroborations": in Mexico, India, and elsewhere, such translational efforts have been crucial to postcolonial nationalist endeavors to incorporate "the indigenous" into the modern nation-state (Prakash 1999; Hayden 2003b; Soto Laveaga 2001). In these terms, the constitution of traditional knowledge as raw material for pharmaceutical development should be (and has been) classic terrain for a postcolonial studies analysis.

But we could also take our inquiries into bioprospecting's representational project in some other directions. We could think about questions of representation here in the dual mode flagged by actor–network theory—the depictive doubling as the political, in an "affirmative" and indeed incredibly explicit sense. Here, I think of the ways in which engaged ethnobotanists and plant chemists have long used their field studies and laboratories precisely like courtrooms; that is as staging grounds for proving the veracity of "indigenous knowledge." In this sense, ethnobotany has been figured by some activist practitioners as a form of what I call "epistemological advocacy"—a commitment to translate "indigenous knowledge" into the language of biochemical efficacy, and to use these assertions of the scientific rationality of indigenous knowledge as an axis of political mobilization and even court-room defense (Hayden 2003b).[7]

It is precisely in the mode of advocacy that some ethnobotanists view bioprospecting as an enterprising mode of representation in the affirmative sense. In the view of prominent U.S. ethnobotanists who have been involved in promoting ethnobotanically-guided drug discovery within U.S. government and corporate sectors, bioprospecting is not just a way to show the (pharmacological) veracity and thus the "value" (to the world) of traditional knowledge. It is also a way to turn this epistemological/biochemical correspondence into a revenue stream for the stewards of traditional knowledge themselves (Plotkin 1995; Balick and Cox 1996). It is in *this* sense—a bid to "include people" (in new idioms of political participation) by "including their knowledge" (in processes of drug discovery)—that some architects and proponents of bioprospecting make strong representational claims themselves.

In the remainder of this discussion, I take my cue from the machinations of one prospecting contract to try and gain some ethnographic purchase on the overly thick question of representation, and some of its many positivities and negativities. If we are interested in the politics of representation we cannot, I suggest, ignore the mechanics of representation: the practices through which bioprospecting contracts aim to animate, activate, construct, and protect the interests of "local people" through the inclusion of their knowledge in the drug discovery process.

## Bioprospecting

Despite the U.S. legislature's continued failure to ratify the 1992 UN Convention on Biological Diversity, an interagency U.S. government program has been among the most prolific sponsors of benefit-sharing agreements in the world. In 1993, the U.S. National Institutes of Health (NIH) began coordinating a bioprospecting initiative called the International Cooperative Biodiversity Groups (ICBG) program. Extending the reach of programs within the NIH (specifically, the National Cancer Institute) that have promoted plant- and ethnobotanical drug discovery, the ICBG program funds U.S. academic researchers to set up royalty-sharing collaborations between developing country communities/researchers and U.S. drug companies (Rosenthal 1997; NSF 1993; Timmermann 1997; Schweitzer et al. 1991; Reid et al. 1993).

One of the first five projects funded under this program in 1993 was the "Bioactive Agents from Dryland Biodiversity of Latin America" initiative (hereafter, the "Latin America ICBG"), based at the University of Arizona and working in Mexico, Chile, and Argentina (Timmermann 1997).

Participating Mexican researchers from the National Autonomous University (UNAM), together with their colleagues in Chile and Argentina, collected medicinal plants and knowledge about their uses, and sent their collections to the pharmaceutical and agricultural divisions of American Cyanamid, a U.S.–based firm. In exchange, the researchers were offered research funds, in order to conduct the required collections, and a percentage of future royalties on any resulting drugs.

The fraught question of "how much" was and remains very much deferred: if a drug is produced at all, it is likely to take 15–20 years (the usual time frame for a new drug to emerge out of the pipeline). In this case, the participating company would share with the University of Arizona between 1 and 5 percent of the net profits. Roughly half of that 1–5 percent would return to the source country, where it would be distributed among participating research institutions and the appropriate "source communities." This ethnobotanically-guided drug discovery initiative has been one of the many collaborations instituted in Mexico and across the world which attempt to bring some degree of "reciprocity" into corporate research and development based on the material and intellectual resources of the south. But as we might surmise from this project's speculative calculus of redistributed royalties, the CBD's notion of "equitable" returns (meaning "just," and not necessarily "equal") is extraordinarily, constitutionally, fragile.

This sense of fragility was compounded in Mexico in the late 1990s, when the legitimacy of plant collections for drug discovery became a deeply unsettled matter. Policymakers, scientists, and indigenous and *campesino* organizations have been grappling not only with the new promises of the CBD but also with the injunctions of the North American Free Trade Agreement (NAFTA), the World Trade Organization (WTO), and their combined imperatives to remove "barriers" to the traffic in biogenetic resources and to introduce stronger intellectual property protection for pharmaceutical and biotechnological products. Intimately entwined with associated shifts toward the "liberalization" of intellectual property, land tenure, and agricultural policy, the question of indigenous political and territorial sovereignty became intensely and violently vexed, most visibly but certainly not exclusively in the southern state of Chiapas (Nash 2001). Meanwhile, national legislation regulating access to genetic resources (required by the CBD) has been pending since the mid-1990s, so that bioprospecting contracts have proceeded on more or less an ad hoc basis. That is, they have gone forth under the watchful eye of the National Institute of Ecology (INE) but without binding national guidelines, legislation—or, many activists argue, legitimacy (Nadal 2000; Hayden 2003b).

## Making and Severing Connections

Whence "legitimacy," then? For ICBG-funded principal investigators, one of the main challenges raised by benefit-sharing agreements is, borrowing an apt turn of phrase from Marilyn Strathern, to ensure that (the appropriate) people are included, and that they are included *well*—that is, that they are not exploited through their very participation (Strathern 2000, 292–294). This is a dilemma that applies to a growing number of arenas in which information/sample collection and property claims promise and threaten to mingle. Prospecting joins genetic databases, tissue banking, and Internet data management as domains in which modes of including people "well" often pull in two opposed directions. First, they must set the ground for potential redistributive or even property claims, but second, they must also "protect" subjects by invoking privacy rights or confidentiality—and thus *cutting off* the relations of identity or provenance that make those redistributive claims possible in the first place (Strathern 2000, 292–294). This seemingly contradictory play of making and severing connections between people and what they "give" proves to be an extraordinarily fitting template for understanding the management of knowledge, plants, and the "interests" of resource providers in the Latin America ICBG project.

Elsewhere, I have discussed in great detail the practices through which the Mexican ethnobotanists working with the Latin America ICBG navigate their new imperative not just to collect plants, but to collect the benefit-recipients who come with those plants—e.g., the people who shall receive a portion of royalties should Mexican plants turn into patented drugs (Hayden 2003a, 2003b). The UNAM ethnobotanists' task is indeed to make connections—not just to identify, but to forge them—between collected resources and the (appropriate) participants. For the sponsoring agency, the U.S. National Institutes of Health, there is a very straightforward logic to the ways in which such connections *should* be made. The NIH stipulates that funded researchers are to sign benefit-sharing contracts (modeled on and serving as informed consent forms) with the people who provide plants and information. These contracts are supposed to produce an ethnobotanical paper trail. They are the crucial documentation that will produce a continuity between "local" people and the plants/ knowledge that might become the drugs that are to become the royalties to be shared (see Hayden 2003a).

But, alongside this imperative to produce continuities lies another move: the deliberate *disconnection* of people from biochemical specimens within the same circuits of exchange.[8] The agreement between UNAM and Arizona has some powerful confidentiality requirements written into it—most notably, the provision that plant extracts traveling from Latin America to participating U.S. companies be stripped of all identifying and ethnobotanical information, and labeled only by a project code. The purpose is to keep valuable identifying information out of the hands of the participating companies, so that they will not be able simply to go elsewhere (e.g., outside of the contractual bounds of this agreement) to resources that look promising. Thus, once back in their own laboratories, the Mexican researchers are charged with making the "connections" between people and their knowledge disappear from public sight, and from the gaze of most other participants in the project.

This appeal to secrecy has some important implications for the work, representational and otherwise, that ethnobotanical knowledge is allowed to do at all as a guide to identifying new drugs. For the incitement to secrecy dramatically undercuts the animation of ethnobotanical knowledge as both a valuable guide for drug discovery and as a conduit for carrying local people into these processes of value production. Rather than keeping people and their knowledge attached to specimens, this project's confidentiality provisions do precisely the opposite—they deliberately anonymize specimens (strip them of identity and connections) as they travel out of Mexico and into the laboratories of the participating companies and U.S. academic institutions. Crucially, in the view of project managers in the U.S., these two divergent practices are enacted toward the same end: *protecting the interests of southern resource providers.*

One of the implications of this move is that this prospecting network is cut off from itself, from within. There is thus no single answer to the seemingly straightforward question which I had hoped to pose here: how much or in what ways does "ethnobotanical knowledge" participate in the drug discovery process? Assessments of the role of ethnobotanical knowledge in the drug discovery process vary dramatically across this rather patchy network—in large part because "it" (ethnobotanical knowledge) does not travel freely among the institutions and actors involved. Ethnobotanical knowledge matters provisionally, or contingently, in the process of drug discovery—according to who is allowed access to it, when, and how.

## On the Uses of Knowledge

If we are to speak of confidentiality here, it is not primarily people but plants whose identities are at stake and under intensified management: much of the management of plants and information in this project involves stripping specimens of their markers of identity. The linchpin in this strategy of information management is the project code that links each extract to its associated data sheet—and thus its crucial identifying information. The information on this form, including plant uses, taxonomic and popular names, and location, is what makes collected samples viable outside of the laboratory. Most importantly, without a species name, no one can go forward with product development.

Neither outsiders such as myself, nor insiders such as the participating companies, are allowed access to the information that connects data about people, knowledge, and place to specific collected plants. Thus, when the Mexican chemists send plant extracts to the participating companies, they are labeled only with the project code. All identifying and supplemental information (i.e., the data sheet) is sent directly to the University of Arizona. Even at annual project meetings, where the Latin American academic researchers, U.S. academic chemists, and corporate representatives gather behind closed doors, the project director at Arizona prohibits participants from revealing sensitive information to each other. Thus, the researchers find themselves treading cautiously and counter-intuitively, trying desperately (for example) not to name the plant whose chemical structure they are discussing on their brand-new slide.

What, though, of those at Arizona who do have access to the information that accompanies each collected plant specimen? In the field of natural products chemistry, there are a number of approaches to the question of how one uses information about plants—e.g., the conditions they treat, their methods of preparation and "administration" (as a tea or a salve, for example), what they might be combined with, when and in what doses they are to be consumed. A nuanced, ethnobotanically-guided drug discovery enterprise is likely to take much of this information into account. Thus, if a plant is used primarily as a tea, a chemist might, at a minimum, start by investigating whether the plant yields bioactive compounds at high temperatures. But this (minimal) attention to the specificities of the local or the "ethno" is not the only approach circulating in the world of natural products chemistry.

For one key participant in this project, ethnobotanical information is valuable not for its specificities but rather for its generalities: that is, as a general beacon of bioactivity, the nuances of which are left for later elucidation. For this Arizona-based natural products chemist, the "uses" attributed to plants are of interest not primarily for their specific content but rather for the sheer fact of their existence. As he explained it to me, the knowledge that, for example, "*gordolobo* is good for coughs" is truly interesting only insofar as there is information that the same plant has some use in other locations—that is, he is looking for signs that the plant has a wide history of use of any sort, which in turn signals a good chance that it contains some active compound. In this view (common to much natural products drug discovery as well as economic botany), the "uses" box on the data sheet might just as well be prompting Latin American researchers to answer "yes" or "no," rather than "gastro-intestinal" or "upper respiratory."

Honing in on the fact, rather than the content, of ethnobotanical knowledge is a well-established approach to natural products drug discovery, but there is room in this project for yet another kind of engagement "with the ethno." As the director of this collaboration, Arizona chemist Barbara Timmermann is perhaps particularly attuned to the politics of participation or inclusion swirling around this endeavor. She assured me that the content of the "uses" box does potentially hold some usefulness in itself. It can matter, in large part, because she, uniquely in this collaboration, knows not just what the uses box contains but is also in a position to relate it to other relevant information. When she is considering which plants to pursue for further analysis, ethnobotanical information serves as one piece of the puzzle, alongside an array of other information. "Future promise" congeals when this information clusters in fortuitous ways—corporate bioassays that suggest a good fit between plant sample and industrial priorities; strong bioactivity readings in the preliminary UNAM tests; and crucially, a literature review that assures that a particular plant or compound is still up for grabs (Timmermann, personal communication, 1997). Too much already-published work on the matter removes a sample from the running, as it ruins both its patent appeal and the Mexican and U.S. chemists' ability to extract some intellectual capital (in the form of publications) from this process (the plant cannot be too useful, then).

From Timmermann's vantage point, ethnobotanical information may further solidify the association of an extract with certain bioactivities; that is, if a plant is supposed to be good for treating wounds, she might lend a particularly careful eye to the results of any anti-microbial screens. Still, the screens continue to proceed according to preestablished company priorities; they do not morph to accommodate the ethnobotanical information to which Timmermann might be privy. Arguably, then, ethnobotanical knowledge holds enough value to help confirm or buttress industrial tests, but it is not nearly weighty enough to contradict these bioassays, much less to direct industrial protocols.

But of course the design of this ICBG project is such that ethnobotanical knowledge is not *permitted* to direct corporate screening protocols. Regardless of the information that the UNAM ethnobotanists may have included on their data sheets, the companies involved subject all of the samples that come to them (identified only as the provenance of one of the three source countries) to the same battery of screens. U.S. government researchers and policymakers affiliated with the ICBG wrote in 1999 that indeed, despite the high profile of ethnomedical and indigenous knowledge in the program, "it is difficult to effectively integrate ethnomedical knowledge into the large-scale high-throughput systems commonly used by the industrial partners . . . Traditional knowledge may be more often useful in academic environments, government labs and with companies that have flexible systems that can be easily customized to take advantage of the information" (Rosenthal et al. 1999, 17).

Yet this approach is *not* the one the corporate liaison to the project says he might have chosen himself. In a 1997 interview, he suggested that the ICBG's approach to using ethnobotanical knowledge—and more explicitly, the blocks placed on its travel—was a bit of a novelty, and not always an effective one at that. From this researcher's perspective, ethnobotanical information just gets "put away." And what if, he asked me (not quite rhetorically), his company does not happen to be screening in an area corresponding to the uses attributed to a particular plant? From his point of view, ethnobotanical information is not emptied of content. It simply disappears.

Importantly, he noted explicitly that this disappearance is a reflection of the interests being represented through plant-based drug discovery. He posed the dilemma this way: wasn't the *requirement* that the company test the extracts *blind* doing "a massive disservice to the researchers who collect this information"? This question brings to the fore the many layers of interest that ethnobotanical knowledge presumably bears here. The explicit doubleness of the term comes to life. If actor–network theory has prepared us well to think about scientists as having an "interest" in "their" knowledge, this suggestion is complicated by the ambiguity of a term which is used just as often in drug discovery and academic circuits to refer to what "local" or "indigenous" people know as to what ethnobotanists produce on data sheets and publications, as experts.

As is suggested by an increasing body of work, postcolonial and otherwise, on natural-historical and botanical knowledge produced in "encounter," the kind of authorship we might identify here is indeed complex (Goonatilake 1984; Grove 1991; Raffles 2000; Pratt 1992). In fact, it is not only rural or "local" participants who are being recruited into the prospecting fold in this agreement. UNAM ethnobotanists and their chemist colleagues are considered participants and potential benefit-recipients as well. For, among the most concrete forms of returns in this exchange are research funds, training for graduate students, promises of equipment and database management tools, and other infrastructural goods destined for the UNAM ethnobotany and chemistry labs (see Hayden 2003b). It is with these provisional returns in mind, perhaps, that the disappearance of "ethnobotanical knowledge" strikes the corporate liaison as problematic—because it seems to remove ethnobotanists themselves from the visible field of participants in prospecting's brand of redistributed value-production.

## Blocks and Flows

Many participating researchers lamented to me and to each other the uncomfortable novelty of the truncated travels of information in this agreement. One of the U.S.-based chemists helping orchestrate the traffic of plant samples in this agreement complained that he is used to being able to collaborate with corporate partners in a much more open way. In this agreement, in contrast, he is not allowed to "share" what he knows in order to help guide the bioassays. What, he asked in exasperation, is the point of this collaboration? Neither the academic scientists (from either side of the salient geopolitical divide) nor the corporate representatives managing their company's involvement in this project were completely sold on the ICBG's injunction that knowledge of plant identities and plant uses shall be so dramatically partitioned and redistributed.

On the surface, such laments might seem to resonate strongly with recent critiques of the quasi-privatization of academic bioscience research. The truncation of information flow is widely seen as one of the most pervasive and negative effects of the increasing prevalence of industry–university partnerships and of patenting as a commonplace part of research in the life sciences (see Soley 1995). Commenting on this situation in the idiom provided by actor–network theory, Marilyn Strathern notes that Latour's famous ever-multiplying networks of knowledge production and interests are indeed interrupted by patent claims. Ownership brings these relations to a stopping point, even if only temporarily (Strathern 1999, 177).

But something distinctive is going on in the case of the ICBG project: it is not ownership itself which blocks the flow of information, but rather the threat of property claims made out of place—at least in the early stages of plant collection, screening, and testing for future promise. Once property claims are ready to be actualized, in the form of a patent, things that were at first kept channeled into very limited avenues of access will start flowing within the agreement. The companies will have to know what the molecule and plant source is. The Latin American researchers will have published articles naming their specimens and perhaps the results of some corporate bioassays, and the potential for royalty-generation is set in motion.

This is not an argument for the liberatory possibilities of intellectual property rights. Certainly, many kinds of claims and types of access stand to be proscribed if a chemical compound from a Mexican plant is patented by the University of Arizona, and licensed to Wyeth-Ayerst. We might point to a long list of examples—the neem tree in India, the enola bean in Mexico, the vines used to make the hallucinogenic beverage ayahuasca in the Amazon basin—of corporate patents taken out by U.S. or European companies on compounds derived from popularly used plants, followed by actions by the companies to sue the long-term users or distributors of these plants for patent infringement. I would suggest that it is precisely the threat of these kinds of "property gone wrong" (or "right," depending on one's perspective) that make provisional sense of the modes of representation on offer in this prospecting agreement.

In other words, the "unconventional" information blocks that most irk participating researchers are not necessarily those imposed by the companies. They are, rather, aimed primarily at upsetting conventional power relationships within industrial–academic, not to mention north–south, collaborations. Secrecy, at least in terms of keeping identifying information out of the hands of corporations, is meant as a form of sanction—a built-in mode of enforcement power within the prospecting contract itself. It is thus a sign of good faith to participating source countries that corporations will not be allowed to circumvent the other "unconventional" aspect of this collaboration—the return of royalty payments to the providers of resources. Thus we arrive at a somewhat paradoxical situation: what could seem like a violation of the spirit of collaborative, ethnobotanically-guided research—virtual erasure of "local" knowledge—is in fact a measure to protect against its undue exploitation.

## Limits of Representation?

What are we to make of these presumed connections and unsettling truncations? The combination produces some notably odd effects. Far from the idea that a continuous chain of value production will be extended from "local people" to ethnobotanical knowledge to a drug and back again, this project coughs up a complex and choppy network of provisional connections and truncations, of short (if not short-circuited) loops of representation and elision. Even in its own terms, the modes of representation on offer in this agreement seem to run into some intriguing self-imposed limits, as the mechanisms in place for asserting and protecting the interests of knowledge providers end up defusing the declared potential (simultaneously biochemical and political) of the resource in question. One consequence of the ICBG's mode of information management is, we might argue, that ethnobotanical knowledge is *marginalized* within the very processes meant to *valorize* it.

I noted at the outset that bioprospecting's self-description seemed eerily suggestive of a world according to Latourian science studies—that is, a world in which people's (and things') interests are spoken for, represented, through their inclusion in processes of knowledge production. But the processes charted above seem to turn that premise inside-out. This bioprospecting agreement and its information- and interest-management practices, on second look, seem to call up a different set of theoretical conversations, leaning more toward the deconstructive denials and negations that inform much postcolonial theory and subaltern studies. Moreiras and other subaltern studies scholars might not be surprised to find the figure of "the indigenous" or of "local people" and their interests constantly invoked in this agreement and yet endlessly deferred. In such circumstances, the inclusion of "people" (and their knowledge) can only be imagined or conducted through their effacement, or their exclusion.

The deferrals enacted in this agreement have a distinctive shape: they come to us, in large part, through recourse to confidentiality, and they highlight the complex representational project that prospecting managers set up for themselves. This is an effort to speak *for* ("local interests") by not necessarily speaking *of* ("local knowledge"). We might argue that if the mechanics of this bioprospecting agreement constitutively refuse to tell us very much (for better or for worse) about the pharmacological "value" of "local knowledge," they do show in vivid detail what a *technics of counter-representation* might look like.

## Efficacy, Transformed . . .

The story I have told here might have been rather different, and certainly much more straightforward. The annals of natural products chemistry, ethnobotany, and medical anthropology are full of rich and nuanced discussions of the ways in which plants and "traditional" medicine have demonstrated their value to drug discovery—and thus, of how, presumably, they can continue to do so. Proponents of natural products chemistry repeatedly point out that "one in four" prescription drugs has been derived from plants (others move the mark up to three-quarters) while still others note that bioprospectors can increase their chances for success from one in 10,000 samples (the skeptical calculus often cited by synthetic chemists) to one in two, by using "ethnobiological" knowledge as their guide (see Rodríguez Stevenson 2000, 1132). We might expect bioprospecting agreements to provide more fodder for such success stories, or (if not), then to help us think about how the complexities of plant-based therapeutics might resist assimilation into pharmacological accountings of efficacy and value. (Examples of such "resistance" are not hard to come by. I think here of the epic difficulty one U.S.–based prospecting company had in synthesizing the plant-based compound that was to lead to its one marketable product. There are also the stories many natural products chemists in Mexico and the U.S. tell about how the process of "isolating" compounds, necessary to produce a patentable pharmaceutical product, often simply does away with all signs of biological activity that plant extracts initially produce.) But, as we saw in the Latin America ICBG project, these always-charged engagements between something imagined as "traditional medicine" and pharmaceutical research have been displaced, even short-circuited. They have become something else altogether.

In the introduction to a thought-provoking 2003 workshop entitled, "Plants, Medicine, and Power: Emerging Social and Medical Relations," historian Gabriela Soto Laveaga argued that in the context of bioprospecting initiatives, articulation between traditional medicine and pharmaceutical research and development has undergone a subtle but important shift. It is, she argues, a relationship that has become thinkable *less* as a matter of the "efficacy" of traditional medicine, and *more* as a matter of the exigencies and complexities of "compensation" (Soto Laveaga, 2003). I would second this formulation and go one half-twist further. With the Latin America ICBG project firmly in mind, I might be tempted to argue that bioprospecting is not primarily about comparative therapeutic efficacies, or "the power of plants" *at all*. It is a strong point, but I say it with the assessments of key project participants in Washington, D.C. firmly in mind. In a 1999 article detailing the contours and accomplishments of the ICBG program up to that point, Joshua Rosenthal, director of the initiative at the NIH, and 21 co-authors from other participating government agencies issued a cautionary note on the relationship between drug discovery and benefit-sharing agreements. In fact, they make clear just how touchy *is* the question of drug discovery within what is, among other things, a natural products (i.e., plant and microbe-based) drug discovery initiative. Seeking to dispel the "popular" conception that bioprospecting's success as a conservation strategy depends on a new drug making it to market (and selling well), Rosenthal and colleagues caution their readers,

> . . . drug discovery is a high risk science. That is, a very small proportion of the research endeavors result in a major drug that will yield financial ben-

efits to the research organizations and their partners. The ICBG program approaches bioprospecting in a more multi-dimensional way, such that progress in any one of the three goals [drug discovery, promoting scientific and economic activity in developing countries, and conservation] ideally strengthens efforts of the other two. By integrating research and development efforts toward all three objectives from the outset, the ICBG aims to make substantial and incremental contributions toward their achievement without pinning all hopes for success on the relatively low probability of producing a major pharmaceutical or agricultural product. (Rosenthal et al. 1999, 7)

Rosenthal and colleagues label the ICBG's challenge as one of "Combining high risk science with ambitious social and economic goals" (to quote the title of their article). I might take their self-diagnosed combination of high ambition and low pharmaceutical expectations and pose a different question about ethnobotanically-based drug discovery as a mode of representing people, knowledge, and "interests." After Soto Laveaga, my question is as follows: how are notions of comparative efficacy being re-calibrated with the promise of benefit-sharing (and its ever-present twin, the accusation of biopiracy) on the horizon? Certainly, the processes I charted above seem to make little sense from the point of view of an effort to turn medicinal plants into drugs. But my argument is that this project has had its sights on the activation of other kinds of value and (counter-)representational projects altogether.

As with many things, the world of bioprospecting does not stand still for long. With one eye on the "difficulties" of eliciting leads out of plant compounds, and another on the fallout from international conflagrations concerning "biopiracy" in Chiapas (following a controversy surrounding a sibling ICBG project), the participating company and the project director at the University of Arizona have decided to dramatically re-tool their collection strategies. These have been orientated away from Mexican plants and the people who come with them, and toward collecting sites that seem to promise more bioactivity and less social-political complexity. As of the fall of 2003, the Mexican team was no longer part of the Latin America ICBG. The UNAM scientists continue to search for new sources of funding for their interdisciplinary plant research, while microbe screening, deep sea bioprospecting (outside of the limits of territorial sovereignty), bioinformatics and combinatorial chemistry, and even pharmacogenomics beckon a new generation of corporate prospecting initiatives, in which neither plants nor their efficacy are very much at the heart of the matter, at all. In its own terms and in a relatively short time-span, the newly inclusive project of bioprospecting has produced its own impossibilities, and is poised to raise some new representational dilemmas altogether.

## Acknowledgments

I would like to extend gratitude, first, to Claudia Castañeda, whose initiative in organizing a panel on Postcolonial Science Studies at the 2000 meetings of the Society for the Social Studies of Science (4S) was the original spark for this essay and for the special issue of *Science as Culture*. For their heroic editorial work thanks go to Maureen McNeil and Les Levidow; for comments and generative conversation, I thank Annelise Riles, Marilyn Strathern, Galen Joseph, and Catherine Alexander; and for their generosity and insights, a final thanks to the participants in the ICBG program, "Bioactive Agents from Dryland Biodiversity of Latin America."

## Notes

1. We might follow George Hartley and frame the "problem" of representation via Marx and Gayatri Spivak (Hartley 2003, 247–252). Hartley notes that in Spivak's famous essay, "Can the subaltern speak?," she works through a critical engagement with the problem of colonialism, marginality, and academic knowledge production in part through an explication of Marx's *Eighteenth Brumaire* (Spivak 1988). As Spivak reads Marx, representation takes on two meanings: as depiction, and as proxy. Representation works in the sense of a depiction, when, for example, peasant proprietors can form a picture of themselves as a class. We might think of representation as "proxy," a mode of speaking for "interests," when these same peasants fail to conceive of themselves as a class, and instead turn to someone else (in Marx's example, Bonaparte) to represent their interests for them (see Hartley 2003, 248). A notion of the fundamental doubleness of representation—as depiction, as proxy, and often as both at the same time—matters a great deal to Spivak's notion of the politics of knowledge production, and, I suggest, it matters a great deal to an analytic project forged in the name of a postcolonial science studies, particularly in the case of bioprospecting.

2. As have, significantly, feminist technoscience studies. Donna Haraway in particular has pointedly remarked upon the gendered and raced notions of agency and identity that underlie Latourian understandings of representation (Haraway 1997, 23–39).

3. As Spivak argues, if the subaltern subject is only knowable through the archives and traces of colonialism, then any effort to speak *for* the (interests of) the subaltern by speaking *of* (depicting) her runs into the fundamental problem that "knowing" is to perpetuate a violence—thus all that is left to the intellectual is to "mime" an act of unknowing (see Spivak 1988; see also Hartley 2003, 235–259).

4. Beverley here draws on an unpublished paper by Alberto Moreiras from 1997, though the point is elaborated in various ways in other conversations in an emerging school of Latin American subaltern and postcolonial studies. See Moreiras's elaborations of a notion of negativity as a critical stance in Moreiras (2001); see also Rabasa (2001).

5. Michel Callon's well-known article on the agency of scallops has been a touchstone for science studies debates over how a "symmetrical" approach to sociological analysis shall distribute attributions of agency among human and non-human entities (Callon 1986).

6. A note on my use of the word "translation" is in order here. A few readers of my work have (mis)understood my argument by assuming that translation must simply mean giving a different name to the "same" substance or knowledge. To the contrary, I draw on work in both science studies and postcolonial and post-structuralist theory which holds that translations are always, necessarily, *transformations*.

7. In a longer discussion of this point (Hayden 2003b), I discuss various modes of epistemological advocacy, and the difference it makes to make truth claims for indigenous knowledges and resource management strategies in the idiom of sustainability (see Posey 1985), a structuralist interest in the fundamental similarities of "scientific" and "traditional" classification structures (see Hunn 1999), or in the language of bioactive chemical compounds and their commercial potential (Schultes and von Reis 1995; Plotkin 1993; Balick and Cox 1996).

8. See Warwick Anderson's fascinating account of the ways in which such moves of attaching and detaching shaped colonial clinical research in the "peripheries" in an era in which the "benefit-recipient" was not an operable category (Anderson 2000).

## References

Adams, V. 2002. "Randomized Controlled Crime: Postcolonial Sciences in Alternative Medicine Research." *Social Studies of Medicine* 32 (5–6): 659–90.

Anderson, W. 2000. "The possession of Kuru: Medicinal Science and Biocolonial Exchange." *Comparative Studies in Society and History* 42 (4): 713–744.

Balick, M. J., and P. Cox. 1996. *Plants, People, and Culture. The Science of Ethnobotany*. New York: Scientific American Library.

Beverley, J. T. 1999. *Subalternity and Representation: Arguments in Cultural Theory*. Durham and London: Duke University Press.

Callon, M. 1986. "Elements of a Sociology of Translation: Domestication of the Scallops and the Fisherman of St. Brieuc Bay," in J. Law, ed., *Power, Action, and Belief: A New Sociology of Knowledge?* 196–233. London: Routledge and Kegan Paul.

Callon, M., and J. Law. 1982. "On Interests and their Transformation: Enrolment and Counter-Enrolment." *Social Studies of Science* 12: 615–25.

Chakrabarty, D. 1999. *Postcolonial Thought and Historical Difference*. Princeton: Princeton University Press.

Goonatilake, S. 1984. *Aborted Discovery: Science and Creativity in the Third World*. London: Zed.

Grove, R. 1991. "The Transfer of Botanical Knowledge between Asia and Europe, 1498–1800." *Journal of the Japan–Netherlands Institute* 3: 160–176.

Haraway, D. 1997. *Modest_Witness@Second Millennium. FemaleMan_Meets_OncoMouse.*™ New York and London: Routledge.

Hartley, G. 2003. *The Abyss of Representation: Marxism and the Postmodern Sublime*. Durham: Duke University Press.

Hayden, C. 2003a. "From Market to Market: Bioprospecting's Idioms of Inclusion." *American Ethnologist* 30 (3): 1–13.

———. 2003b. *When Nature Goes Public*. Princeton: Princeton University Press.

Hunn, E. S. 1999. "Ethnobiology in Court: the Paradoxes of Relativism, Authenticity, and Advocacy," in T. Gragson and B. Blount, eds., *Ethnoecology: Knowledge, Resources, and Rights*. 1–11. Athens and London: University of Georgia Press.

ISE (International Society of Ethnobiology). 1988. *Declaration of Belém*. (available at http://users.ox.ac.uk/~wgtrr/belem.htm).

Kloppenburg, J. 1988. *First the Seed: The Political Economy of Plant Biotechnology*. Cambridge, UK: Cambridge University Press.

Latour, B. 1993. *We Have Never Been Modern*, trans. Catherine Porter. Cambridge: Harvard University Press.

Lozoya, X. 1984. "La Herbolaria Medicinal de México," in X. Lozoya and C. Zolla, eds., *La Medicina Invisible: Introducción al Estudio de la Medicina Tradicional de México*. Mexico, DF: Folios.

Lozoya, X., and C. Zolla, eds. 1984. *La Medicina Invisible: Introducción al Estudio de la Medicina Traditional de México*. Mexico, DF: Folios.

Moreiras, A. 2001. "A Storm Blowing from Paradise: Negative Globality and Critical Regionalism," in I. Rodríguez, ed., *The Latin American Subaltern Studies Reader*. 81–110. Durham and London: Duke University Press.

Nadal, A. 2000. "Biopiratería: El Debate Politico." *La Jornada*. Mexico City: 13 September, 22.

Nash, J. 2001. *Mayan Visions: The Quest for Autonomy in an Age of Globalization*. New York and London: Routledge.

National Science Foundation. 1993. Request for Proposals, International Cooperative Biodiversity Groups Program, Washington, D.C.

Plotkin, M. 1993. *Tales of a Shaman's Apprentice: An Ethnobotanist Searches for New Medicines in the Amazon Rainforest*. New York: Viking.

———. 1995. "The Importance of Ethnobotany for Tropical Forest Conservation," in R. Schultes and S. von Reis, eds., *Ethnobotany: Evolution of a Discipline*. 349–361. Portland: Dioscorides Press.

Posey, D. 1985. "Indigenous Management of Tropical Forest Ecosystems: The Case of the Kayapó Indians of the Brazilian Amazon." *Agroforestry Systems* 3:139–158.

Prakash, G. 1999. *Another Reason: Science and the Imagination of Modern India*. Princeton: Princeton University Press.

Pratt, M. L. 1992. *Imperial Eyes: Travel Writing and Transculturation*. New York and London: Routledge.

Rabasa, J. 2001. "Beyond Representation? The Impossibility of the Local (Notes on Subaltern Studies in Light of a Rebellion in Tepotzlán, Morelos)," in I. Rodriguez, ed., *The Latin American Subaltern Studies Reader*. 191–210. Durham and London: Duke University Press.

Raffles, H. 2000. "The Uses of Butterflies: Henry Walter Bates and the Ambiguities of Victorian Natural History." *American Ethnologist* 28 (3): 513–48.

Reid, W., et al., eds. 1993. *Biodiversity Prospecting: Using Genetic Resources for Sustainable Development*. Washington, DC: World Resources Institute; Instituto Nacional de Biodiversidad; Rainforest Alliance; African Centre for Technology Studies.

Rodríguez Stevenson, G. 2000. "Trade Secrets: The Secret to Protecting Indigenous Ethnobiological (Medicinal) Knowledge." *New York University Journal of International Law and Politics* 32: 1119–74.

Rosenthal, J. 1997. "Integrating Drug Discovery, Biodiversity Conservation, and Economic Development: Early Lessons from the International Cooperative Biodiversity Groups," in F. Grifo and J. Rosenthal, eds., *Biodiversity and Human Health*. 281–301. Washington, DC: Island Press.

Rosenthal, J., et al. 1999. "Combining High Risk Science with Ambitious Social and Economic Goals." *Pharmaceutical Biology* 37 (Supp): 6–21.

Schultes, R., and S. von Reis, eds. 1995. *Ethnobotany: Evolution of a Discipline*. Portland: Dioscorides Press.

Schweitzer, J., et al. 1991. "Summary of the Workshop on Drug Development, Biological Diversity, and Economic Growth." *Journal of the National Cancer Institute* 83: 1294–98.

Soley, L. 1995. *Leasing the Ivory Tower: The Corporate Takeover of Academia*. Boston: South End Press.

Soto Laveaga, G. 2001. "Root of Discord: Steroid Hormones, a Wild Yam, 'Peasants' and State Formation in Mexico (1941–1986)." PhD Thesis, Department of History, University of California, San Diego.

———. 2003. "Plants, Medicine, and Power: Emerging Social and Medical Relations, Workshop Introduction." University of California, Berkeley, 14 March.

Spivak, G. 1988. "Can the Subaltern Speak?" in C. Nelson and L. Grossberg, eds., *Marxism and the Interpretation of Culture*. 271–313. Urbana: University of Illinois Press.

Strathern, M. 1999. "Potential Property: Intellectual Rights and Property in Persons," in M. Strathern, *Property, Substance, & Effect: Anthropological Essays on Persons and Things*. 161–178. London: Athlone Press.

———. 2000. "Accountability . . . and Ethnography," in M. Strathern, ed., *Audit Cultures: Anthropological Studies in Accountability, Ethics, and the Academy*. 279–304. London: Routledge.

Timmermann, B. 1997. "Biodiversity Prospecting and Models for Collections Resources: The NIH/NSF/USAID Model," in K. E. Hoagland and A. Y. Rossman, eds., *Global Genetic Resources: Access, Ownership, and Intellectual Property Rights*. 219–34. Association of Systematics Collections.

# IV. Moving Forward: Possible Pathways

I think first of all that the Western theoretical establishment should take a moratorium on producing a global solution. . . . I think in the language of commercials, one would say: Try it, you might like it. Try to behave as if you are part of the margin, try to unlearn your privilege.—GAYATRI SPIVAK, "The Post-modern Condition"

Only when science and technology evolve from the ethos and cultural milieu of Third World societies will they become meaningful for our needs and requirements, and express our true creativity and genius. Third World science and technology can evolve only through a reliance on indigenous categories, idioms and traditions in all spheres of thought and action.
—THIRD WORLD NETWORK, *Modern Science in Crisis*

The Indian civilization today, because it straddles two cultures, has the capacity to reverse the usual one-way procedure of enriching modern science by integrating within it significant elements from all other sciences—premodern, non modern, and postmodern—as a further proof of the universality and syncretism of modern science. Instead of using an edited version of modern science for Indian purposes, India can use an edited version of its traditional sciences for contemporary purposes.—ASHIS NANDY, "Science as a Reason of State"

The strength of social studies of science is its claim to show that what we accept as science and technology could be other than it is; its great weakness is the general failure to grasp the political nature of the enterprise and to work toward change. With some exceptions it has had a quietist tendency to adopt the neutral analyst's stance that it devotes so much time to criticizing in scientists. One way of capitalizing on the strength of social studies of science, and of avoiding the reflexive dilemma, is to devise ways in which alternative knowledge systems can be made to interrogate each other.—HELEN WATSON-VERRAN AND DAVID TURNBULL, "Science and Other Indigenous Knowledge Systems"

Recently, many historians of colonial medicine and technology have followed anthropologists in mapping out a distinct, complex set of engagements with the "modern." . . . Yet laboratory science, defetishized at its "origins," still moves around the globe as a fetish, with its social relations conveniently erased. It seems to arrive with capitalism, "like a ship," then magically arrive else-where, just as powerful, packaged, and intact. We remain attached to the "Marie Celeste" model of scientific travel. . . . We sometimes hear from scholars in science studies that postcolonial perspectives are not relevant to their work because they are not examining science in temporally postcolonial locations—we hear this even (perhaps especially) from those working on science in the United

States!—but this seems to us to echo the assertion of an earlier generation of scholars who argued that because they did not study *failed* science, sociology was irrelevant to their work.—WARWICK ANDERSON AND VINCANNE ADAMS, "Pramoedya's Chickens: Postcolonial Studies of Technoscience"

PESSIMISM CAN OFTEN BE THE FIRST RESPONSE by Western readers when they begin to recognize the long histories, extensive reach, and continuing practices of Western sciences' complicity with colonial and imperial projects. It can seem that inaction is the only safe policy when faced with the horrible consequences of earlier generations' attempts to do the right thing by the standards of their day. Yet no societies can afford the luxury of inaction when faced with the kinds of environmental, health, economic, social, and political disasters that confront people around the world today. Moreover, in the preceding parts, we have already begun to encounter proposals for improving conceptual frameworks, theories, policies, and practices. The authors in part IV focus on several different sites at which to engage in transformations: Third World societies, a "white settler" society such as South Africa, the U.S. federal government, sciences in the North, and grassroots citizen organizations committed to changing scientific priorities and practices. Three questions can illuminate some of the most important challenges here.

## Who Decides?

The weight of opinion in the preceding parts supports the view that it cannot be left only to scientists and engineers to determine the direction of science and technology research and policy, though of course their expert advice must always be weighed in any such decisions. However, to many Western-trained scientists and engineers, denying them such authority seems to invite the insertion of uninformed and partisan politics into scientific and technological policy—precisely what the commitment to autonomy and value neutrality had sought to block. Indeed, many think that federal science policy in the United States in the last decade has been far too happy to adopt research agendas and policies shaped by social interests (in this case, of right-wing Christian groups), by the Department of Defense, and by corporate interests in pharmaceutical products, fossil fuels, ranching, and agribusiness, among others. It has all too often ignored advice from scientific institutions and agencies. The only reliable solution to such insertions of ignorance and politics into science policy, according to such critics, is to leave decisions about scientific research in the hands of the leaders of the scientific community and to scientists who strictly follow ideals of value-free objectivity.

Yet the authors in this collection observe how again and again bad consequences follow when what are fundamentally political and ethical issues get transformed into only technical ones by governments and science and engineering experts. Experts are trained to make only technical decisions. As part of their commitment to value neutrality, often they are trained even to refuse to consider any possible social or political consequences of their work.

Several cautions are helpful here. No one group should be making policy decisions. Rather, as Jenny Reardon's essay earlier argued, all those who will be affected by such decisions should participate fully in scientific and technical projects. This is a fundamental principle of any democratic ethic. But how early in a scientific project should such participation occur? What constitutes full participation? And how should the structure and priorities of science and technology institutions be reorganized to maximize such end-user participation from the beginning of research projects? Is it only the science and technology institutions that need transformation to ensure full participation? What are the obstacles to full participation by women and poor people? Identifying effective responses to such questions, let alone transforming policies and practices, is not an easy task, as the essays here point out. These authors call for a prioritization of ethical and political over intellectual goals in the selection and design of scientific and technical projects. Thus they call for "civic science" (Bäckstrand), "negotiated science," and "epistemic modernization" (Hess). Of course, such proposals will be controversial. But do any reasonable alternatives exist? To whom should they be reasonable?

Fortunately, we do not have to reinvent science or its governance from scratch. These authors identify a number of research fields and ways of proceeding in which such goals are already being implemented. Valuable transformations are already under way, and those interested can join in and promote them.

## What Are Non-Western Priorities?

Science and technology theorists working in non-Western societies have their own priorities, as earlier essays have pointed out. They too have mostly been resisting temptations either to demonize or to romanticize their own cultures' scientific and technological traditions, as Sardar's account of Islamic science and Hoppers's of South African scientific legacies explore. Many different and often conflicting interests and preferences exist among non-Western scientists, policymakers, and informed citizens. Yet they often call for grounding scientific and technological projects in the cultural legacies that have been most highly valued in non-Western societies rather than in legacies of Western societies such as the Enlightenment and its exceptionalist and triumphalist assumptions. How could this work in a world in which transnational cooperation is required to solve many kinds of problems, such as pandemics, climate change, the management of oceans and of space, global economic and financial relations, governments that oppress their own peoples, militarism, and immigrants and refugees? How are such projects already under way in such contexts?

Part IV opens with debates among Third World science and technology theorists. The Sri Lankan engineer Susantha Goonatilake argues that it is unrealistic for any social movement to imagine that it can today start over from scratch and create a scientific and technological tradition radically different from modern Western sciences, contrary to the perception of the Third World Network quoted at the beginning of this introduction. It is a delusion, Goonatilake claims, to imagine that indigenous knowledge can be preserved in situ, as the anti-biopiracy activists hope. The particular environments, knowledge legacies, and cultures on which such societies depend are rapidly disappearing, he points out, as they are transformed by unstoppable processes of globalization and its environmental effects. Thus indigenous groups have no effective choice except to try to "suture" into modern Western sciences the kinds of projects that better serve them. Modern Western sciences are too powerful to resist. This view is also the dominant one among Western scientists, as well as among leading historians and philosophers of science. Do such critics overlook ways that such different scientific traditions can be bridged, integrated, or blended without prioritizing only modern Western scientific ideals? Is seeking to nourish "other ways of knowing" really only a romantic fantasy? Or is it a realistic strategy that not only will strengthen both sciences and democratic social relations but is already well established in ways not visible to the critics?

Other Third World intellectuals, such as Sardar, Hoppers, and the Third World Network, assess the possibilities for retaining what they think is the best of both Western and indigenous legacies. Often these proposals make ample use of the modern scientific and technological expertise in which many of the authors were trained. Yet the projects they propose are to be sutured into the cultural values and interests of their own societies. Notice that both modern Western sciences and indigenous knowledge systems are here conceptualized not as fully integrated, coherent bodies of knowledge but as collections—assemblages—of ontologies, epistemologies, ethics, and cultural preferences and practices, practical theories, inquiry methods and techniques, empirical results, and more, which can be disaggregated and their elements selectively used within other culturally specific knowledge systems (see Watson-Verran and Turnbull 1995). Hasn't this been just how Western traditions have borrowed from other cultures? Note that where Goonatilake would incorporate non-Western elements into Western knowledge systems, this proposal recommends the incorporation of elements of Western knowledge systems into non-Western ones—a very different matter. The direction of integration does make a difference, it turns out. In such a context, what will philosophies of science that embrace and theorize such integrations look like?

## What Should Western Priorities Be?

How can those of us working in Western contexts, and often educated in Western traditions, make ourselves fit to engage in democratic negotiations with peoples around the globe who have borne a disproportionate share of the costs and received relatively few of the benefits of Western scientific and technological agendas? As readings in the preceding parts have indicated, we need more realistic understandings of the strengths and limitations of both Western and the many non-Western science and technology traditions in history. We need to develop philosophies of science and technology that recognize that modern Western sciences are just one tradition (or, rather, one loosely connected set of traditions) among many viable ones, albeit much too powerful and self-ignorant ones today. What should we save, and what should we reject, of the Western legacies? While we must learn how not to romanticize them, as the dominant exceptionalist and triumphalist accounts do, demonizing them is an equally unfruitful and inappropriate strategy. We cannot just walk away, leaving them in the hands of those who are happy to benefit from the egregious ways that they often function around the globe today. Moreover, these Western sciences and their philosophies have great strengths that need to be transformed to meet the kinds of democratic goals articulated by authors here and in earlier parts of the book.

Many of the chapters in this collection explicitly or implicitly propose changes in the powerful regulative ideals of Northern scientific research— such as objectivity, rationality, empirical reliability, comprehensiveness, and others. A number of feminist philosophers of science, sociologists of knowledge, and political philosophers have focused directly on transforming those ideals to increase their competence and effectiveness (Haraway 1991; Harding 1986; Keller 1984; Longino 1990). One such approach is standpoint theory, developed as a research methodology and epistemology in the 1970s and 1980s, as discussed in the book's introduction. Yet many researchers and scholars from around the world implicitly use a standpoint logic when they insist on thinking about any situation, social or natural, from the lives of the most economically, politically, and so-cially vulnerable peoples. The "strong objectivity" project discussed in the book's introduction has been one contribution to a more adequate phi-losophy of science that has emerged from most (perhaps all) grassroots science and technology activist projects.

Daniel Sarewitz, David Hess, and Karin Bäckstrand approach from three different directions strategies for transforming the epistemologies, policies, and practices of Northern sciences. Sarewitz focuses on needed changes in U.S. government policies. Sciences search for certainty, pre-diction, and control of nature, yet both democracy and nature itself flourish only without such constraints. Sarewitz proposes some policy changes. Hess explores different ways in which scientists have commit-ted their research to social justice priorities. Bäckstrand examines how ordinary citizens have organized and effectively participated in the pro-duction of scientific knowledge by convincing scientists to change the priorities and practices of research and, sometimes, undertaking part of such research themselves. Hess and Bäckstrand provide many concrete examples of how scientific and technological practices are already under-going social justice transformations. Clearly science education must also change to prepare citizens and experts for working in these transformed environments.

## References

Amin, Samir. 1985. *Delinking*. London: Zed.

Anderson, Warwick, and Vincanne Adams. 2007. "Pramoedya's Chickens: Post-colonial Studies of Technoscience." In *The Handbook of Science and Technol-ogy Studies*, 3rd ed., ed. Edward J. Hackett et al., 181–204. Cambridge: MIT Press.

Haraway, Donna. 1991. "Situated Knowledges: The Science Question in Fem-inism and the Privilege of Partial Perspectives." In *Simians, Cyborgs, and Women*, 183–202. New York: Routledge.

Harding, Sandra. 1986. *The Science Question in Feminism*. Ithaca: Cornell Uni-versity Press.

———. 1991. *Whose Science? Whose Knowledge? Thinking from Women's Lives*. Ithaca: Cornell University Press.

———, ed. 2004. *The Feminist Standpoint Theory Reader: Intellectual and Politi-cal Controversies*. New York: Routledge.

Keller, Evelyn Fox. 1984. *Reflections on Gender and Science*. New Haven: Yale University Press.

Longino, Helen. 1990. *Science as Social Knowledge*. Princeton: Princeton Uni-versity Press.

Nandy, Ashis. 1990. "Science as a Reason of State." In *Science, Hegemony, and Violence: A Requiem for Modernity*, ed. Ashis Nandy, 1–23. Delhi: Oxford.

Spivak, Gayatri. 1987. "The Postmodern Condition." In *In Other Worlds*. New York: Methuen.

Third World Network. 1993. "Modern Science in Crisis: A Third World Response." In *The Racial Economy of Science*, ed. Sandra Harding, 484–518. Bloomington: Indiana University Press.

Watson-Verran, Helen, and David Turnbull. 1995. "Science and Other Indigenous Knowledge Systems." In *Handbook of Science and Technology Studies*, ed. Sheila Jasanoff et al., 115–39. Thousand Oaks, Calif.: Sage.

ZIAUDDIN SARDAR

# 20. Islamic Science

## *The Contemporary Debate*

THE DEBATE ON THE MEANING, nature, and characteristics of a contemporary Islamic science first emerged during the late seventies. The rise of OPEC, the Iranian revolution, and a growing consciousness in Muslim societies of their cultural identity led many scientists and academics as well as institutions to emphasize the distinctive scientific heritage of Islam. The reflections on the history of Islamic science generated a question of contemporary relevance: how could modern Muslim societies rediscover the spirit of Islamic science as it was practiced and developed in history? A number of international seminars and conferences—most notably the series of seminars on "Science and Technology in Islam and the West: A Synthesis" held under the auspices of the International Federation of Institutes of Advanced Studies in Stockholm in 1981 and Granada in 1982; the "International Conference on Science in Islamic Polity—Its Past, Present and Future" backed by the Organization of Islamic Conference and held in Islamabad, Pakistan in 1983; and "The Quest for a New Science" Conference organized by the Muslim Association for the Advancement of Science (MAAS), in Aligarh, India in 1984—were held to explore the question. The conferences provide a launchpad for the debate and revealed a great deal of confusion around the whole notion of Islamic science and its meaning and relevance to contemporary times.

During the last decade, as the debate spread to all parts of the Muslim world, a number of different and distinct approaches to Islamic science have come to the fore. A considerable amount of literature, ranging from the sublime to the mediocre, has been produced. So, what we now have is an embryonic, diffused, sociology-of-knowledge type of discipline which roughly defines a vigorously contested territory. The proponents of various schools of thought, each deeply entrenched in its position, regularly publish in the discipline's core journal: MAAS *Journal of Islamic Science.*

The discussions have centered around two basic questions: what is Islamic science? Also, how does it differ from the practice of conventional science? Five points of view can be distinguished on these questions. The first is what we may call western-type *scientific fundamentalism*: it totally rejects the whole idea of Islamic science. The second position represents the other end of the spectrum and takes us toward deep subjectivity: *mystical fundamentalism*. The third and fourth viewpoints maintain that

Islam has a great deal to say about science, but are divided on where exactly Islam enters the picture. The third position is mainly concerned in reading the verses of the *Qur'ān* in the theories, advances, and discoveries of modern science. This tendency has been labeled Bucaillism, after the French surgeon who initiated the movement. The fourth, *science in Islamic polity*, school is the official, establishment position in that it is backed and supported by the Islamic Secretariat of the Organization of Islamic Conference, the international body that brings the Muslim world together on a political platform. The fifth viewpoint argues that Islamic science is something quite different from science as it is practiced today. It is attributed largely to the *Ijmalis*, a group of independent scholars and thinkers who have championed a future oriented critique of contemporary Muslim thought; and to the *Aligarh* school, which has evolved around the Center for Studies on Science in Aligarh, India.

## Scientific Fundamentalism

Those who reject the notion of Islamic science argue from the conventional, positivist perspective on science. Science, it is argued, is neither Western nor Eastern, Christian nor Islamic. It is neutral, value-free, and universal. While advances in science may raise ethical issues, scientific knowledge itself has nothing to do with values. At best, the proponents of this position argue, value judgment can be applied in giving emphasis and pursuing a particular area of scientific research. However, basic scientific research in all areas must be pursued for its own sake; religion should be kept firmly away from science.

Initially, this approach, which ignores recent advances in the philosophy of science and sociology of knowledge, was the dominant view among rank and file Muslim scientists. However, recent debates on eugenics, efforts in theoretical physics to produce a Theory of Everything that can be proudly displayed on a T-shirt, and the emergence of complexity, have seriously undermined the position of scientific fundamentalists. The positivist view of science is being increasingly questioned in the Muslim world.

## Mystical Fundamentalism

In this perspective, Islamic science becomes the study of the nature of things in an ontological sense. The material universe is studied as an integral and subordinate part of the higher levels of existence, consciousness, and modes of knowing. Thus we are talking about science not as a problem solving enterprise, but more as a mystical quest for understanding the

Absolute. In this universe, conjecture and hypothesis have no real place; all inquiry must be subordinate to the mystical experience. This school of thought has a very specific position on Islamic science in history. All science in the Muslim civilization was "sacred science," product of a particular mystical tradition—namely the tradition of gnosis, stripped of its sectarian connotations and going back to the Greek neoplatonists. "Traditional science" does not necessarily mean science as it has existed in Muslim tradition and history, but products produced within the tradition of Islamic mysticism or Sufism. Traditional science is *science sacra*, the Science of Ultimate Reality, as taught by Sufi masters and mystics of other traditions. The goal of Islamic science today, they argue, is to rediscover the classical Islamic traditions and their sacred nature. This perspective is represented by a small and influential group.

## Bucaillism

This is a combination of religious and scientific fundamentalism. Bucaillists try to legitimize modern science by equating it with the *Qur'an* or to prove the divine origins of the *Qur'an* by showing that it contains scientifically valid facts. Bucaillism grew out of *The Bible, the Qur'an and Science* by Maurice Bucaille published in 1976. Bucaille, a French surgeon, examines the holy scriptures in the light of modern science to discover what they have to say about astronomy, the earth, and the animal and vegetable kingdoms. He finds that the Bible does not meet the stringent criteria of modern knowledge. The *Qur'an*, on the other hand, does not contain a single proposition at variance with the most firmly established modern knowledge, nor does it contain any of the ideas current at the time on the subjects it describes. Furthermore, the *Qur'an* contains a large number of facts which were not discovered until modern times. The book, translated into almost every Muslim language from the original French, has spouted a whole genre of literature looking at the scientific content of the *Qur'an*. Subjects ranging from relativity, quantum mechanics, and the big bang theory to the entire field of embryology and much of modern geology have been discovered in the *Qur'an*. Conversely, experiments have been devised to discover what is mentioned in the *Qur'an* but not known to science—for example, the program to harness the energy of the jinn! Bucaillism takes the reverence of science to a new level: Bucaillists do not just accept all science as Good and True, but attack anyone who shows a critical or skeptical attitude toward science and defend their own faith as "scientific," "objective," and "rational." This is the most popular version of Islamic science.

## Science in Islamic Polity

In this perspective, science is seen in similar terms to those of the Western paradigm as neutral and universal. But, it is argued, we can approach science with a secular or an Islamic attitude. The Islamic approach to science is to recognize the limitations of human reason and acknowledge that all knowledge comes from God. Within an Islamic polity—that is, an idealized "Islamic state"—the principles and injunctions of Islam which are the basis of the state would automatically guide science in the direction of Islamic values. The individual Muslim scientists would also bring their own values to bear on their work. The "Statement on Scientific Knowledge seen from Islamic Perspective," issued after the first International Conference on Science in Islamic Polity in 1983, states that science is one way humanity seeks "to serve the Supreme Being by studying, knowing, preserving and beautifying His creation." The Islamic framework seeks a "unifying perspective, combining the pursuit of science and the pursuit of virtue in one and the same individual." The "Islamabad Declaration" also called for "the creation of the Islamic science and technology system" by the end of the century. However, the emphasis on Islamic values in this perspective has remained largely at the level of rhetoric. Much of the work done at the national and international level within the framework of the "Islamic Conference Standing Committee on Scientific and Technological Cooperation" (COMSTECH) has been very conventional and concerned largely with nuclear physics, biotechnology, and electronics.

## The Ijmali Position and the Aligarh School

Both the Ijmalis and the Aligarh school argue that while Western science claims the norm of neutrality for itself, it is in fact a value-laden and culturally biased enterprise. The Ijmalis emphasize the "repulsive facade" of the metaphysical trappings of Western science, the arrogance and violence inherent in its methodology, and the ideology of domination and control which has become its hallmark. These things are inherent both in the assumptions of Western science as well as its methodology. Thus attempts to rediscover Islamic science must begin by a rejection of both the axioms about nature, universe, time, and humanity as well as the goals and direction of Western science and the methodology which has made meaningless reductionism, objectification of nature, and torture of animals its basic approach. But science in this framework is not an attempt to reinvent the wheel; it amounts to a careful delineation of norms and values within which scientific research and activity is undertaken. At the Stockholm Seminar in 1981, Muslim scientists identified a set of fundamental concepts of Islam which should shape the science policies and scientific activity of Muslim societies. The concepts generate the basic values of Islamic culture and form a parameter within which an ideal Islamic society progresses. There are ten such concepts, four standing alone and three opposing pairs: *tawhīd* (unity), *khalīfa* (trusteeship), *ʿibādat* (worship), *ʿilm* (knowledge), *halal* (praiseworthy) and *haram* (blameworthy), *adl* (social justice) and *zulm* (tyranny) and *istislah* (public interest) and *dhiya* (waste). When translated into values, this system of concepts embraces the nature of scientific inquiry in its totality: it integrates facts and values and institutionalizes a system of knowing that is based on accountability and social responsibility. How do these values shape scientific and technological activity? Usually, the concept of *tawhīd* is translated as unity of God. It becomes an all-embracing value when this unity is asserted in the unity of humanity, unity of person and nature, and the unity of knowledge and values. From *tawhīd* emerges the concept of *khalīfa*: that mortals are not independent of God but are responsible and accountable to God for their scientific and technological activities. The trusteeship implies that "man" has no exclusive right to anything and that he is responsible for maintaining and preserving the integrity of the abode of his terrestrial journey. But just because knowledge cannot be sought for the outright exploitation of nature, one is not reduced to being a passive observer. On the contrary, contemplation (*ʿibādat*) is an obligation, for it leads to an awareness of *tawhīd* and *khalīfa*, and it is this contemplation that serves as an integrating factor for scientific activity and a system of Islamic values. *ʿIbādat*, or the contemplation of the unity of God, has many manifestations, of which the pursuit of knowledge is the major one. If scientific enterprise is an act of contemplation, a form of worship, it goes without saying that it cannot involve any acts of violence toward nature or the creation nor, indeed, could it lead to waste (*dhiya*), any form of violence, oppression, or tyranny (*zulm*) or be pursued for unworthy goals (*haram*); it could only be based on praiseworthy goals (*halal*) on behalf of public good (*istislah*) and overall promotion of social, economic, and cultural justice (*adl*). Such a framework, argue the Ijmalis, propelled Islamic science in history toward its zenith without restricting freedom of inquiry or producing adverse effects on society. When scientific activity was guided by the conceptual matrix of Islam, it generated a unique blend of ethics and knowledge. It is this blend—which produces a distinctive philosophy and methodology of science—that distinguishes Islamic science from other scientific endeavors. Rediscovering a contemporary Islamic science, argue the Ijmalis, requires using the conceptual framework to shape science policies, develop methodologies, and identify and prioritize areas for research and development.

While accepting the position of the Ijmalis, the Aligarh School has added a number of other concepts to their framework of inquiry into Islamic science. However, both schools maintain that the practice of science within the conceptual framework of Islam involves a change in methods and direction from the dominant style and practice of science. The difference in emphasis and priorities of Islamic science, together with its assumptions and methods, generates a scientific enterprise that is radically different from Western science. Imagine a biology without vivisection or animal experimentation, or physics based on synthesis rather than reduction. The Aligarh school focuses on such methodological issues in an attempt to bring Islamic values to the level of the laboratory. The Ijmalis, on the other hand, are interested in developing a futuristic framework within which science policies and research and development activities in the Muslim World could be shifted gradually toward a generally accepted, and consensus oriented, framework of Islamic science. The two approaches are complementary and attempt to produce a viable alternative while recognizing the complexity of the issues and the difficulties involved in solving the problems generated by Western science.

## Outlook

Given the fragmentary nature of the debate on Islamic science, little of pragmatic and practical value has emerged so far. This debate, it appears, has largely followed the same process as the attempts to develop a "science for the people" in the early 1970s. The rhetoric has become more and more extreme while the practical dimension has, on the whole, been ignored. However, unlike the "science for the people" movement, the Islamic science debate has produced a theoretical framework which is evolving toward a consensus. While explorations of the theoretical framework for Islamic science will continue, it will be the pragmatic policy and methodological work that will determine and shape its future.

## Further Reading

Ahmad, Rais, and S. Naseem Ahmad. *Quest for A New Science*. Aligarh: Centre for Studies on Science. 1984.

Al-Attas, Syed Muhammad Naquib. "Islam and Philosophy of Science." MAAS *Journal of Islamic Science* 6 (1): 59–78. 1990.

Anees, Munawar Ahmad. "What Islamic Science Is Not." MAAS *Journal of Islamic Science* 2 (1): 9–20. 1986.

———. *Islam and Biological Futures*. London: Mansell. 1989.

Anees, Munawar Ahmad, and Merryl Wyn Davies. "Islamic Science: Current, Thinking and Future Directions." In *The Revenge of Athena: Science, Exploitation and the Third World*. Ed. Ziauddin Sardar. London: Mansell. 1988. 249–260.

Bakr, Osman. *Taweed and Science*. Kuala Lumpur: Secretariat for Islamic Philosophy and Science. 1991.

Bucaille, Maurice. *The Bible, the Qur'ān and Science*. Paris: Seghers. 1976.

Bull, Nasim. *Science and Muslim Societies*. London: Grey Seal. 1991.

Hoodbhoy, Pervez. *Islam and Science*. London: Zed Books. 1991.

Jamison, Andrew. "Western Science in Perspective and the Search for Alternatives." In *The Uncertain Quest: Science, Technology and Development*. Ed. Jean-Jacques Saloman et al. Tokyo: United Nations Press. 1994. 131–67.

Kirmani, Zaki. "Islamic Science: Moving Towards a New Paradigm." In *An Early Crescent: The Future of Knowledge and the Environment in Islam*. Ed. Ziauddin Sardar. London: Mansell. 1989. 140–62.

———. "An Outline of an Islamic Framework for a Contemporary Science." *MAAS Journal of Islamic Science* 8 (2): 55–76. 1992.

Manzoor, S. P. "The Unthought of Islamic Science." *MAAS Journal of Islamic Science* 5 (2): 49–64. 1989.

Nasr, Seyyed Hossein. *The Need for A Sacred Science*. Richmond, Surrey: Curzon Press. 1993.

Ravetz, J. R. "Prospects for an Islamic Science." *Futures* 23 (3): 262–72. 1991.

Salam, Abdus. "Islam and Science." *MAAS Journal of Islamic Science* 2 (1): 21–46. 1986.

Sardar, Ziauddin. "A Revival for Islam, A Boost for Science." *Nature* 282: 354–7. 1979.

———. "Can Science Come Back to Islam?" *New Scientist* 88: 212–16. 1980.

Sardar, Ziauddin, ed. *The Touch of Midas: Science, Values and the Environment in Islam and the West*. Manchester: Manchester University Press. 1982.

Sardar, Ziauddin. *Explorations in Islamic Science*. London: Mansell. 1989.

SUSANTHA GOONATILAKE

## 21. Mining Civilizational Knowledge

PRESCRIBING PROGRAMS FOR SCIENCE can lead to misperceptions of one's position. It is perhaps a hazardous task. So *Toward a Global Science* is both a disclaimer of positions that may be attributed to me as well as a statement of what I see as possibilities in expanding scientific knowledge.

First, to state my position clearer, I hope I am allowed a biographical note. Before I started a second career in the social sciences, I had practiced in the physical sciences. Trained as an engineer in my own country (in a system cloned from the British) as well as in Britain and Germany, I have designed, constructed, operated, and repaired or helped perform projects in electrical engineering and the cement industry. In the case of a power outage in my Third World country of Sri Lanka, it would be my duty—with consumers screaming curses—to get the system going again or, if a factory burned down, to get it functioning in the shortest time. So I am fully wedded to science-based technologies (and their constituent sciences) that work. If something does not work, I should know why, so that I can deliver a satisfactory product or service. And if something goes wrong due to my ignorance, I will be held accountable, ultimately in a court of law: if I were a civil engineer and designed a building that collapsed and killed people, or if I were a medical practitioner whose patient died due to my malpractice, I would be again held accountable.

As this aside indicates, modern science's applications work, and their efficacy is held to be valid in the realm not only of the "laws" of nature, but also of human law. Modern science and many of its applied fields, like engineering and medicine, are based on testable laws and are expected to be governed rigorously. But as a student of the sociology of science and technology, I also know that sciences came into their present form through a set of particular social pressures.

Yet science, whatever its social, political, psychological, or philosophical roots, is ultimately "that which works." To simplify a bit, it is a black box into which one puts a question—and out comes a reasonable and testable answer. Whatever the trajectory of the present delivery system of science, this is the ultimate test. Consequently, all knowledge systems cannot be equivalent. I do not believe that a forest dweller's knowledge of laws of motion comes anywhere near that of a Galileo, Copernicus, Kepler, or Newton. To hold such a position is, at the most, romantic ignorance. But I do believe that the forest dweller has a deep knowledge of the forest, its flora, and fauna. I believe, with many anthropologists who have studied the problem, that the intellectualism that drives the forest dweller is in an ultimate sense not different from that of a Newton. He or she happens to have faced a different set of problems, a different set of historical givens, and come to different positions. And I also believe that if Newton were to be magically transported to the forest, he would learn much systematic, testable knowledge of plants from the local groups, and he would have little to contribute on the subject in return. On the other hand, if we were to transport a Linnaeus to the forest, there would indeed arise some enlivened talk on categorizing plants.

But if we shift these hypothetical meetings to civilizational entities, where formal systems for knowledge gathering and transmission exist, there would be deeper discussions. If Leibniz, Newton's contemporary and rival, were to be transported to twelfth-century China, he would find much in common with his concepts of monads and point-moments, as Joseph Needham has documented. Similarly, if Leibniz were to be transported to fifth-century Sri Lanka, where atomic theories of time were elaborated in great detail, he would again have invigorating discussions on the nature of time. Or, returning to more modern times, Mach, the philosopher who influenced Einstein the most, would recognize South Asian positions on epistemology—as indeed he discovered on his own. Cognitive scientists, evolutionists, or computer specialists would find familiar ideas in the South Asian scene, as some have indeed already found out.

In a recent public statement, U.S. scientists warn of a creeping relativism in dealing with sciences whereby all knowledge systems are held to be equivalent; the statement also complains of a parallel return to obscurantism through multicultural courses and postmodernism.[1] Being a cosmopolitan person, speaking several languages and having lived, studied, and worked in different cultural milieus, I am fully wedded to multiculturalism. But there are shades of differences in the use of the word "multicultural," especially when applied to the sciences. Let me digress.

I have visited, say, Athens, Angkor Wat, and the Museum of Modern Art in New York. I found the Greek sculpture boring in its naturalness, while works of art at the other two locales struck deep responsive chords. I found especially congenial some of the West African–influenced pieces by a Picasso or a Braque at MOMA. Later, in West Africa itself, I saw some of the types of masks that influenced Cubism. But this does not mean that given a choice of mathematics, I would select a West African system (which writers like Zaslavasky have documented) instead of the Euclidean system to do my scientific problems. In another setting, if a sophisticated non-Euclidean geometry were to emerge from these West African ideas, and, if it were consonant with the scientific task at hand, I would choose it. So multiculturalism in science and multiculturalism in the humanities are two different things. In the case of sciences, every-day reality is always the reference point: it should work. Art does not have to "work"; it does not solve problems about nature, except in an indirect, allusory sort of way.

Further, my reading of postmodernism is not just as a system (or more appropriately a nonsystem) of thought. To me, postmodernism is a re-flection of the zeitgeist of today's Western intellectual life, which is seem-ingly exhausted. The initial thrust of the Enlightenment has lost both its vigor and its intellectual certainty. Current problems about the limits of tinkering with the environment have laid to rest some of the simplistic readings of Bacon's writings on science as the torturer of nature. Current epistemological problems in several fields—including the seeming foun-tainhead of them all, physics—have questioned the Cartesian dichotomy of subject and object. And the project of mathematizing the "true knowl-edge" of science and completing the rational project at a full foundational level has also collapsed, because mathematics, after Gödel, has lost its earlier assumed certainty.

So the Enlightenment and modern project is floundering, not because of postmodernist relativism, but on the terms set by the project itself. As for postmodernism, I find the writings insightful, but at times the authors are boring and trivial, seeming to waste hundreds of pages to prove some rather simple points. If more serious stuff in that philosophical genre is needed, required reading would be, among others, Nagarjuna (first cen-tury), Aryadeva (first century) and Bhartuhari (eighth century)—three South Asians. Compared to some of the trivialities of postmodernism, here lies real meat in the same broad genre.

So we have to read a different message into these postmodern times of the Western psyche: the modern agenda has run out of steam, and postmodernism is both a symptom and a reflection of this exhaustion. It is not an agenda to replace modernism, which would be a contradiction in terms. The new agenda, instead, has to come from the remnants of the earlier certainties that still have validity and the new certainties that can come from other cultures, including other civilizational spheres.

In addition, on the threshold of the twenty-first century, we are faced with a set of new and different challenges in the scientific sphere. Science is not what it seemed at the beginning of its journey in Newton's time. Since then—taking into consideration the number of scientists working and the number of scientific journals or other fora for the dissemina-tion of discoveries—science has been growing exponentially, doubling its population of practitioners and output every few years.[2] If science pro-ceeds at this pace, scientists would soon outnumber the rest of the human population, and the weight of all the scientific papers ever published would be greater than that of the earth.[3] The result is that in terms of present practice, certain quantitative growth limits will be reached in science within the next generation or so. These quantitative changes will require qualitative shifts in the nature of science. Twenty-first-century science is not going to be seventeenth-century science.

Part of the solution to this quantitative problem is to let the machine-computer sphere take over some of the work now done in the human sphere. This is part of a trend in human problem-solving, which has always sought shortcuts, replacing once cumbersome procedures with simpler, more powerful ones. Thus there were changes in scientific productivity from the Greco-Roman system of calculation, to the place-value arithmetic of the Renaissance, through logarithms, through the slide rule and to the computer. All the calculations that Newton and others of his time took years to do by hand are today done at the touch of a button by my teenaged nephew's astronomy program: almost instantaneously, he can see the conjunctions and outcomes of the planetary system for eons to come, from any vantage point in space or time —by means of a software that in a developed country today costs the same as a meal in a restaurant. Or to take another example, at the click of a button, a whole set of individual equations and problems—say, the floor design of a multi-storied building, or expected results from a particle accelerator—can be produced. In a similar fashion, the use of automation in laboratories is extending experimentation to the machine sphere; Galileo is no longer dropping objects from the tower of Pisa. Library research is also changing, with a tremendous increase in productivity because of the availability of extensive databases.

In the eighteenth century, Diderot and the Encyclopedists had the then farfetched dream of recording all knowledge.[4] Speaking of his encyclopedia, Denis Diderot wished:

> to collect all the knowledge scattered over the face of the earth, to present its general outlines and structure to the men with whom we live, and to transmit this to those who will come after us, so that the work of the past centuries may be useful to the following centuries, that our children, by becoming more educated, may at the same time become more virtuous and happier, and that we may not die without having deserved well of the human race.[5]

Today, the technology potentially exists that would allow almost total access to the entire formal written information system. Thus the dreams of Diderot, according to information specialists, may now be on the verge of realization, as far as Western written cultural knowledge is concerned.[6] So, this is one other instance where the Enlightenment Project is running out of its agenda.

Because of this knowledge explosion, the means of accessing and comprehending a mass of data have become central to the scientific enterprise. This automation has already been extended to search for theory. In fact, a decade or so ago, a program interestingly called "Bacon" after the Scientific Revolution's theorist of knowledge, "rediscovered" several laws in physics, such as Ohm's Law and Snell's Law.[7] Current experimental techniques like those involved in the Human Genome Project and the Hubble Telescope are spewing out so much data that automated processing at a higher level, using artificial intelligence techniques that mimic some aspects of human problem-solving, have become essential.[8] Such programs see patterns in the data, analyze them, and come to conclusions.

Serious research is also being conducted on what have been called "knowbots" and "discovery machines," which will swim in a sea of data and come out with discoveries.[9] But such artifacts do not exist in a vacuum. They are distillations, in hardware and software, of theories and debates on the cognitive process. Here discussions of psychology, the mind, perception, knowledge, and so on become of indirect—yet sometimes central—concern to the scientific enterprise.

*Toward a Global Science* does not take the position, for instance, that the ancient Egyptians knew of modern batteries, a charge leveled in the U.S. scientists' statement (cited above) as an example of the untruths some multicultural courses are teaching. Or for that matter, I do not hold that ancient Sri Lankans knew how to fly simply because the Sanskrit poetic epic *Ramayana* mentions that its villain Ravana, the king of Lanka, used an alleged flying contraption. In *Toward a Global Science*, I have rested on formally accepted fact in the essentially Western realm of discourse. Also, although science has many incongruities in its growth, as Feyerabend has pointed out, I do not take his position that "anything goes" in science and that there is no method.[10] There are *methods*, and different sciences use different approaches.

In that sense, all systems of knowledge are not equivalent. There are also hierarchies of explanations possible for many phenomena. Some explanations are more parsimonious than others, explaining many phenomena with a smaller number of variables; others predict better; and in still others, the results of scientific enquiry into a phenomenon can be replicated. There are also phenomena that cannot, by definition, be repeated, like the Big Bang or the beginning of life on earth. One can only simulate such events.

There is also an incommensurableness that extends to different disciplinary approaches. Thus a physicist would use Newton's equations for a calculation of a planetary orbit, and then he could well run the system in reverse, as time in the Newtonian world is reversible. At the same time, that physicist, in designing an efficient heat engine, would work with the field of thermodynamics, which has an arrow of time. The research program of molecular biologists is not the same as that of systems theorists, or for that matter that of biological evolutionists, or botanical taxonomists. Their views of their fields, definitions of problems, and approaches to potential solutions are all different.

So a variety of epistemological and even ontological positions (such as appear in quantum physics) exist in today's sciences. Hence the choices for potential interventions in the enlarged sciences of the twenty-first century are much wider than those normally understood as scientific methodology. As the shift to increasing automation of knowledge gathering must necessarily gather momentum, so these other deeper dimensions in science become important. With automatic laboratories, knowbots, and discovery machines, the seventeenth-century Scientific Revolution's agenda can be partly left to machines. This situation begs the need to broaden scientific horizons: the new sciences must incorporate into the emerging mixture results in ontology, epistemology, logic—results already arrived at in disciplines outside the boundaries of traditional Western scientific endeavor. Thus the enlarged science project of the twenty-first century must incorporate other cultural elements.

These cultural elements include those values that consciously or unconsciously seep into the scientific project. The decision to emphasize the Manhattan Project and the intense interest in particle physics was such a value decision, fueled by the Second World War and the Cold War. Now other values prevail, and so the big particle accelerator is dropped, blocking, for the time being, any new discoveries to be revealed through its massive particle-smashing power. Questions revolving around values also dog such subjects as biotechnology and biomedicine.

Often, values in science are incorporated unconsciously. Can we incorporate them consciously? Yes, if our lives or other important matters are at stake. Thus the perception of an endangered environment has stopped particular scientific and technological developments. And the values of Green movements are reflected in the globally binding Rio Declaration. These restraining factors have already explicitly guided human knowledge.

But how far can we consciously bring in elements from outside traditional Western science? Can one have a Hindu science, a Buddhist science, a Christian science, a Marxist science, or for that matter an Islamic science? With the recognition of the social basis of science, especially its capitalist underpinnings, there were attempts to develop a socialist science in the former Soviet Union and at times in China. Some of these efforts led to the crudities of Lysenkoism and the anti-Intellectualism of the Cultural Revolution. These failures prove that completely totalizing changes are no longer possible in science. Science is like the biological tree of evolution that already has its own rigidities. Changing it completely is an impossible task, as impossible as starting a new biological system, replacing the 4,000 million years old existing one. The time for total revision has passed; the existing system has enough entropic and other rigidities. The existing science may be capitalistic, Eurocentric, patriarchal, and/or class-based. But to grow a new one wholesale is no longer possible. One can only graft elements to the existing tree, and such grafts only take if there is some compatibility. One could make piecemeal adjustments, however, and some of these could have a major impact if they are grounded in deep epistemological or ontological changes. But all such changes have to be governed within the paradigmatic and other needs of a given discipline. So there are limits to, say, an Islamic or any other religion-based science.

More importantly in this porous world, fundamentalist projects based only on a priori assumptions are doomed. Witness the failed attempts in the Soviet Union and in China. The outside has a tendency to come streaming in, upsetting the most antiseptic of cultural enclaves. Thus, a Khomeini could use a Sony tape recorder as a tool for disseminating his message (as he did); but behind every Sony and its tape there is a world of technical and scientific culture providing the scientists and technicians among Khomeini's flock with a different set of cultural messages. Khomeini's message becomes "only" a surface resting on a deeper structure of a given scientific culture.

In this sense, unlike in the remote past, searches for absolute fundamentalist sovereignty in scientific epistemology are doomed. Today, one cannot without contradictions "build socialism in one country" or a regime of pure Islam. Eastern Europe, China, and Cambodia all have in this sense imploded from their earlier searches for purity, because of the dynamics of the globalized system. The enemy is no longer across the border, it is within, it is part of oneself; we are now exposed to more messages than any individual or country can contain. . . .

I do not want to replace the Western Doctor of Philosophy with the witch doctor. I believe, however, that there are elements that the witch doctor—or in our more sophisticated cases, "civilizational knowledge carrier"—can contribute to the knowledge base of doctors in the different sciences. Some of these contributions may yield completely fresh and sophisticated approaches, as in the case of civilizational knowledge. Thus there may be strict limits attainable through a solely Mullah-driven science, but on the other hand there would be many elements in the great Islamic scientific traditions that could still be drawn upon. Very probably there are important nuggets that did not get translated to Latin from Arabic in the manuscripts in Cordoba and elsewhere, concepts still worth examining. Similarly, in the other great civilizational areas like East Asia, and lesser ones like the pre-Columbian Americas, there are mines of knowledge yet to be adequately explored.

If it sounds like I am accepting the "totalizing" hegemony of modern science, I am. I want to enlarge it if possible, not destroy it. I want to reach beyond the Enlightenment and the modern projects and some of their Eurocentric limitations. But the modern sciences, when taken individually, are not monolithic, ontologically and epistemologically totalizing projects. There are too many differences and even contradictions in the approaches of the different disciplines as to methodology, epistemology, and at times ontology. So science as a totalizing project is totalizing only to the extent that it is an organized skeptical attempt to gather valid knowledge. With that pursuit I am perfectly comfortable. I want only to increase the skepticism, to make it more valid, and to enlarge the catchment area.

## Notes

1. I. Brown, Malcolm W., "Scientists Deplore Flight from Reason," *New York Times*, June 6, 1995.

2. Price, D. J. de Solla, *Little Science, Big Science* (London, Macmillan, 1963).

3. Science Foundation Course Team, *Science and Society*, Science Foundation Course Units 33–34 (The Open University Press, 1971).

4. Dizardt, Wilson P., *The Coming Information Age* (New York and London, Longman, 1982).

5. Quoted in Horton, Allan, "Electronic Access to Information: Its Impact on Scholarship and Research," *Interdisciplinary Science Reviews*, 1983, Vol. 8. No. 1, 67.

6. Ibid.

7. Langley, P. M., "Rediscovering Physics with Bacon 3," *Proceedings of the International Joint Conference on Artificial Intelligence* 6, 1977, 505–7.

8. Waldrop, M. Mitchell, "Learning to Drink from a Fire Hose," *Science*, 11 (May 1990) 248:674–5.

9. Ibid.

10. Feyerabend, P., *Against Method: Outline of an Anarchist Theory of Knowledge* (London, New Left Books, 1975).

CATHERINE A. ODORA HOPPERS

## 22. Towards the Integration of Knowledge Systems

*Challenges to Thought and Practice*

### Indigenous Knowledge Systems: A Profound Challenge to Human Development in the Twenty-first Century

SOUTH AFRICA'S DRIVE for the development, promotion, and protection of indigenous knowledge systems (IKS) within which this initiative is located, comes at a time when major "winds of change" are blowing in the country. On the one hand, there are major transformation and democratisation processes being implemented under the new dispensation. Several macro-level policies provide frameworks for understanding the equity, empowerment and development thrusts in government policies—for example, the Reconstruction and Development Programme (RDP), the policy of Growth, Equity and Reconstruction (GEAR), the National System of Innovation, and the African Renaissance.

The latter, in particular, sets forth an agenda that combines identity reconstruction and innovation, human rights, sustainable development and democratisation in South Africa and throughout the African continent. The African Renaissance aims at building a deeper understanding of Africa, its languages, and its methods of development. It is a project that includes the rewriting of major tenets of history, both past and contemporary. IKSs thus posit tremendous challenges for the reconstruction and development of strategies in South Africa.

Several global imperatives also underpin the need for renewed attention to IKSs. To begin with, the world stands at a crossroad in search of new, human-centered visions of development in health, in preserving and conserving biodiversity, in human rights, and in the alleviation of poverty. All the agencies of the United Nations are seeking to promote paradigms of sustainable human development that build on knowledge resources that exist in communities. As a continent, Africa is seeking its own renaissance and seeking to establish the terms of its development. Despite the affluence that globalisation has brought to a small minority, we live in a world in which subjection, suffering, dispossession, and contempt for human dignity and the sanctity of life are at the centre of human existence.

Emotional dislocation, moral sickness, and individual helplessness remain ubiquitous features of our time. Moreover, for a great majority of the population of Africa, the loss of cultural reference points has culminated in the fundamental breakdown of African societies with dire consequences for the social and human development project as a whole. Finally, globalisation is threatening the appropriation of the collective knowledge of non-Western societies into proprietary knowledge for the profit of a few.

At the international level, key agencies of the United Nations, such as the World Health Organisation (WHO), the United Nations Environmental Programme (UNEP), the United Nations Development Programme (UNDP), and the United Nations Educational, Scientific and Cultural Organisation (UNESCO), have conventions or mandates of various kinds on the issue of diversity, indigenous and traditional knowledge, and of the rights of indigenous people worldwide. The issue of diversity, and especially biodiversity, and the role of indigenous communities in the protection and utilisation of the natural products around them, were elucidated during the 1992 United Nations Conference on Environment and Development (UNCED) held in Rio de Janeiro.

Also involved, playing an active though ethically unclear role, are the Trans-National Corporations (TNCs) with immense financial power, but with short-term profit motives keen to exploit genetic and other resources from the so-called Third World countries. The TNCs are protected by the agreements initiated at the Uruguay Round, which led to the formation of the World Trade Organisation (WTO). The WTO is the global trade "police" interested in eliminating all trade barriers worldwide. However, it is also the WTO that legislates the idea of intellectual property rights (IPRs) in a manner that assumes individual ownership (a Western phenomenon) as a universal phenomenon, and excludes the notion of "collectivity" that characterises ownership of IPRs in many societies in the Third World.

The WTO is attempting to establish new regimes of intellectual property rights at a time when Third World countries are not at all prepared for this. Many developing countries still need to invest in the development of skills to manage IPRs and leverage its influence for the benefit of the poor. As it stands, too many existing IPR systems are focused on private ownership, and are at odds with indigenous cultures that emphasise collective creation and ownership of knowledge. Present IPR systems encourage the appropriation of traditional knowledge for commercial use; moreover, without the fair sharing of benefits with the holders of knowledge. Current IPRs also violate indigenous cultural precepts by encouraging the commodification of such knowledge. There is, therefore, a serious need for innovation in the intellectual property regime itself (Mashelkar 2000).

At the level of science, a recent cornerstone cue for change emanates from the UNESCO *Declaration on Science and the Science Agenda Framework* adopted in Budapest, Hungary (1999). This declaration urged member countries to define a strategy to ensure that science responds better to society's needs and aspirations in the twenty-first century. It reiterated the need for political commitment to the scientific endeavour and, especially, to make science *more responsive* and *more inclusive*. Science should become more accountable, more communicative, and more dialogical. The vision of science in the twenty-first century should be a science that can appreciate interconnectedness and interdependence; a science that can decipher the meaning of words like "responsibility" and "ethics" in the use of scientific knowledge; a science that can comprehend the fact that science is a product of culture, or cultures, and that its diverse manifestations must be recognised; and a science that can be seen by all to be a shared asset.

This declaration called for a drastic change in the attitude, methods, and approach in the scientific field, and to the problems of development, especially in the social and human dimension. Science, the document emphasized, must be put to work for sustainable development, and must thus transform itself and become inclusive of women and other forms of knowledge in terms of its culture of admission and operation. But most of all, it must take a stand on issues affecting global development, such as pollution, depletion of natural resources, poverty, and the widening disparities in well-being.

## Institutional Development Imperatives

It is with this in mind that South Africa needs to develop a corpus of academics and scientists who can act as catalysts and agents of change. Within their ranks, the process of the gradual transformation of scientific ethos, ethics, and practice should emerge, while developing a strong and committed system of protocols for developing and protecting indigenous knowledge systems, biodiversity, and especially, the IPRS of local communities. It is crucial for institutions that engage with IKSS to develop an understanding of the relationship between science and the series of suppressed "Others" and the role of science in such suppression. Part of the legacy of colonialism, and the science that accompanied it, that still lingers in academic practice in general, is that non-Western societies and the knowledges that sustained them are taken as obsolete. In the rush towards modernity, we, "the newly modernized," have not wanted to give those on whom we have imposed the signifier of obsolescence a voice. In fact, as a group, victims are never part of scientific history or discourse.

There is thus a need to open up new moral and cognitive spaces within which constructive dialogue between people, and between knowledge systems, can occur (Visvanathan 1997).

The starting point for this is a realisation that new directions in the philosophy and sociology of science are emerging today, not from the academia, but from questions raised by grassroots movements. The grassroots movements represent dissenting organic academia and raise issues that the universities are reluctant to confront. It is the grassroots movements and their determination of new directions in the philosophy and sociology of science that have constantly demonstrated that knowledge is an intrinsic part of democratic politics. It is through such voices that we have come to understand, not only policy frameworks and the political economy of science, but also how the cosmology within which modern Western science was embedded could be a source of violence. It is by undertaking this work of reconnecting systems that questions of cosmology and political economy can be linked to democratic theory (Visvanathan 1997).

This initiative is intended, therefore, to facilitate the development of community-conscious scientists and researchers across the natural and social sciences on the one hand, and help reshape the content of curricula at universities in South Africa on the other. Central to the endeavour will be to work towards improving the lives of holders of knowledge in the rural areas who have been subjected to epistemological disenfranchisement by the combination of colonial and apartheid practices buttressed by the attitudes, ethos, and practices of the scientific community. This should lead to high-level capacity-building for critical analytical work, knowledge development, and strategic policy advice on crucial development issues in Africa and internationally.

Even more urgent is the fact that some universities, university faculties, and individual academics are beginning to question the eschewed knowledge production formats that undergird the curriculum of tertiary institutions. The relations between universities and communities are also coming under the spotlight, giving rise to the question of how universities perceive local communities. It is also quite apparent that the knowledge that has generally informed policy development in South Africa and most of the developing countries excludes IKS, thereby continuing to render people in rural areas in a deficit mode with acutely disempowering consequences.

Formal research institutions should, therefore, demonstrate their readiness for work to the advantage of the holders of knowledge in rural communities, and to facilitate their recognition. Where it is identified that particular communities are "resource-rich but economically poor," concerted research and development interventions grounded in an IKS perspective should help to rapidly highlight the situation at community level, and offer advice as to how government could and should act.

## Challenges to Research and Development Strategies and Perspectives

From an IKS perspective, research as we know it is intricately linked to European imperialism and colonialism. In many local communities, the word "research" conjures up silence and bad memories, and raises smiles that are knowing and distrustful. The ways in which scientific research is implicated in the worst excesses of colonialism remains a powerfully remembered history for many of the world's colonised people in whose domain indigenous knowledge systems repose (Smith 1999, 1). It is a history that still offends our deepest sense of our humanity. It galls indigenous communities that Western researchers and intellectuals can so glibly assume to know all that it is possible to know of local people and systems on the basis of shallow altruism or brief encounters with some individuals.

The Maori intellectual, Linda Smith, writes: "It appals us that the West can desire, extract and claim ownership of our ways of knowing, our imagery, the things we create and produce, and then simultaneously reject the people who created and developed those ideas, and seek to deny them further opportunities to be creators of their own culture . . . , (and) deny them the validity of their own knowledge . . ." (Smith 1999, 1).

This collective memory of imperialism has been perpetuated by the ways in which knowledge about indigenous people was collected, classified, and then represented to Western audiences, and then, through the eyes of the West represented back to those who have been colonised. It is this circularity that is still perpetuated by most formal research institutions in South Africa. It is a Western discourse about the "other" that is supported by institutions, vocabulary, scholarship, doctrines, and research methods. It is here, where formal scholarly pursuits of knowledge are intertwined with the anecdotal constructions of the "Other," that research becomes a significant battleground for the struggle between the interests and ways of knowing of the West, and the interests and ways of knowing of indigenous people and communities that, for so long, could only survive in the underground trenches of epistemological resistance.

In a decolonising framework, therefore, deconstruction has a larger intent. It consists of addressing social issues within a wider framework of self-determination and social justice. The simultaneous project is one of establishing new research protocols that are acceptable for those research had proscribed as its "objects," and excavating the subjugated knowledges with a view to making them available in real time for present and succeeding generations. Research is thus not some innocent or distant academic exercise, but is recognised as an activity with intent. This intent must be made overt and explicit to those whom we have turned into the rats in our scientific cages.

## Recognising Innovation and Creativity at Community Level

As Gupta (1999) emphasises, a higher consciousness about what exactly is at stake here must develop. At the material level, it is clear, for instance, that there is a gross asymmetry in the rights and responsibilities of those who produce knowledge, particularly in the informal sector, and those who go about valorising it in the formal sector. This brings to light the issue of the ethics of extraction and responsibility.

It is also strongly felt that the notion is not true that research and development by small-scale firms or individual community-based scientists cannot generate globally valuable intellectual property. At the same time, it is important to note that while no claims exist that all problems can be solved better by local communities, some problems that they solve in a creative manner, through their own genius, indicate that there are niches that mainstream science and technology institutions have either failed to fill, or have not even noticed. In that regard, there exists tremendous scope for complementarity between the knowledge systems (i.e., formal and informal) (Gupta 1999), and for the reciprocal valorisation of knowledge systems (Hountondji 1977).

It is also recognised that a major threat to the sustainability of natural resources is *the erosion of people's knowledge*, and the basic reason for this erosion is *the low value attached to it*. The erosion of people's knowledge associated with natural resources is under greater threat than the erosion of the natural resources themselves. Questions therefore need to be asked as to why the most disadvantaged people have to carry the heaviest burden of maintaining genetic diversity for future generations. Indigenous herbalists, veterinary experts, and pastoralists know a lot about the habitats and life cycles of plants and animals, and various other aspects of other resources. Yet efforts to build upon knowledge systems of people who have maintained their natural resources are, so far, quite inadequate.

Language is another problem in revitalising local communities. Writing in English, for example, has been recognised as valuable for connecting internationally, but there is an awareness that it alienates locally. The goal here is not only to feed back what (as scientists) we learn from local communities and individual innovators in their own language, but also to share with them what we learn from others (i.e., enhancing the public understanding of scientific developments).

A strong view, therefore, needs to be taken of the unprofessional conduct on the part of scientists, especially the tradition of making local communities anonymous in scientific practice, even when these communities have divulged key information regarding certain plants or drugs. This must be brought to a rapid halt, and a change in the ethics of scientific practice should gradually impact on the basic tenor of academic discourse on local knowledge (Society for Research and Initiatives for Sustainable Technologies and Institutions, n.d.), and on IKS as a whole.

A development imperative to emerge from these considerations is the issue of *value addition*. Value addition to indigenous knowledge will help local communities to co-exist with biodiversity resources by reducing primary extraction and generating long-term benefits. From this perspective, local communities are termed "knowledge-rich, but economically poor." Any dynamic knowledge system therefore has to evolve through the continuance of traditional knowledge and contemporary innovations, and this should be pursued by individuals as well as communities. The aim is to connect creative people engaged in generating local solutions that are *authentic* and *accountable*, thus facilitating people-to-people learning. As scientific institutions, we should be seriously concerned with the ethics of knowledge extraction, its documentation, dissemination, and its incorporation into real life products and services or policy framework. The search is, therefore, for a middle-way in the dialogue and linkage between IKS and formal knowledge systems.

At the same time, the discussion on biodiversity can only become authentic if we probe deep enough into the knowledge traditions of each part of the world to discover the roots of *sustainable ethics*. But this discovery requires preparing our minds for visions that collide with the dominant materialistic worldview. The existing epistemology relies excessively on the language and ethos of the elite, whose record of sharing their rent with providers of knowledge is not very honourable. When they plead for reform, they advocate action by everyone else in the world, ranging from local governments to world bodies, but their own conduct and practices remain untouched by the logic of their appeal. With this in mind, it is urgent that a code of conduct be developed (Society for Research and Initiatives for Sustainable Technologies and Institutions, n.d.).

The scientific community must, therefore, transform their knowledge legitimation and accreditation cultures in order to build linkages between excellence in formal scientific systems and innovations in informal knowledge systems, and thus create an inclusive knowledge network to link various stakeholders through applications of information technologies (The National Innovations Fund, India).

## IKS: Towards a Conceptual and Analytical Framework

Knowledge is a universal heritage and a universal resource. It is diverse and varied. The acquisition of Western knowledge has been and still is invaluable to all, but on its own, it has been incapable of responding adequately in the face of massive and intensifying disparities, untrammelled exploitation of pharmacological and other genetic resources, and rapid depletion of the earth's natural resources. For its part, IKSS represent both a national heritage and a national resource that should be protected, promoted, developed and, where appropriate, conserved. But it is also a resource that should be put at the service of the present and succeeding generations.

By way of a definition, the word *indigenous* refers to the root, something natural or innate (to). It is an integral part of culture. *Indigenous knowledge systems* refer to the combination of knowledge systems encompassing technology, social, economic and philosophical learning, or educational, legal and governance systems. It is knowledge relating to the technological, social institutional, scientifics, and developmental, including those used in the liberation struggles (Odora Hoppers and Makhale-Mahlangu 1998).

The idea of indigenous knowledge as espoused within this framework, for instance, is not just about woven baskets, handicraft for tourists or traditional dances per se. Rather, it is about excavating the technologies behind those practices and artefacts: the looms, textile, jewellery, and brass-work manufacture; exploring indigenous technological knowledge in agriculture, fishing, forest resource exploitation, atmospheric and climatological knowledge, and management techniques (Dah-Lokonon 1997), indigenous learning and knowledge transmission systems (Doussou 1997), architecture, medicine and pharmacology, and recasting the potentialities they represent in a context of democratic, equitable participation for community, national, and global development *in real time*.

## The Questions and Challenges to Practice

The issue of IKSS posits profound challenges to contemporary practice, a few of which are listed below.

The first relates to the *knowledge generation and legitimation processes*, such as the type of knowledge being generated in scientific institutions;

the type of research questions being asked, and the existing rules and regulations governing legitimation and accreditation of scientific knowledge.

The second relates to *social and economic survival of "resource-rich, but economically poor" local communities* (Gupta 1999). How can the study and validation of ıĸs assist directly in the economic and socio-cultural empowerment of the communities?

A third agenda that the integration of knowledge systems highlights, is the need to explore deeper the interface between epistemology, diversity, and democracy, to explore the potential for true exchange and what Hountondji refers to as the *"reciprocal valorisation among knowledge systems"* (Hountondji 1997, 13). Central to this is the need to establish knowledge as an intrinsic part of democratic politics (Visvanathan 1997, 2).

Fourthly, internal to the ıĸss, there is a separate need to engage in its critical evaluation and careful validation, while recognising its inner truths and coherence in order *to facilitate its active re-appropriation and authentication into current, living research work* (Hountondji 1997, 15).

A fifth point, one of great importance for many Africans, is that it is strongly felt that the time has come to subject to *direct interrogation, the historical, scientific and colonial discourses* behind the semantic shift that turned the illiterate from someone ignorant of the alphabet, to *an absolute ignorant*; pitting what is not written as thoughtless, as a weakness, and, at its limit, as primitivism (Hountondji 1997, 33–34), which has been central to the strategic disempowerment of African societies since the advent of colonialism.

Sixth, in realising the fundamental intolerance of modern science towards the legitimacy of folk or ethnic knowledges, coupled with our increasing inability to develop an ecologically coded society, engaging with ıĸs enables us to re-open crucial files that were summarily closed somewhere in the chaos and violence of colonialism (Visvanathan 1997, 38–40).

Colonialism remains a factor insofar as it provided the framework for the organised subjugation of the cultural, scientific, and economic life of many on the African continent and the Third World (Mugo 1998, 6). This subjugation extended in a spectrum from people's "way of seeing," their "way of being," their way of negotiating life processes in different environments, their survival techniques, to technologies for ecologically sensitive exploitation of natural resources. All these knowledges were, en masse, rendered irrelevant to their use as millions of people became transmogrified by the combined advent of modern science and colonialism, into an inverted mirror of Western identity—a mirror that belittled them and sent them to the back of the queue (Esteva 1992, 6, 7).

Seventh, IKSS enable us to *move the frontiers of discourse and understanding* in the sciences as a whole, and to *open new moral and cognitive spaces* within which constructive dialogue and engagement for sustainable development and collective emancipation can begin. In effect, it makes it possible for us to "clear spaces" in order to *enable new issues* in science development to be generated and fostered, and thus determine new directions for the philosophy and sociology, as well as political economy of the sciences (Visvanathan 1997, 7, 8).

Finally, IKSS enable us to re-establish science as the story of all animals, and not just of the lion; to develop a clearer sense of the ethical and judicial domain within which science works, and to begin to understand the political economy of "othering." More importantly, IKSS humanise our practice, and enable those silent witnesses of marginalisation (i.e., those regarded as *refractory to the scientific gaze*) to become part of an empowering process, and strengthen their capacity to take an active part in questioning the competence and ethics of the professional expert (Visvanathan 1997, 9–13), at the same time working to forge genuine partnerships and informed alliances for development.

IKSS are characterised by their embeddedness in the cultural web and history of a people, including their civilisation, and form the backbone of the social, economic, scientific, and technological identity of such people. It is these knowledges that are referred to as "indigenous knowledge." They consist of tangible and intangible aspects that, in contemporary contexts, can be identified as those that:

—have exchange value and that, with support, can be transformed into enterprises or industries;
—perpetuate social, cultural, scientific, philosophical, and technological knowledge, that can provide the basis for an integrated and inclusive knowledge framework for a country's development;
—represent major sociocultural institutions and organisational systems.

These knowledges are rich and varied, ranging from soil and plant taxonomy, cultural and genetic information, animal husbandry, medicine and pharmacology, ecology, climatology, zoology, music, arts, architecture, social welfare, governance, conflict management, and many others (Odora Hoppers and Makhale-Mahlangu 1998, 5). Their intrinsic efficiency and efficacy as tools for personal, societal, and global development must, therefore, be identified and accredited as necessary. It is this recovery of indigenous knowledges, and the systems intricately woven around them, that will enable the move towards a critical but resolute reappropriation of the practical and cognitive heritage of millions of people around the continent (Hountondji 1997, 35) and elsewhere in the world.

It is, in turn, the re-appropriation of this heritage that may provide new clues and directions as to the visions of human society, human relations, sustainable development, poverty reduction, and scientific development, all of which cannot be resolved using the existing ethos of the Western framework alone.

A focus on IPRS within IKSS will also enable the indigenous authorities and communities to publicly and legally lay claim to IPRS, and copyright to the wide range of artistic, pharmacological, and other products currently being extracted largely without recompense. The development of new protocols for benefit sharing, value addition, and new ethics of extraction, will further lead to the strategic revisiting of the adequacy of existing legal, educational, industrial, commercial, and other sectoral provisions currently under implementation, with a view to questioning the extent to which they are oriented to serving, promoting, developing, and protecting *all* sources of knowledge, and putting them to use for the benefit of all.

In methodological terms, engaging with IKSS implies sensitisation, empowerment, and restoration of holism and ethical practices, including spirituality, for individuals, system, and institutions. It means going beyond the appraisals of the work of individual scientists, beyond the output of particular research teams and the competitive acumen of individual research institutions, and reaching the point where it is possible to ask questions that can serve to re-centre Africa and the Third World (Hountondji 1997). . . .

## The Relations between Knowledge Systems

In seeking to move towards an IKS research agenda, the core strategy at the level of epistemology is to seek the best of both. Both the Western system and the IKSS represent national resources. The local contextual expertise and technologies that indigenous knowledge frames offer can complement some of the mechanical and technical precision capabilities of the Western knowledge systems to generate forms of creativity that benefit and empower everyone. But for this to happen, power must shift.

Accordingly, the initiative seeks at a deep level to contribute to the emancipation, development, integration, and protection of IKSS. A process of dialogue through research should be central to such a methodology, in order to clarify the role of IKSS within human development. The *emancipation* of IKSS is a necessary condition for development; *development* is crucial if they are to be integrated with other knowledge systems to achieve a holistic framework; *integration* is crucial if IKSS are to fulfil their potential for contributing to human and social development in a rapidly changing global context; and *protection* is a strategy of vigilance against exploitation by dominant world forces.

We also need to "clear space" in institutional and other policy arrangements for these diverse knowledges to exist and participate in unfolding modernisation processes. By bringing in both the developing and developed countries, the intention is to signal the seriousness of the commitment to the agenda for dialogue on the issue of the transformation of the knowledge-generating fields at the global level, knowing, uncomfortably so, that "having insisted on the authority of the Western 'tradition' throughout the colonial period in Africa, Europeans must join in and contribute to the relativisation of its legitimacy" (Professor Devisch, in personal communication with the author in 1999).

Within South Africa, continentally and internationally, there is a need to establish new research protocols and codes of conduct for scientific work, by initiating a critical reflection on the ethical questions surrounding research in the human and natural sciences, and to raise new research questions. In the years to come, such interventions should inform the overall framework of the National Innovations Fund and the National System of Innovations that have been established in South Africa.

We need to generate and present debates and analyses on the internal and external characters of both IKSS and Western knowledge systems, separately and in relation to one another with the objective of promoting strategies and terms under which their integration can be achieved. The contributions that Western knowledge has made to the development of science and technology are acknowledged. However, it is also recognised that, despite their internal fragmentation, Western knowledge systems have achieved world hegemony under subjugative colonialism and imperial relations. Conversely, indigenous knowledge systems have undergone mutations, have been subjugated, sometimes underdeveloped and exploited, but retain the potential for human and social development.

## Strategies for Achieving the Circular Empowerment Objective

These strategies will be achieved by positing public policy dialogue as dialogue between fields that problematises the relationships between knowledge, power, and human development. The basic perspective follows from Gramsci, Foucault, and Freire, and has been articulated in the African context by Odora Hoppers (1997) as both a methodology and strategy for ensuring informed participation and thus empowerment. This empowerment strategy involves at least three constituencies (be they at national or international levels): individuals and organisations in civil society; the scientific, especially the academic community; and policy makers.

The scientific community is meant to include individuals based at universities, the academia in general, and science institutions. In the South African context, this would include science councils. But this also includes autonomous, authoritative, and influential persons in the academic or intellectual world who are able to contribute global or African perspectives on IKSS. The role of this constituency will be to interrogate and explicate the epistemological foundations of knowledge systems, and the processes of knowledge generation that take place within these institutions. . . .

The notion of comparison in this initiative, therefore, seeks to bring perspectives, insights, and experiences from different contexts into a critical reflective dialogue on the tenets of thinking and practice, and the potential for something qualitatively better to emerge. The cases that are referred to in this volume are introduced to reflect diversity and plurality. Participants in the initiative take cognisance of the limitations and dangers posited by the reductionist project of Descartes, in which objects of study are arbitrarily isolated from their natural surroundings and relationship with fellows, *which in itself reflects a political choice aimed at controlling nature and the exclusion of other ways of knowing.* Sandra Harding has characterised this Cartesian tendency as the "contemporary alliance of perverse knowledge claims with the perversity of dominating power" (cited in Shiva 1989, 29).

The comparative aspect of this initiative would aim at examining the manner in which the exclusion of other traditions of knowledge by reductionist science is itself part of the problem at the ontological level (in that properties of other knowledges are *simply not taken note of*); at the epistemological level (in that other ways of perceiving are simply *not recognised,* even where they should); and at the sociological level (in that the non-specialist, the non-expert is *deprived of the right of access to knowledge and to judging claims made on its behalf*).

How, then, can modern science be regarded as signalling advance for humanity, when it was achieved at the cost of tremendous silencing, parochial legitimation procedures and, most of all, the deterioration in social status for most of humanity, including women and non-Western cultures? (Shiva 1989, 29–31) . . .

## Note

I am extremely grateful for the invaluable and sharp insights that came out of the four days of intensive engagement with Mr. Daryl McLean in late 1998. The discussions led to the crystallisation of the conceptual models that will be used in this research initiative.

# References

Capra, F. 1982. *The Turning Point. Science, Society and the Rising Culture.* Simon and Schuster: New York.

Chambers, R. 1983. *Rural Development. Putting the Last First.* Longman: London.

Dah-Lokonon, G. B. 1997. "Rainmakers: Myth and Knowledge in Traditional Atmospheric Management Techniques," in P. Hountondji, ed. *Endogenous Knowledge: Research Trails.* CODESRIA: Oxford.

*Declaration on Science and the Use of Scientific Knowledge and the Science Agenda Framework for Action.* 30C/15. UNESCO General Conference, 30th Session, Paris 1999.

Doussou, F. C. 1997. "Writing and Oral Tradition in the Transmission of Knowledge," in P. Hountondji, ed. *Endogenous Knowledge: Research Trails.* Oxford: CODESRIA.

Esteva, G. 1992. "Development," in W. Sachs, ed. *The Development Dictionary: A Guide to Knowledge as Power.* Zed Books: London.

Freire, P. 1972. *Pedagogy of the Oppressed.* Penguin Books: London.

Gupta, A. 1999. *Compensating Local Communities for Conserving Bio-diversity: How Much, Who Will, How, and When?* Indian Institute of Management: Ahmedabad, India.

Heisenberg, W. 1988. In F. Capra. *Uncommon Wisdom, Conversations with Remarkable People.* Flamingo: London.

Hountondji, P. 1977. *Endogeneous Knowledge: Research Trails.* CODESRIA: Dakar.

Kenway, J. 1996. "The Information Superhighway and Post Modernity: The Social Promise and the Social Price," in *Comparative Education,* Volume 32, Number 2, June 1996.

Makhale-Mahlangu, P. 1998. "Discussions on the Issue of Indigenous Knowledges." HSRC.

Mashelkar, R. A. 2000. "The Role of Intellectual Property in Building Capacity for Innovation for Development: A Developing World Perspective." Paper prepared for the World Intellectual Property Organisation (WIPO). CSIR. Government of India.

Mugo, M. 1998. "Culture and Education." Paper presented at the African Rennaisance Conference. Johannesburg, 28, 29 September 1998.

Nash, P. 1977. "A Humanistic Gift from Europe: Robert Ulich's Contribution to Comparative Education," in *Comparative Education Review,* Volume 21, Numbers 2 & 3, June/October 1977.

Odora Hoppers, C. 1997. *Public Policy Dialogue: Its Role in the Policy Process.* Johannesburg: Centre for Education Policy Development.

Odora Hoppers, C., and Makhale-Mahlangu, P. 1988. *A Comparative Study of the Development, Integration and Protection of Indigenous Systems in the Third World. An Analytical Framework*. Document. HSRC.

Rahman, A. M. D. 1993. *People's Self Development. Perspectives on Participatory Action Research*. Zed Books: London.

Shiva, V. 1989. *Staying Alive. Women, Ecology, and Development*. Zed Books: London.

Smith, L. T. 1999. *Decolonizing Methodologies. Research and Indigenous Peoples*. Zed Books: London.

"Social Sciences and the Challenges of Globalisation in Africa." Introduction to the International Symposium. Johannesburg. 14–18 September 1998.

Society for Research and Initiatives for Sustainable Technologies and Institutions (SRISTI). N.d. Knowledge Networks for Rewarding Creativity. Honey Bee Network. Indian. Institute of Management: Ahmedabad.

The National Innovations Fund. N.d. CSIR & IIM India: India.

Visvanathan, S. 1997. *A Carnival for Science: Essays on Science, Technology and Development*. Oxford University Press: Calcutta.

DANIEL SAREWITZ

# 23. Human Well-Being and Federal Science

## *What's the Connection?*

THIS BRIEF CHAPTER CONSIDERS how the organizational roots of federally funded science influence the capacity of the nation's research enterprise to contribute to human well-being. I take well-being (individual and collective) to have several components, including: (1) the fulfillment of all elemental needs necessary for survival; (2) the achievement and preservation of human dignity; and (3) the capacity to act on a more or less level civic and moral playing field.[1] The essence of my argument is that the Cold War origins of the federal research enterprise, and the philosophical foundations on which this enterprise rests, are implicated in a range of tensions and challenges to human well-being that the enterprise—as currently organized—cannot address coherently. These tensions arise in large part from the fact that science aims at delivering benefits to society through the achievement of predictive certainty and technological control, while the vitality of both nature and democracy derive from a lack of predictability and controllability. My discussion starts with the organization of federal (U.S.) science, but considers human well-being in a global context. This connection is reasonable because the U.S. is by far the most scientifically productive nation, and because the impacts of science on society are global in character.

## Cold War Roots

The role of the military in organizing the nation's current science and technology enterprise cannot be overstated. From the end of World War II until the launch of Sputnik by the Soviet Union in 1957, 80 percent or more of all federally funded research was justified in terms of national security needs. The creation of the American research university and the explosion of technology-intensive industries that lay at the core of the nation's economic growth were strongly and directly catalyzed by funding from the Department of Defense. Moreover, when Sputnik stimulated a highly politicized call for an increased national commitment to civilian research, the lion's share of resources during the subsequent decade went to the manned space program, which in many ways was simply a technological adjunct to the Cold War defense effort. For example, many of the information management, advanced materials, and navigation and control technologies necessary for space travel were also applicable to—or borrowed from—the nation's high-technology defense system.[2]

In the early 1950s, as the scale and complexity of America's Cold War geopolitical commitments became clear, research and development came increasingly to be viewed in the Defense Department and among leading science administrators not simply as the provider of particular weapons systems, but as a continual source of new knowledge, innovation, and technical expertise that would preserve American military preeminence across a diverse range of potential national defense applications. The scientific-military nexus entrained—and sustained—all sectors of the post-War research enterprise. From the perspective of those who designed and built this enterprise, the important functional distinction in science was not between basic and applied, but between classified and unclassified. The knowledge-production process was viewed not in terms of particular disciplines of basic science, but specified outcomes for military needs. The university was viewed not as an ivory tower, but as a vital cog in the national defense machine that included private industry and a range of government agencies. The role of the scientist was not as maverick roaming the frontier of knowledge, but as an interactive member of a multi-talented research group that often included theorists, experimentalists, and engineers. While research tools such as particle accelerators were used to carry out what might be termed *pure* research on fundamental physics, they were paid for by the Defense Department and the Atomic Energy Commission in large part because they were valuable test beds for technologies with military applications, and because they were the training ground for scientists who would help to create the coming generation of Cold War weaponry.[3]

The research organization that flowered from these Cold War roots was thus dominated by physical science, justified in terms of its role in technology development, and characterized by a dependency relationship between scientists, be they governmental or not, and their sponsoring federal agencies. The persistence of this organization can be seen in the continued dominance of three agencies—the Department of Defense, NASA, and the succession of energy research agencies—which peaked at nearly 90 percent of the federal R&D budget at the height of the Apollo program in 1965, and today still constitutes 66 percent of all federal research and development spending. New political momentum to develop missile defense systems, at expenditure levels that dwarf most other research programs, demonstrates that these patterns will continue for the foreseeable future.[4] Even in academia, many important fields, such as electrical engineering, computer science, and materials science, are today strongly supported by Defense Department funds, while physics continues to derive much of its support from the Department of Energy, which is a direct descendent of the Atomic Energy Commission (Wulf 1998, 1803).

One consequence of this organizational heritage is that the tools at hand are applied to emerging issues and problems, even if they are arguably inappropriate. For example, the massive U.S. Global Climate Change Research Program (USGCRP), which was established in law in 1990, the year after the Berlin Wall fell, is dominated by NASA space technology programs for data acquisition, and physical science approaches to modeling and interpreting atmospheric (and, to a lesser extent, oceanic) processes and evolution (cf. Subcommittee on Global Change Research 1997). One could easily imagine an alternative (and less expensive) program that placed a considerably greater emphasis on the life and social sciences—indeed, a growing sentiment for such a reprioritization is now beginning to emerge in some quarters—aimed at understanding the dynamics of ecosystems and social systems in a changing global and policy environment (cf. Lawler 1998). The organizational and political basis for such an effort, however, was not in place at the time the USGCRP was being planned. Our national approach to climate change thus strongly reflects the organization of Cold War science.

While the Cold War justified a top-down approach to setting science priorities, the ideology of basic research called for a bottom-up arrangement where scientists themselves would determine the most fruitful directions for fundamental investigation, based on their expert judgment as exercised through peer review and other mechanisms. This ideology has been most successfully implemented through research funded by the National Science Foundation and the National Institutes of Health. To a very great extent, of course, the ideal of an autonomous scientist exploring the frontiers of nature is strongly buffered by bureaucratic and policy decisions about how and where to allocate money, by the organization of research institutions such as universities and federal laboratories, and by the disciplinary organization of science itself, but within these constraints there is no question that the state of fundamental knowledge about nature has been spectacularly advanced by federally funded scientists acting with individual autonomy. All the same, individual autonomy can be exercised—and science advanced—in settings that are severely bounded. For example, during the Cold War, the conduct of classified military research on universities' campuses was commonly justified by the argument that scientists and engineers had to be free to pursue whatever research they chose, regardless of whether or not it was subject to the strictures of military secrecy (Lowen 1997, 144). An even more extreme case is illustrated by the Soviet Union under Stalin, where scientists somehow managed to conduct fruitful and sometimes world-class fundamental research programs, even as they were subjected to severe political persecution that often included prison and torture (Graham 1998).

## The Enlightenment Program

Given this combination of top-down organization motivated by the Cold War, and bottom-up research trajectories determined in part by individual scientists, how is human well-being introduced into the equation? After all, the promise of science to fulfill human needs is perhaps the principal political justification for public funding of civilian research. Over much of the past 50 years, this question has often been sidestepped by the assumption and assertion that the benefits of science flow automatically to society, deriving spontaneously from the progressive increase in the reservoir of fundamental knowledge, the development of innovative and beneficial new technologies, and the operation of the market economy that introduces the products of science and technology into society. In furthering this view, one can reasonably argue that the *program* for science (to use a term favored by social scientists) that was articulated and embraced by great Enlightenment thinkers from Bacon and Descartes to Jefferson and Voltaire, and reframed in a modern guise by Vannevar Bush's famous report, *Science, The Endless Frontier* (1960/1945), has been almost inconceivably successful. This program prescribed the linking of scientific knowledge about the laws of nature to the technological control of nature itself for the benefit and progress of humanity; it was implemented in its most comprehensive and successful form by the Cold War organization of American science; and it is internalized today at every level of the diverse and complex modern research enterprise, and throughout industrialized society as a whole. One can choose one's symbol of the culmination of this program—the polio vaccine; the hydrogen bomb; the invention of the transistor; the cloning of Dolly the sheep or the defeat of world chess champion Garry Kasparov by Deep Blue the computer—but the overall point seems unavoidable: human affairs are now mediated at every level and on every scale by science-based technology and the economic activity that it engenders. There are optimists who view this achievement as the salvation of the species (and even of nature [Ausubel, 1996]) and pessimists who portray it as a disaster-in-progress, but few if any would argue with the observation itself.

Neither would many suggest that we have arrived at utopia. But the idea that some desirable type of steady-state if metastable condition for humanity may actually be within reach has in fact taken hold of optimists and pessimists alike, as embodied in the environmental and economic concept of sustainability, the political ideal of "the end of history," and the technologist vision of nature as infinite cornucopia and the human life span as infinitely extendable. That such ambitions can be taken seriously and exist simultaneously in our jaded and hypersophisticated age is strong confirmation that the Enlightenment program for science has fulfilled its ambitions.

Yet the reality of both nature and democratic governance reveal crucial tensions embodied in the Enlightenment program for science. In terms of nature, the central paradox is that while the scale of control afforded by science and technology continues to increase, so does the domain of uncertainty and potential risk. "Knowledge of the [natural] system that we deal with is always incomplete," writes the ecologist C. S. Holling. "Surprise is inevitable . . . there is an inherent unknowability, as well as unpredictability. . . . The essential point is that evolving systems require policies and actions that not only satisfy social objectives [i.e., control for human benefit] but, at the same time, also achieve continually modified understanding of the evolving conditions and provide flexibility for adaptation and surprises" (Holling 1995, 13–14). Whereas science-based technological control of nature for human benefit has been the hallmark of the Enlightenment program, this control has aimed at severing the ties between society and nature, of escaping the threats and uncertainties imposed by those ties, of overcoming the natural limits on human endeavor in every activity, from food production to cognition. Holling's insight is that this goal—freedom from natural caprice—is unachievable because the very act of controlling natural systems introduces new variables that increase the unpredictability of the systems' dynamics. This insight has been borne out repeatedly in failed efforts to "manage" ecosystems. Protecting and preserving complex natural systems—systems on which human survival depends—thus requires that the expectation of control be abandoned, and replaced by an awareness that we cannot dictate the consequences of our actions in nature. The completely unanticipated discovery of the Antarctic ozone hole in 1985 is a stark example of this problem. In exercising control over the local environment through the use of chlorofluorocarbon refrigerants, we also perturbed the component of the stratosphere that protects the earth's surface from ultraviolet radiation. As the scale of human activity increases, we should expect surprises such as this to become more common.

Less recognized and appreciated is a similar relation between democracy and the Enlightenment program for science. It is often said that the price of democracy is "eternal vigilance." This means that democratic society must constantly guard against monopoly in the competition among ideas, principles, and morals. The competition itself is crucial to the existence of democracy. In its absence is oligarchy and authoritarianism. Human civilization has been marked by an ongoing discourse about the essential attributes that characterize a "good" society and its citizens—freedom, justice, equality, wisdom, mercy, tolerance, restraint, sharing—but there will never be an equilibrium equation describing the perfect, utopian balance among these attributes. Conflicts are inevitable and necessary, trade-offs must constantly be made—justice must be *tempered* with mercy; freedom with restraint. This struggle proceeds through a succession of social and political consensuses that must be hashed out in a process that is reasoned but not strictly rational, practical but not pretty. In a civil society, "there can be no ultimate closure," writes the social theorist Philip Selznick, "because values reflect existential conditions, which are always subject to change. . . . Reason takes into account the temptations and limitations of human conduct; therefore it is self-critical and self-limiting. This moderating outcome is also a source of indeterminacy. . . . Certainty is sacrificed on the altar of reason" (1992, 61–2).

The fact that indeterminacy is not only inevitable but essential to democracy—something to be embraced rather than overcome—does not comport well with a scientific worldview whose most legitimating measures of success are predictive certainty and control of nature. Having created the material welfare and technological infrastructure on which democracy has now come to depend, the significance of Enlightenment science for the democratic process itself seems murky. The issue here is not only that science and technology constantly transform the structure of society (usually without the consent of the governed), but also that over the past half-century or so, they have become tools that are called on to assist in the explicit improvement of democratic process—helping to resolve political dispute, set priorities for action, and manage social change. Through scientific eyes the mechanisms that enable democracy—politics, laws, bureaucracy—may look not just messy and irrational but subject to scientific correction. But the quest for scientifically based certainty and control in the political realm can conflict with democratic ideals by demanding, and fueling a demand for, that which is fundamentally incompatible with civil society: closure.

## Eight Problems

The scientific ideals of precision, determinacy, and control may thus lie in profound tension with society's struggle to comprehend and manage complex and evolving systems, be they natural or democratic. It can be no surprise, then, that while the Enlightenment program for science, catalyzed by an organizational structure that grew out of the Cold War, has helped to create heretofore unimaginable breadth of knowledge, degrees of control and levels of affluence, it has also revealed and provoked new challenges, contradictions and conflicts. Perhaps most prominent among these is that nature has gone global. Issues such as climate variability (El Niño, global warming), disease migration (AIDS, ebola), and invasive species (kudzu, killer bees) reflect the interaction of a global industrialized economy and a closed but infinitely complex earth system. The range of direct if uncertain threats to human well-being is daunting. These threats demonstrate that a science and technology program focused on ceding control over the environment to individuals acting quasi-independently yields unanticipated complications at higher organizational levels of nature and society. Ozone depletion has already been mentioned as one such complication. Continued sea level rise is another, with impacts exacerbated by the ongoing migration of populations to coastal regions. The cumulative effects of global ecosystem destruction loom as a threat of unknown but potentially disastrous proportions. And such problems are greatly complicated by fundamental issues of justice and equity. Obviously, the adverse impacts of global environmental problems are experienced disproportionately by poor people and poor nations, who have fewer resources and less flexibility in responding to changing environmental conditions (cf. Sachs 1996). Science is not organized to integrate such considerations of equity into its research priorities.

Yet even as these types of disparities grow more severe, a second problem can be recognized. As the affluence of industrialized nations of the world continues to grow, the priorities and capabilities of science and technology (fueled by the market incentives that promote commercial application of science) become increasingly divorced from the basic needs of those people—in rich nations as well as poor—who have not proportionately benefited from the products of the Enlightenment program. The clearest example is biomedical research, which in the U.S. focuses its formidable resources on fundamental investigations into molecular function and the search for high technology cures for diseases of affluence and old age. These priorities largely neglect a broad range of low-cost (and low-profit) opportunities in public health, such as nutrition research; they also fail to meaningfully address the debilitating health care problems of the developing world (cf. World Health Organization 1996). Even when a problem is of a global scale, the benefits of science may be disproportionately appropriated by affluent societies. This moral dynamic is strikingly illustrated in the case of AIDS, where high-technology and high-priced drug therapies are having a remarkable measure of success in reducing mortality among AIDS-sufferers who live in developed nations. In the developing world, the story is entirely different, and in many nations where advanced drug therapies are unaffordable, AIDS has become not only an overwhelming public health problem, but a tangible threat to economic and social prospects as well (Warrick 1998, A2). Understanding and controlling the HIV virus at the molecular level is the very stuff of the Enlightenment program: high-prestige science that attracts money, peer approval, and Nobel prizes. Again, the organization that encourages and rewards such science includes no provision to change research trajectories in response to a moral imperative.

Third, rapid progress in science and technology is transforming the fundamental institutions of civil society—including the structure of community, and the democratic process itself—in ways that are neither well understood nor easily controlled. Moreover, the speed of scientific and technological progress is such that profound and often wrenching societal transformations appear with greater frequency than ever before in history. In recent years, the erosion of civic community in America has been a subject of much concern to intellectuals, politicians, and pundits of every stripe. While the causes of any such decline must be complex, the proliferation of one transforming technology after another—telegraph, telephone, automobile, radio, television, air conditioner, home computer, Internet—introduces a pervasive instability into the structure and function of community. Similarly, within the period of a generation or two, the Green Revolution made the idea of the small, family-owned farm both economically and technologically obsolete (in the absence of government subsidy, at least) across much of the globe. The Enlightenment program supports a social consensus that accepts such change as the unavoidable price of progress. The program simply does not accommodate the idea that criteria of societal well-being are a legitimate guide for and metric of scientific progress.

Fourth, an overlay on this transformation process seems to be that technological progress exacerbates inequitable distribution of wealth, both within and between nations. A nation with many scientists and engineers, many telephones, computers, universities, and high technology companies will generate more ideas, more opportunities, more productivity and economic growth, than a nation that lacks these assets. This kind of growth is self-perpetuating, and has resulted in, among other things, a considerable increase in concentration of wealth among the world's industrialized nations over the past three decades—despite the remarkable gains made by a few East Asian nations (cf. United Nations Development Programme 1992). This phenomenon contradicts a basic tenet of the Enlightenment program—that new knowledge is cosmopolitan in its benefits—and demonstrates that scientific knowledge and technological control are appropriable commodities. Indeed, the Cold War organization of science was in its essence a quest to generate national advantage through the appropriation of knowledge and innovation.

Fifth, scientific uncertainty has become an increasingly common cause of political gridlock, especially in controversies related to the environment and natural resources. Global climate change is the archetypical example of this trend. The technical debate over the scientific validity of global warming has become a surrogate for a value debate about the preservation of the environment and the distribution of the benefits of industrialization. A sixth, possibly related issue is the overwhelming increase in the volume and availability of technical information relevant to human decision making at every level of society, unaccompanied by indications that this trend is in fact leading to greater wisdom or better decisions in the public sphere. These two trends reflect the Enlightenment confidence that more scientific information is in itself sufficient to drive political solutions to a range of societal problems. Scientific debate thus becomes a surrogate for underlying value debate necessary to make decisions, while the demand for "more information" displaces the demand for effective decision making.

Seventh, and partly as a reflection of its own success, the research enterprise is increasingly caught up in vexing and divisive ethical questions, such as those surrounding the cloning of higher organisms, technological erosion of privacy, patenting of genetic material, and clinical testing of new medicines in different cultures. Such questions often reflect a collision of scientific progress, commercial incentive, and ethical norms. For example, personal privacy may be threatened by the availability of genetic information that could be used as a basis for offering or denying medical or life insurance coverage to individuals (Hudson, Rothenburg, Andrews, Kahn, and Collins 1995, 391–393). In cases such as this, ethical conflict is an inevitable byproduct of the success of the Enlightenment program. The proliferation of new knowledge and techniques for the control of both nature and societal activity lies in profound tension with the desire of democratic society to exercise control through political processes. This conflict is exacerbated by the operation of the open market, which seeks to introduce the products of science into the economy, and resists any efforts to restrict its ability to do so.

Finally, the research community itself increasingly reports on a breakdown in confidence, optimism, and morale, especially in academia. Explanations of this phenomenon include loss of autonomy due to increased demand for scientific accountability to the public; ever-increasing degrees of specialization that alienate scientists from the real world; and the dissolution of community due to competition among peers for recognition and funding (cf. Pollack 1997). These explanations suggest an increasing tension between the evolving role of scientists in society, and some key tenets of Enlightenment science, for example, that scientists are accountable to society only through the quality of their science; that the sure path to comprehending nature is reductionism; that individual productivity is the key to scientific progress.

Now it may not be unreasonable to pronounce the foregoing issues to be political, sociological, and economic in nature, and thus largely beyond the capacity of science to address. One might also simply observe that while social progress is often slow, and politics irrational and difficult, more science and technology—a continuation of the Enlightenment program—cannot help but overcome these hurdles and continue to move things in the right direction, as they have done in the past.

But what is crucial here is not simply that the building and shaping of the entire portfolio of federal science and technology activities in the early post-War years took place *without regard* to any of these problems (many of which, of course, either did not exist or were not recognized), but as well that the organizational structure and knowledge products of today's enterprise are often not suited to addressing them productively. While the explicit question of how best to connect the enterprise to human well-being has raised its head from time to time in political debate over how science and technology should be organized, the idea that such beneficial connections will arise automatically through implementation of the Enlightenment program has always, in the end, prevailed.

Thus, recent consideration of how federal science should be connected to human well-being is usually framed in terms of how to enhance the performance of the Enlightenment program, not in terms of questioning the program itself. For example, the science community and its many advocates focus strongly on the need for "better communication" between scientists and the public, in order to ensure continued public support for research funding. The ideal of the "civic scientist" or "citizen scientist" has begun to get some attention from such scientific leaders as the director of the National Science Foundation (and current Presidential Science Advisor) (Lane 1996). While this ideal is certainly laudatory, as commonly articulated it is also rooted in the assumption that the feeding tube that delivers information from the scientific community to the public just needs to be widened, and it neglects the perhaps more fundamental issues of what type of nourishment, exactly, is being provided, and what the public might have to offer the research enterprise in terms of wisdom and guidance. "Sound science" is also commonly invoked as a cure for the political battles that often emerge over science- and technology-related issues such as global environmental change, biodiversity preservation, energy policy, and nuclear waste disposal. Technocratic approaches such as risk assessment are often prescribed to help rationalize the political decision-making process (House Committee on Science 1998). Again, these types of prescriptions, while not without some potential application, accept the received, Cold War organization of the science and technology enterprise as a starting point. I am arguing instead that the fundamental operational realities of democracy and nature, which dictate an essential unpredictability and uncontrollability, render the continued implementation of the Enlightenment program through the mechanisms of the Cold War inherently problematical. New organizational tenets and models may be necessary to confront these difficulties.

## New Links between Science and Well-Being

If the organization of science is understood to significantly reflect social, political, and economic processes—especially the very specific if multifaceted priorities dictated by the conflict between the United States and the Soviet Union—then the eight problems mentioned above can be seen as defining a reality within which new links between science and human needs can potentially be forged. The shape of some of these new links is gradually becoming apparent. New philosophical approaches now compete with the key insights of the great Enlightenment thinkers. The most conspicuous and controversial of these has arisen from the field of social studies of science. This approach sheds important light on the social and political context within which scientists are working and scientific problems are defined and confronted. But it has been reviled and rejected by most mainstream scientists—and not a few social scientists—for its assertion that scientific knowledge is socially constructed. In my view, this assertion is simply trivial as a critique of science per se: In the real world, the success and impact of science is argument enough for the validity of its method and philosophical underpinnings—socially and politically constructed as they may be. As David Hull writes: "No amount of debunking can detract from the fact that scientists do precisely what they claim to do" (1988, 31). This success, however, is necessarily and appropriately defined within the context of the Enlightenment program, and especially the application of scientific knowledge to the technological control of nature. A more fruitful question, then, would address the extent to which the Enlightenment program is appropriate for and compatible with the types of challenges facing society today. The social studies of science have helped to position science where it belongs—in the heart of society, rather than as an insular satellite—and even through the rancor it stimulates, brings attention to this question, and thus raises the possibility of alternative programs for the future.[5]

As a more practical matter, mechanisms to better connect democratic process to the establishment of scientific priorities and practices are fitfully beginning to develop, although often over the strenuous objections of the scientific community. In Europe, citizens conferences that give communities the opportunity to make decisions about scientific and technological choices are gradually becoming recognized as an effective tool, while in the United States, environmental stakeholder groups, often organized at the scale of a local watershed, now increasingly replace, augment, or subsume expert technical debate in the effort to resolve environmental dilemmas. The National Institutes of Health has begun placing patients on a few of its peer review panels to help introduce broader perspectives into the process of allocating biomedical research funds. And community-based research is a nascent but highly promising mechanism for linking scientists to local people and problems in a manner that was never envisioned or allowed for in the Cold War organization of science (cf. Sclove 1995; Landy, Susman, and Knopman 1999; Agnew 1999; Sclove, Scammell, and Holland 1998).

Efforts to achieve a more synthetic view of nature by breaching the barriers that separate traditional scientific disciplines are occurring with variable success in such disparate areas as cognitive neuroscience and environmental science. The emerging and well-publicized field of complexity science acknowledges the intrinsic limitations of traditional approaches to understanding and describing nonlinear systems such as consciousness or economies. Throughout the sciences, there is an increasing recognition that nature is simply not comprehensible in terms of narrow, disciplinary, reductionist investigation (cf. Cornwell 1995).

The concept of sustainability is giving birth to new measures of progress and new agendas for research that explicitly link science to a moral framework rooted in the tenet of intergenerational equity. Sustainability has become a guide for research in diverse fields, such as agriculture, economics, ecology, and public policy. The idea of adaptive management of complex systems acknowledges that such systems (natural and social) are not predictable, and that both science and policy are thereby always subject to error and amenable to correction. Adaptive management recognizes that values are usually less malleable than science, and thus prescribes for science the role of assessing and monitoring the impacts of policy decisions that have already been made. Science becomes a tool for correcting and improving the incremental democratic policy process by providing insight, rather than dictating policy by providing predictions. Industrial ecology views manufacturing and energy use as cycles, rather than independent streams of production, consumption, and waste, and thus defines entirely new criteria for judging the viability of technologies (cf. Lee 1993; Graedel and Allenby 1995). Sustainability is a goal; adaptive management is a policy process for moving toward the goal; industrial ecology is a technical perspective that supports the policy process. These linked concepts view nature and democracy as models to be emulated and supported, not obstacles to be overcome.

Progress in these and related directions is important and promising, but the relevant scientific activities remain a marginally small proportion of the total federal science enterprise. Typically, it seems, initiatives along these lines are undertaken in isolation, often as a result of the action of individuals with vision and energy. There are few institutional structures within which successful experiments can grow and propagate. Funding sources and reward systems still militate against those who would seek to define stronger connections between science and human well-being.

One problem, of course, is that the organizational inertia of the Cold War is difficult to displace. Doing so will take time and persistence. Another problem is that the mental model of the Enlightenment program is so simple and elegant—more science and more control always yield more well-being, in essence—that it resists being replaced by something more nuanced. But a few things can be said about the components of an alternative organization for science that seeks to relieve the tensions created by the Enlightenment program. This new organization will certainly direct us toward new types of institutions that promote, as a foundation for determining research priorities, meaningful interaction between scientists and the people who scientists are serving. It will portray complex, real-world problems, rather than scientific disciplines, as organizing foci for research programs. Artificial taxonomies, such as basic versus applied research, will be abandoned as meaningless, and the boundaries between social and natural science will become increasingly permeable. Metrics of scientific excellence will focus as much on social outcomes as on scientific ones. Scientists and science administrators will internalize the idea that complex and indeterminate structures of nature and democracy must be a basis for such new organizational approaches to research as adaptive management and industrial ecology.

The transition to an organization of science characterized by these and related attributes is a matter, for the most part, of political vision and will. It is worth emphasizing that whereas the mythologies of the "golden age" of Cold War science tell a story of abundant funds available to individual scientists who freely pursued exciting new knowledge wherever it might lead, the broader reality underlying this Elysium was that the Department of Defense created a huge, integrated knowledge production enterprise aimed at achieving a particular desired outcome—victory over the Soviet Union. Similarly, the creation of stronger linkages between science and human well-being can be framed as an organizational challenge that requires a clear definition of the outcomes desired, and a mobilization of intellectual activity aimed at achieving these outcomes. If the resources and institutional structures are put in place, the science will happily follow.

## Notes

1. I recognize that the precise meaning of the second and— especially—the third components have been and will remain infinitely contestable, yet this contesting is usually a matter of balance among agreed upon variables, rather than competition between mutually exclusive concepts.

2. For information on science budgets from the early Cold War years, see Office of Management and Budget, *The Budget of the United States Government, Fiscal Year 1999*, Historical Table 9.7 (Washington, D.C.: Government Printing Office, 1998); and National Science Foundation, *Federal Funds for Research and Development: Detailed Historical Tables, Fiscal Years 1951–1998*, at www.nsf.gov/sbe/srs/nsf98328/start.htm.

There are many accounts of the political history of Cold War science and technology. Some of the better ones include: Leslie (1993), Zachary (1997), Norberg and O'Neill (1996), Lowen (1997), Sapolsky (1990), McDougall (1986), Kevles (1987), and Sherry (1977).

3. Ibid.

4. Biomedical and health-related research is the second largest component of the research portfolio, making up 22 percent of federal expenditures in FY 2000 (and 46 percent of nondefense research). For post-1960 research and development data, see National Science Board, *Science and Engineering Indicators* (Washington, D.C.: National Science Foundation), published biennially; and Intersociety Working Group, *Research and Development* (American Association for the Advancement of Science), published annually. Also see James Glanz, "Missile Defense Rides Again," *Science* 284 (April 16, 1999): 416–20.

5. For an introduction to some of these ideas, Jasanoff, Markle, Petersen, and Pinch (1995).

## References

Agnew, B. (1999, March 26). "NIH Invites Activists into the Inner Sanctum." *Science* 283:1999–2001.

Ausubel, J. H. (1996, Summer). "The Liberation of the Environment." *Daedalus* 125:1–7.

Bush, V. (1960). *Science, the Endless Frontier*. Washington, DC: Office of Scientific Research and Development (Original work published 1945).

Cornwell, J. (Ed.). (1995). *Nature's imagination*. New York: Oxford University Press.

Glanz, J. (1999, April 16). "Missile Defense Rides again." *Science* 284: 416–20.

Graedel, T. E., and B. R. Allenby. (1995). *Industrial ecology*. New York: Prentice Hall.

Graham, L. R. (1998). *What Have we Learned about Science and Technology from the Russian Experience?* Stanford: Stanford University Press.

Holling, C. S. (1995). "What Barriers? What Bridges?" In L. H. Gunderson, C. S. Holling, and S. S. Light (Eds.), *Barriers and Bridges to the Renewal of Ecosystems and Institutions*. New York: Columbia University Press. 13–14.

House Committee on Science. (1998, September 24). *Unlocking Our Future: Toward a New National Science Policy. A Report to Congress*. Washington, DC: House Committee on Science.

Hudson, K. L., K. H. Rothenberg, L. B. Andrews, M. J. Ellis Khan, and F. S. Collins. (1995, October 20). "Genetic Discrimination and Health Insurance: An Urgent Need for Reform." *Science* 270:391–93.

Hull, D. L. (1988). *Science as a Process: An Evolutionary Account of the Social and Conceptual Development of Science*. Chicago: University of Chicago Press.

Jasanoff, S., G. E. Markle, J. Petersen, and T. Pinch (Eds.). (1995). *Handbook of Science and Technology Studies*. London: Sage Publications.

Kevles, D. J. (1987). *The Physicists: The History of a Scientific Community in Modern America*. Cambridge: Harvard University Press.

Landy, M. K., M. M. Susman, D. S. Knopman. (1999). *Civic Environmentalism in Action: A Field Guide to Regional and Local Activities*. Washington, DC: Progressive Policy Institute.

Lane, N. (1996, February 9). "Science and the American Dream: Healthy or History." Speech presented at the Annual Meeting of the American Association for the Advancement of Science, Baltimore, MD. Available on-line: www.nsf.gov/od/lpa/forum/lane/slaaa.htm.

Lawler, A. (1998, June 12). "Global Change Fights off a Chill." *Science* 280: 1683–85.

Lee, K. N. (1993). *Compass and Gyroscope: Integrating Science and Politics for the Environment*. Washington, DC: Island Press.

Leslie, S. W. (1993). *The Cold War and American Science*. New York: Columbia University Press.

Lowen, R. S. (1997). *Creating the Cold War University: The Transformation of Stanford*. Berkeley: University of California Press.

McDougall, W. A. (1986). *. . . the Heavens and the Earth: A Political History of the Space Age*. New York: Basic Books.

National Science Foundation. (1998). *Federal Funds for Research and Development: Detailed Historical Tables, Fiscal Years 1951–1998*, at: www.nsf.gov/sbe/srs/nsf98328/start.htm.

Norberg, A. L., and J. E. O'Neill (1996). *Transforming Computer Technology: Information Processing for the Pentagon, 1962–1986*. Baltimore: Johns Hopkins University Press.

Office of Management and Budget. (1998). *The Budget of the United States Government, Fiscal Year 1999*. Washington, DC: Government Printing Office.

Pollack, R. (1997, August). "Hard Days on the Endless Frontier." *The FASEB Journal* 11: 725–31.

Sachs, A. (1996). "Upholding Human Rights and Environmental Justice." In L. R. Brown and C. Flavin (Eds.), *State of the World 1996*. New York: W. W. Norton and Company. 133–51.

Sapolsky, H. M. (1990). *Science and the Navy: The History of the Office of Naval Research*. Princeton: Princeton University Press.

Sclove, R. E. (1995). *Democracy and Technology*. New York: Guilford Press.

Sclove, R. E., M. L. Scammel, and B. Holland. (1998). *Community-Based Research in the United States: An Introductory Reconnaissance, Including Twelve Organizational Case Studies and Comparison with the Dutch Science Shops and the Mainstream American Research System*. Amherst, MA: The Loka Institute.

Selznick, P. (1992). *The Moral Commonwealth: Social Theory and the Promise of Community*. Berkeley: University of California Press.

Sherry, M. S. (1977). *Preparing for the Next War*. New Haven: Yale University Press.

Subcommittee on Global Change Research. (1997). *Our Changing Planet: The FY 1998 U.S. Global Change Research Program*. Washington, DC: Office of Science and Technology Policy.

United Nations Development Programme. (1992). *Human Development Report 1992*. New York: Oxford University Press.

Warrick, J. (1998, October 28). "AIDS's Long Shadow Cools Global Population Forecast." *The Washington Post*, A2.

World Health Organization. (1996). *World Health Report 1996*. Washington, DC: World Health Organization.

Wulf, W. A. (1998, September). "Balancing the Research Portfolio." *Science* 28: 1803.

Zachary, G. P. (1997). *Endless Frontier: Vannevar Bush, Engineer of the American Century*. New York: The Free Press.

DAVID J. HESS

# 24. Science in an Era of Globalization:
# Alternative Pathways

HOW HAS THE SCIENTIFIC FIELD changed during the era of globalization, that is, roughly since the 1970s, when substantial changes in the global
economy brought about trade liberalization and increased competition
among national economies? How have the changes opened and closed
opportunities for scientific research to be more responsive to the world's
poor, to historically excluded groups, and to environmental problems?
Although government and industry have long prized scientific knowledge as a source of technological innovation, STS researchers have also
developed an increasingly precise knowledge about how the institutions
of science have changed during the late twentieth century and the early
twenty-first. Whether one calls the transition mode-two knowledge production, academic capitalism, audit culture, the triple helix, asymmetric
convergence, or, more poetically, degrees of compromise and impure cultures, there is considerable documentation of the ways in which universities, academic disciplines, industrial research and development, laboratory organization, funding sources, intellectual property rules, and other
fundamental aspects of the scientific field have changed since the 1970s.
Established disciplines have, in some cases, lost ground to new interdisciplinary enterprises that are aligned with military and industrial priorities,
such as the "T" fields of IT (information technology), BT (biotechnology), NT (nanotechnology), and increasingly CT (clean technology). As
scientific research has become increasingly central to national competitiveness goals, the alignment of scientific research agendas with industrial priorities makes it more difficult for researchers to wrap themselves
in the flag of academic freedom, pursue research programs that conflict
with the strategic goals of government and industry, and, in short, operate outside the mainstream of their fields and departments. Yet such work
does proceed. As I will argue, the changes in scientific research in an era
of globalization cannot be reduced to the increasing influence of military
and industrial research priorities over scientific research. A countervailing process has also emerged.[1]

## Epistemic Modernization

The globalized academy with its new research and development centers
constitutes one way in which the degrees of freedom for scientists are

constrained from the outside, but it by no means represents the whole story. Just as there is increasing scrutiny of the selection of research programs from above, there is also increasing scrutiny of science from below. I call this second change the epistemic modernization of science, a term that is borrowed loosely from the concepts of reflexive modernization and ecological modernization. Such general concepts tend to be vague and imprecise rubrics; to combat the tendency, I will sketch in some detail what topics of study are brought into focus through the concept.[2]

Epistemic modernization refers to the process by which the agendas, concepts, and methods of scientific research are opened up to the scrutiny, influence, and participation of users, patients, nongovernmental organizations (NGOs), social movements, ethnic minority groups, women, and other social groups that represent perspectives on knowledge that may be different from those of economic and political elites and those of mainstream scientists. In a sense, the change represents a return, but under very different historical circumstances, to the conditions of early modern science. In the history of early modern science, an epistemic primitive accumulation occurred when Western explorers and scientists traveled around the world and brought home the diverse local knowledges of plants, animals, landscapes, languages, medicines, and social institutions. Modern science as we know it today was built up from the interaction with, and codification of, lay and non-Western knowledges. As science became increasingly denaturalized and laboratory based, lay and local knowledges became less important to, and less valued in, most scientific fields. However, in the late twentieth century, various criticisms of science reopened the doors to greater public interaction. Epistemic modernization has echoes of an earlier era but is also quite different.

The epistemic modernization of science and technology occurs as a result of various institutional changes. In the simplest form, the social composition of science is undergoing differentiation both within societies, as the doors are opened to previously excluded social groups, and internationally, as university-based research and education become ubiquitous in countries throughout the world. Another form of epistemic modernization is the development of community-oriented research projects that involve laypeople in the tasks of agenda setting, problem definition, research design, and implementation. A third form involves the growth of interaction with civil society and social movement organizations that have increasingly challenged the epistemic authority of science. Finally, a fourth form comes internally, from dissident scientists who have broken ranks with consensus opinion over agenda-related issues. The dissidents have sometimes coalesced into NGOs of socially responsible scientists, formed coalitions with NGOS and social movement organizations to provide them with counterexpertise, and contributed to the development of alternative research fields.

The epistemic modernization of science often involves a conflict over the framing of science with respect to a broader public interest. Groups that claim to represent the public interest from within science, such as leaders of research fields who serve as spokespersons for good science, tend to adopt the frame of paternalistic progressivism; that is, they wrap themselves in the empiricist flag of gradual scientific progress and methodological neutrality and reject efforts to shift research agendas by dissident scientists, laypeople, social movements, or reform movements in science that are aligned with social movements. In contrast, those who support alternative research agendas, such as civil society organizations and reformist scientists, often adopt a corresponding frame or discourse of scientific devolutionism, that is, a historical narrative of a fall from grace in which scientific research has been captured by corporate profit motives and corrupted by cultural bias. The emergent institutions of epistemic modernization represent a third approach, which attempts to engage the loss of confidence in the equation of "Science" with a general or public interest. Rather than accept the devolution frame or attempt to counter it with the frame of paternalistic progressivism, research under conditions of epistemic modernization accepts the challenges mounted by dissident experts and civil society organizations, but incorporates and transforms them through an emergent institutional structure that opens scientific inquiry, and to some degree technological design, to participation by nonexperts and historically excluded groups. Rather than reject all epistemic challenges as unscientific, the approach selects some challenges for research and some challengers for research funding. In the sections that follow, I will consider four dimensions of epistemic modernization: the increasing diversification of the social composition of science, community-based research, the rise of the interactive model of public communication, and alternative pathways in science.

## Diversification of Social Composition

In its simplest and most straightforward form, epistemic modernization results from the effects that the increasing universalism of the membership policies of science as an institution has on its research agendas. Especially since the mid-twentieth century, there has been growth in the opportunities for women, African Americans, Latinos, gays and lesbians, and other historically excluded social groups of people who have gained access to education and jobs in science. At the same time, science has become more internationalized, so that graduate programs and major universities exist throughout the world and in many postcolonial societies, where there is now the option to pursue graduate work at home rather than in a former colonial center.

One might argue that because scientific knowledge production is (or should be) demographically neutral, such demographic changes should have no impact on the content of scientific knowledge. The argument is defensible up to a point, and at an individual level the science of new members of formerly excluded social categories may appear to be indistinguishable from that of white males in North America and Europe. The argument can also be taken to considerable length when applied to methods and the knowledge-vetting process, even across international communities. American scientists in some research fields sometimes dismiss the research of foreign scientists as methodologically unsophisticated; however, when international controversies heat up, arguments that allude to the demographic address of a scientist have the same illegitimate status in the rhetoric of science that ad hominem arguments have. Demographically based prejudices must be translated into methodological and evidentiary arguments to count as legitimate.

However, when one understands that scientific research takes place in fields of conflict and cooperation, then one also realizes that the theories, concepts, methods, and problem areas of a specific research field are not universally shared but instead contested between dominant and subordinate (or nondominant) networks. In some cases the subordinate networks may not be demographically different from the dominant networks, but one would also expect to find cases where the social address of challenger scientists is closely linked to controversies that have emerged. As Donna Haraway has shown for primatologists and Sharon Traweek for physicists, social differences can translate into epistemic differences regarding preferences for methods, problem areas, concepts, and even equipment design. Furthermore, when one examines how the diversification of the social address of scientists has affected the agenda-setting politics of scientific knowledge (that is, the choices of which research fields to develop and which ones to leave undone), the issue of demographic change is hardly trivial. When the doors of science open to a broader social composition, the answer to the question of what counts in science as an important problem area, especially for health and environmental research, depends a great deal on whom one asks. Debates over different agendas and methods benefit science in the sense of making visible unseen biases that have previously passed as unquestioned neutrality, and they result in improved research methods as well as better allocation of resources to problem areas. To use Sandra Harding's phrase, diversity leads to "stronger objectivity."[3]

A parallel process occurs in the technological and design fields (such as engineering, marketing, and the design professions), where technology and product development teams have also opened design and innovation teams to diverse social and occupational addresses. By increasing the social diversity of the design team, new opportunities emerge for seeing the problem from previously excluded and invisible perspectives. The diversification of the social composition of professional design teams can be enhanced through methods such as participatory design, which are analogous to the institutions of lay interaction in science to be discussed in the next sections. As a result of both increased professional diversity and the incorporation of lay perspectives in the design process, design teams are better able to check their own views of what users want and how users interact with the prototypes. The innovations that follow are also likely to be more universal. Universal design is to technology as strong objectivity is to science.

In following the research of feminist and multicultural science studies scholars—who argue that changes in the social composition of science lead to the modernization of science in the sense of making possible more robust and universalistic conceptual systems, methods, and prioritization of problem areas—it is important not to overstate the case and thereby discredit it with essentialism. There is not necessarily an identifiable woman's perspective or African American perspective on every scientific issue and method. Likewise, to some degree the transnational nature of scientific conferences and journals may have reduced some of the more dramatic cultural differences in science, such as the differences between French and English styles of physics that Pierre Duhem noted at the beginning of the twentieth century. However, as the scientific field has become more diverse socially, there are many instances of networks of scientists who belong to historically excluded groups and bring a new sensitivity to what counts as science. In doing so, they contribute to the modernization of science in the sense of the ongoing critique of cultural baggage that has previously passed as neutral and universal knowledge, methods, and priorities for research.[4]

## Community-Oriented Research

In the second dimension of epistemic modernization, the scientist-layperson relationship is brought into close contact through local, community-oriented research projects that involve lay participation. One example is pro bono, community-oriented research, where the researcher undertakes the project without extramural funding and with the intention of benefiting a local community. Self-funding allows the researcher great flexibility in defining the extent of public participation, but in most research fields the absence of funding also restricts the choice of methods and scope of the project.

The pro bono type of community-oriented research was institutionalized in Europe through the development of science shops—small offices that are funded by the university to serve as conduits between the research needs of the surrounding community and the resources of the university. Under the science shop model, a community-based organization comes to the university office with a research problem, and the office attempts to find a faculty member who is willing to work on the problem or to supervise students who work on the problem. The university generally provides some limited funding to staff the office, such as by paying for a part-time staff person, but the research is conducted by faculty or students on a pro bono basis. The science-shop movement (as it was sometimes called) grew during the 1970s and 1980s but suffered retrenchment and cutbacks under the budgetary pressures and neoliberal reorientation of the universities during and after the 1990s.[5]

Somewhat analogous to science shops in the United States, community-based research may originate from the community or the university. An example of the former is the emergence of what Phil Brown calls "popular epidemiology," in which residents in a community identify a new disease or a cluster of cases of a known disease and may also identify potential causes. In the case of the cancer cluster that developed in Woburn, Massachusetts, the community members also sought out and developed a partnership with researchers at Harvard University's School of Public Health. In a somewhat different example, researchers at the University of Pennsylvania developed partnerships with community groups in the impoverished surrounding community of West Philadelphia. The university provided support for faculty and students to engage in participatory action projects, many of which took place through the local schools, such as a nutrition education project.[6]

To the extent that the university defines its core mission as an engine of technological innovation oriented to regional industries, it will be increasingly difficult to find a place for pro bono research projects that involve the participation of community organizations and social movement organizations. There will be exceptions, such as the case of universities that are located in low-income neighborhoods and find it in their enlightened self-interest to support community-development projects and partnerships with community organizations. Even in those cases, the models of research associated with business-oriented economic development (the "right hand" of the university) are likely to receive much more attention and greater resources than those driven by participatory action. The growth of service learning programs may represent one opportunity for pro bono community-based research, but the quality of research that is possible in short-term internships may not serve community needs for collaborative research and expertise.

One opportunity for growth for community-based research, and in many ways the model for this type of epistemic modernization in the United States, is public health research oriented toward low-income, ethnic minority, and rural populations. The history of clashes between researchers and community groups regarding research agendas, methods, and bias has motivated health and environmental funders to be more concerned with disadvantaged populations and much more sensitive to the problems of access and community participation. For example, in 2001 the Agency for Healthcare Research and Quality, in collaboration with the Kellogg Foundation, held a conference that developed recommendations on community-based participatory research. Likewise, in 2005 the Environmental Protection Agency supported a similar conference on community-based participatory research and environmental justice. Those changes represent recognition by mainstream governmental and funding organizations of the value of enhancing community access to research agendas and research (see Agency for Healthcare Research and Quality 2001; U.S. Environmental Protection Agency 2004).

## The Interactive Model of Public Communication

Another dimension of epistemic modernization has been the shift in the public communication models of scientists, engineers, medical researchers, and other expert groups. The transmission model, which was associated with the assumption of the value of scientific autonomy for a democracy, defined communication between scientists and the public as a one-way process through the media and educational system. In turn, feedback generally came not through the media and educational system but through passive public support of research funding either as authorized by elected public officials or through public contributions to nonprofit research organizations.

In contrast, a growing body of literature on the public understanding of science has shown the remarkable ability of lay publics to acquire expertise and fluency when necessary, as in the case of communities that face problems of environmental justice or patients who battle chronic disease.

426   *David J. Hess*

The literature documents the ability of publics to reconstruct knowledges and reappropriate technologies for their own purposes. Furthermore, it tends to articulate a concept of the public that is more differentiated, with pockets of literacy and illiteracy that are strategically based on a need to know. The pockets may take the form of geographically based local knowledge, experience-based knowledge (such as that of chronic disease patients), or combinations of the two (such as communities that are suffering the health effects of toxic exposure). Although the transmission model dismisses their knowledge as scientifically groundless, they operate from an alternative epistemic authority grounded in personal experience. Their own versions of *cogito ergo sum*—"I am sick; therefore I doubt," "I can smell the pollution in the water; therefore I doubt," and so on—provide the basis for their confidence in questioning expert authority and their need to engage it in detail (Allen 2003; Brown and Mikkelsen 1990; Eglash et al. 2004; Hess 1999, 2005; Irwin and Wynne 1996).

Under conditions of epistemic modernization, the assumptions behind the transmission model are put into question, and an interactive model emerges in its place. Research communities come to recognize civil society groups, such as environmentalists and disease-based patient advocacy groups, as bearers of legitimate questions rather than merely misinformation and lack of knowledge, and they recognize lay knowledge as complementary to scientific knowledge rather than merely inferior to it. Scientists who embrace epistemic modernization replace the older communication policy of improved transmission to the lay masses and stepped-up suppression of the insider challengers with the institutionalized incorporation of selected epistemic challenges. The challenges can be converted into researchable knowledge claims, just as some challengers can be converted into institutionalized participants in the research and policymaking process.

The changes are especially evident in the incorporation of health and environmental social movements. What once took a social movement to open up research policymaking to public debate (AIDS, breast cancer, alternative cancer therapies, antitoxics, etc.) is now increasingly institutionalized through the conversion of social movement organizations into insider advocacy organizations with their own affiliated research fields. As research undergoes epistemic modernization, the public involvement in, and shaping of, science not only becomes more prevalent but is recognized by the leaders of the scientific community as a legitimate part of the agenda-setting process. As Phil Brown has noted, citizen-scientist alliances also form to challenge the accepted scientific knowledge. To the extent that the change becomes widespread and exercises a significant influence on research agendas, we can say that the knowledge-making process has undergone epistemic modernization. Again, the question can be studied empirically across different scientific fields (Brown 2007; Hess 2005).

The interactive model is emerging in two major forms. In the indirect form, interaction involves contributions by lay advocates and activists to processes that set overall agendas. Usually the work involves splitting the social movement or advocacy groups into some people who have developed the appropriate literacy to engage the legislative appropriation and public funding agencies; in other words, they have undergone an "expertification" process described by Steven Epstein. An example can be found in the advisory boards of the National Institutes of Health, where patient advocacy groups have won seats at the table of funding decisions. Another example is the Danish consensus conference, which enables a lay "jury" to deliberate and give advice on a policy issue such as whether a research field should be pursued, and if so, what regulatory guidelines should be put in place.[7]

In the direct form of the interactive model, activists and advocates make the transition from contributors to general agenda setting and formulation of regulatory issues to the status of contributors to scientific research fields. In my historical and ethnographic research, three basic processes became evident: conversion (activists target scientists who are working in a dominant research program and attempt to convince them to shift to a research program aligned with the activists' goals), biographical transformation (the change of activists and advocates into researchers through additional formal education and credentialing as scientists), and network assemblage (the lay activist or advocate does not necessarily acquire the formal credentials or expertise to contribute to a specific scientific field but instead serves as a catalyst for research by obtaining funding or by leading research projects but delegating the more technical work to experts).[8]

In summary, the demise of the transmission model means that the leaders of science recognize that a new kind of relationship with the public is possible, one that is analogous to the transition in the media from broadcast to interactive communication. Unlike community-based research, where the interaction with laypeople is more localized through community organizations on specific research problems, the interactive model is less geographically localized and more oriented toward general research agendas. The lay activists or advocates can find places on funding panels and conferences, help scientists to convert to new research programs, undertake further education to become experts themselves, arrange for funding and research opportunities for sympathetic scientists, and help orchestrate research projects.

## Alternative Pathways in Science

The fourth major dimension of epistemic modernization involves scientists who directly incorporate into their research programs the goals of social change associated with social movements. Usually the dissident scientists use the repertoires of action available within the scientific field; consequently, in the terminology that I have developed, this type of mobilization is best described as a professional reform movement within science rather than a protest-based social movement. However, the scientific reform movements often exist alongside and in alliance with broader social movements, so there is considerable overlap between reform movements within the scientific field and social-movement mobilizations. This section discusses some of the alternative pathways in science that emerge in two types of reform movements within science: those with a goal of stopping a field of research and those with a goal of developing an alternative research field.

One way of engaging the politics of research agendas is to call for a moratorium on certain areas of research. In the United States since the mid-twentieth century, two examples of research moratoria are scientist-driven mobilizations to stop some kinds of recombinant DNA and weapons research. In the case of recombinant DNA research, the intention of the scientists was never to halt the development of a research field but only to halt it temporarily to put in place safety guidelines and some restrictions on the riskiest research. When the research moratorium did emerge, it came from outside the scientific community in the form of temporary moratoria imposed by local and state governments. In the case of weapons research, the drive for a moratorium came more from the scientists themselves and took the form of a pledge not to engage in a type of research. In effect, scientists had called a research strike. The pledge or research strike represents only the organized end of a continuum of responses to research programs that are controversial to segments of a research community. At the other end of the continuum is the stigmatization of research programs that benefit the military or a specific industry on the grounds that the research would generate technologies that create undue risks and dangers for the broader society. For example, scientists who work on genetically modified food for agribusiness, on nuclear or fossil fuel energy, on chlorine-based chemistry, and on related topics that are environmentally controversial find that their research programs can be the target of criticism from environmental groups as well as stigmatization from some of their colleagues.[9]

The organized pledge and stigmatized research problem area are examples of how broader public concerns, as articulated by social movement or advocacy organizations, can be translated into the scientific field in the form of debates and mobilizations by scientists in opposition to some research agendas. They provide another example of epistemic modernization, in that there is a contestation of the contours of a research field as shaped by political and economic elites. Furthermore, oppositional politics within the scientific field can also coincide with a second type of scientific mobilization: the development of new and alternative research fields. For example, the radical science movement of the 1970s diversified from opposition to war-related research and technology to various projects in support of "people's science," which included agricultural, computing, and scientific assistance for North Vietnam and Nicaragua as well as for underserved segments of the population in the United States. The experiments in people's science shifted activism in science away from the oppositional politics of the research moratorium toward new research programs that were geared to the needs of the world's poor and toward the idea of developing alternative knowledges and technologies. In turn, the goal of developing sciences for the people drew on and supported the appropriate technology movement, a grassroots research and development movement that emerged in nonprofit organizations largely outside the major universities. The appropriate technology movement drew attention to the need to develop new energy and agricultural technologies that today would be recognized under the rubric of sustainable design, particularly strands that involve local ownership and control. Although there are many types of alternative research fields, I focus here on three that are relevant to environmental issues: renewable energy, sustainable agriculture, and green chemistry.[10]

In all cases, the alternative scientific fields oriented toward more environmentally sustainable science show significant but limited gains since the 1960s. Over time, support from the federal government has opened up for each of the three fields. For example, research on renewable energy achieved recognition during the 1970s in the wake of the rise in oil prices and the decline of public confidence in nuclear energy. At that time, the federal government earmarked some funding for research on "soft path" alternatives such as solar energy. In the 1980s, the federal government launched what was later named the Sustainable Agriculture Research and Education program. Finally, in the 1990s the Environmental Protection Agency began supporting research initiatives in green chemistry.

In each case the fields can point to some success stories. Renewable energy has become increasingly mainstreamed through industrial growth in areas such as wind and solar energy. As the industry has developed and government-based funding has become available, there has been corresponding growth of scientific journals and research networks as well as support for research and development in the private sector. Even in times of retrenchment of federal research spending, such as during the presidency of George W. Bush, renewable energy research has received some funding and policy support. Likewise, sustainable agriculture research can point to some growth of acres under organic cultivation at research universities and to the development of sustainable agriculture research centers in some of the major agricultural schools. Green chemistry can also point to the growth of research panels at mainstream chemistry conferences and to the emergence of research networks dedicated to "green chemistry."[11]

However, when set against the broader background of funding levels in comparison fields, the growth of the three research fields needs to be understood as having occurred in a context of low levels of relative funding. For example, federal funding for renewable energy research and development peaked at about $1.6 billion in 1979, declined throughout the 1980s, and in real dollars never returned to the level of support in the 1970s. Across both Democratic and Republican administrations since the 1970s, nuclear energy and fossil fuels have each consistently received more federal funding support than renewable energy in the U.S. Department of Energy budget. Furthermore, funding for renewable energy has been highly selective; in the first decade of the twenty-first century, the Bush administration favored roadmaps for a long-term transition to the hydrogen economy and ethanol fuels but fired, at least temporarily, a number of researchers at the National Renewable Energy Laboratory. Funding levels for renewable energy improved during the Obama presidency, but in the U.S. Department of Energy budget, nuclear and fossil fuel funding each remained much higher than renewable energy funding such as solar.[12]

A similar pattern of relative lack of research funding can also be found for organic agriculture and green chemistry. A search of the U.S. Department of Agriculture's database of research projects in 1997 found that only thirty-four projects out of thirty thousand were explicitly directed toward organic farming systems and methods (about 0.1 percent of the database and of Department of Agriculture funding). Likewise, a study in 2003 found that organic agriculture research acreage was only about 0.13 percent of total research acreage. If anything, the studies overestimated the total proportion of research on organic farming because they focused on government-sponsored research; industrial research funding has overwhelmingly favored biotechnology and other types of conventional agriculture. Another sign of the weakness of the research field is the fact that researchers interested in career advancement tended not to select sustainable agriculture as a research field. Just as most federal spending for energy research is devoted to fossil fuels and nuclear energy, most agriculture research is dedicated to biotechnology and the problem of increasing monoculture production through industrial agriculture. A similar pattern of marginal funding plagues green chemistry. No major university's chemistry or chemical engineering department is dedicated to it, and green chemistry research programs tend to be viewed with disdain because they are seen as "too applied." Federal funding for research in this area has been on the order of 1 to 2 percent of the total funding of $300 million for chemistry research.[13]

From the optimistic perspective, research on renewable energy, organic agriculture, and green chemistry are examples of growing alternative research fields that attempt to fill the gaps of undone science. However, to the extent that scientists who sympathize with environmental movements wish to undertake research in the three areas, they will be joining subordinate research networks and facing the attendant career risks. The risks are probably greater for agricultural scientists who opt to work on organic research and chemists who opt to study green chemistry than for engineers who want to work on solar power or other forms of patentable renewable-energy technology. Since the late 1990s, renewable energy has become a growth industry, and in the first decade of the twenty-first century, a wave of solar energy start-ups recalled the wave of biotechnology start-ups in the 1990s. As a result, the research field has undergone a transition from the alternative status to alignment with competitiveness and industrial development goals. Individual researchers who make a decision to develop a patentable product and become an entrepreneur have the option of becoming wealthy, but in essence they are also leaving the prestige game of reputation building through publication.

Whatever the shortcomings of the alternative research fields, the fact that they exist and have shown some growth over time is, like the existence of attempts by some scientists to end some categories of weapons research, another indication of the permeability of the scientific field to shifts in agendas that are consistent with the goals of activists and advocates such as environmentalists. The alternative pathways in science, and their linkages to alternative pathways in industry, represent another dimension along which we can track the epistemic modernization of science.

## Conclusions

Although conflict and controversy among networks in a research field are inherent in the scientific field, two historical changes associated with globalization enhance the level of conflict. First we see an increasing emphasis on mission-based funding oriented toward technology transfer and industrial innovation, especially when geared to national industrial priorities and regional industrial clusters. As a result, the selection of research agendas becomes a policy problem to be addressed from the value perspective of the competitiveness of industrial (and military) innovation. Second there is a countervailing trend toward epistemic modernization, which involves opening the content of scientific research fields to greater public participation and influence. In some cases, conflicts within a scientific field between two research networks are associated with general societal conflicts, where two articulations of a public benefit (one defined by military-industrial organizations and one defined by civil society organizations) are in conflict. Yet because some degree of autonomy always exists in the scientific field, the relations of cooperation and conflict among networks in science do not always map neatly onto broader social conflicts and divisions.

Even where the alignments with broader social divisions are not readily visible, funding priorities shape network dominance. Because the priorities of funding sources reflect, however imperfectly, the interaction of the negotiated priorities of scientists and those of economic and political elites, the dominant networks of a research field tend to align with the interests of the elites. Researchers who are developing transferable and licensable technology will tend to win huge helpings of funding served on elegant platters, whereas those who wish to explore the health and environmental effects of such technologies may end up being sent to the kitchen to beg for the scraps off the table of the funding system. Because dominant networks tend to control access to the means of disciplinary reproduction (top journals, leading departments, best graduate students, and positions on funding panels), they can afford to ignore the subordinate networks and let them wither on the vine of inattention. No conspiracy theory is needed to explain the alignments that occur; one needs only to understand that the fields of science are not completely autonomous regarding the self-determination of the broad priorities of research agendas. To some degree, they never were: from the seventeenth century through the twentieth, scientists have always fought to maintain a degree of autonomy from extrascientific intervention. Attention to the specific institutional changes that have occurred in the era of globalization makes it possible to understand how the scientific field is increasingly a site where general societal conflicts play themselves out.

In view of the ongoing influence that industrial and political elites exercise over research agendas, not to mention the industrial processes for technological innovation, epistemic modernization could be viewed as little more than a strategy to colonize lay knowledge, co-opt civil society challengers, and quell internal dissidents and reform efforts. In discussing the phenomenon of epistemic modernization, one should not forget that the dominant changes in scientific fields will be driven by the dominant groups in society. Although the quasi autonomy of the scientific field also makes possible some exceptions, this fundamental "law" in the sociology of knowledge has been recognized since its discovery by a marginalized social scientist over a century and a half ago, and it is unlikely to be repealed anytime soon. However, the concept of epistemic modernization draws attention to another dimension of a nonautonomist approach to the scientific field in an era of globalization: the role of pressure and participation from below. As with reflexive modernization and ecological modernization, the question of the political significance or lack of significance of epistemic modernization should be left open to empirical research. Research on the topic is likely to result in nuanced determinations about occasions where the processes of epistemic modernization, including alternative pathways in science, are more and less significant.[14]

## Notes

This selection is based on an edited condensation by the author of chapter 2 of *Alternative Pathways in Science and Industry* (MIT Press).

1. On mode-two knowledge production, see Nowotny, Scott, and Gibbons 2001; on academic capitalism, see Slaughter and Leslie 1997; on audit culture, see Strathern 2000; on the triple helix, see Etzkowitz and Leydesdorff 1997; on asymmetric convergence, see Kleinman and Vallas 2001; on degrees of compromise, see Croissant and Restivo 2001; and on impure cultures, see Kleinman 2003.

2. On reflexive modernization, see Beck 1992; and on ecological modernization, see Mol and Spaargaren 2000.

3. On the field perspective, see Bourdieu 2001 and other chapters from Hess 2007. On changes in the theory, methods, and research-problem-area priorities of primatology due to internationalization and increasing participation by women, see Haraway 1989; on American and Japanese physicists, see Traweek 1992; and on strong objectivity, see Harding 1992.

4. On French and English physics, see Duhem 1982. On other examples, see Harding 1998, 1993; and Hess 1995, chap. 2.

5. Farkas (1999, 2002) found retrenchment happening during her dissertation fieldwork in the Dutch science shops during the early years of the twenty-first century. See also Leydesdorff and Ward 2005; Walchelder 2003.

6. On types of participatory research, especially the continuum between professionally initiated and lay-initiated research, see Moore 2006; on popular epidemiology, see Brown and Mikkelsen 1990; on community-based research

and the university, see Strand et al. 2003; and on the Philadelphia projects, see Puckett and Harkavy 1999.

7. On expertification, see Epstein 1996. On various types of citizen participation, including the consensus conference model, see Fischer 2000.

8. See Hess 1999, 2005. In Collins's terms (2000), biographical transformation involves the transition from participatory to contributory expertise.

9. On scientists and the Vietnam War, see Moore 2008 and Moore and Hala 2002. On later antiweapons activism, see Gusterson 1996. On the recombinant DNA controversy, see Krimsky 1982, 1992.

10. On Science for the People, see Beckwith 1986. On the history of the appropriate technology movement, see Kleiman 2000. The development of alternative research fields might be compared with "scientific and intellectual movements" (Frickel and Gross 2005) and countermovements in the sciences (Nowotny and Rose 1979). Alternative pathways in science have similar features to scientific and intellectual movements, but alternative pathways may tend to remain more marginalized than scientific and intellectual movements because they lack high-status leaders.

11. On the absolute growth of sustainable agriculture, see Organic Farming Research Foundation 2003. Hassanein (1999) charts the growth of sustainable agriculture research through the 1990s, including the development of research centers, and provides an analysis of an alternative form of research even within this alternative pathway: a more farmer-oriented research model based on farmer networks that occurs in what I call localist pathways in the food and agricultural field. On green chemistry, see Guterman 2000; Woodhouse 2006 and Woodhouse and Breyman 2005.

12. On the suppression of researchers at the National Renewable Research Laboratory, see Carman 2006. However, given that the presidential administration was also courting midwestern states through the ethanol initiative of 2006, the fired researchers were later rehired. See Holdren 1998; Renner 2001; and Sissine 1999. On the funding levels during the first year of the Obama presidency, see U.S. Department of Energy 2009 and Power Engineering International 2009.

13. On the database study and 0.1 percent figure for funding for organic agricultural research, see Lipson 1997. For a similar study, see also Anderson 1995. On acreage, see Organic Farming Research Foundation 2003. On career considerations in the selection of agricultural research fields, see Busch and Lacy 1981 and Goldberg 2001. On the industrial orientation and funding of agricultural research, see Busch 1994. The figure of 1 to 2 percent of $300 million (for National Science Foundation chemistry research), the absence of research university chemistry departments, and the perception that green chemistry proposals are seen as having a do-not-fund stigma are based on discussions with my colleague Woodhouse, who has researched the topic extensively. See especially Woodhouse 2006.

14. With respect to the nineteenth-century social scientist, see "Theses on Feuerbach" in Marx and Engels 1973.

## References

Agency for Healthcare Research and Quality. 2001. "Community-Based Participatory Research: Conference Summary." http://www.ahrq.gov/about/cpcr/cbpr.

Allen, Barbara. 2003. *Uneasy Alchemy: Citizens and Experts in Louisiana's Chemical Corridor Disputes*. Cambridge: MIT Press.

Anderson, Molly. 1995. "The Life Cycle of Alternative Agricultural Research." *American Journal of Alternative Agriculture* 10 (1): 3–9.

Beck, Ulrich. 1992. *The Risk Society: Towards a New Modernity*. Newbury Park, Calif.: Sage.

Beckwith, Jon. 1986. "The Radical Science Movement in the United States." *Monthly Review* 38 (1): 118–28.

Bourdieu, Pierre. 2001. *Science of Science and Reflexivity*. Chicago: University of Chicago Press.

Brown, Phil. 2007. *Toxic Exposures: Contested Illnesses and the Environmental Health Movement*. New York: Columbia University Press.

Brown, Phil, and Edwin Mikkelsen. 1990. *No Safe Place: Toxic Waste, Leukemia, and Community Action*. Berkeley: University of California Press.

Busch, Lawrence. 1994. "The State of Agricultural Science and the Agricultural Science of the State." In *From Columbus to ConAgra*, ed. Alessandro Bonanno, Lawrence Busch, William Friedland, Lourdes Gouveia, and Enzo Mingione, 69–84. Lawrence: University Press of Kansas.

Busch, Lawrence, and William Lacy. 1981. "Sources of Influence on Problem Choice in the Agricultural Sciences: The New Atlantis Revisited." In *Science and Agricultural Development*, ed. Lawrence Busch, 113–28. Montclair, N.J.: Allanheld, Osmun.

Carman, Diane. 2006. "Plug Pulled on Renewable Energy Gurus." DenverPost .com, February 14.

Collins, Harry. 2000. "Surviving Closure: Post-rejection Adaptation and Plurality in Science." *American Sociological Review* 65 (6): 824–25.

Croissant, Jennifer, and Sal Restivo, eds. 2001. *Degrees of Compromise: Industrial Interests and Academic Values*. Albany: SUNY Press.

Duhem, Pierre. 1982. *The Aim and Structure of Physical Theory*. Princeton: Princeton University Press.

Eglash, Ron, Jen Croissant, Giovanna Di Chiro, and Ray Fouche, eds. 2004. *Appropriating Technology: Vernacular Science and Social Power*. Minneapolis: University of Minnesota Press.

Epstein, Steven. 1996. *Impure Science: AIDS, Activism, and the Politics of Knowledge*. Berkeley: University of California Press.

Etzkowitz, Henry, and Loet Leydesdorff, eds. 1997. *Universities and the Global Knowledge Economy: A Triple Helix of University-Industry-Government Relations*. New York: Pinter.

Farkas, Nicole. 1999. "Dutch Science Shops: Matching Community Needs with University R&D." *Science Studies* 12 (2): 33–47.

———. 2002. "Bread, Cheese, and Expertise: Dutch Science Shops and Democratic Institutions." Ph.D. diss., Science and Technology Studies Department, Rensselaer Polytechnic Institute. http://www.scienceshops.org.

Fischer, Frank. 2000. *Citizens, Experts, and the Environment: The Politics of Local Knowledge*. Durham: Duke University Press.

Frickel, Scott, and Neil Gross. 2005. "A General Theory of Scientific/Intellectual Movements." *American Sociological Review* 70:204–32.

Goldberg, Jessica. 2001. "Research Orientation and Sources of Influence: Agricultural Scientists in the U.S. Land-Grant System." *Rural Sociology* 66 (1): 69–92.

Gusterson, Hugh. 1996. *Nuclear Rites: A Weapons Laboratory at the End of the Cold War*. Berkeley: University of California Press.

Guterman, Lila. 2000. "'Green Chemistry' Movement Seeks to Reduce Hazardous Byproducts of Chemical Processes." *Chronicle of Higher Education*, August 4, A17–A18.

Haraway, Donna. 1989. *Primate Visions: Gender, Race, and Nature in the World of Modern Science*. New York: Routledge.

Harding, Sandra. 1992. "After the Neutrality Idea: Science, Politics, and 'Strong Objectivity.'" *Social Research* 59 (3): 567–87.

———, ed. 1993. *The "Racial" Economy of Science: Toward a Democratic Future*. Bloomington: Indiana University Press.

———. 1998. *Is Science Multicultural? Postcolonialisms, Feminisms, and Epistemologies*. Bloomington: Indiana University Press.

Hassanein, Neva. 1999. *Changing the Way America Farms: Knowledge and Community in the Sustainable Agriculture Movement*. Lincoln: University of Nebraska Press.

Hess, David. 1995. *Science and Technology in a Multicultural World: The Cultural Politics of Facts and Artifacts*. New York: Columbia University Press.

———. 1999. *Evaluating Alternative Cancer Therapies: A Guide to the Science and Politics of an Emerging Medical Field*. New Brunswick, N.J.: Rutgers University Press.

———. 2005. "Medical Modernisation, Scientific Research Fields, and the Epistemic Politics of Health Social Movements." In *Social Movements in Health*, ed. Phil Brown and Stephen Zavestoski, 17–30. Malden, Mass.: Blackwell.

———. 2007. *Alternative Pathways in Science and Industry: Activism, Innovation, and the Environment in an Era of Globalization*. Cambridge: MIT Press.

Holdren, John. 1998. "Federal Energy Research and Development for the Challenges of the 21st Century." In *Investing in Innovation*, ed. Lewis Branscomb and James Keller, 299–335. Cambridge: MIT Press.

Irwin, Alan, and Brian Wynne, eds. 1996. *Misunderstanding Science? The Public Reconstruction of Science and Technology*. Cambridge: Cambridge University Press.

Kleiman, Jordan. 2000. "The Appropriate Technology Movement in American Political Culture." Ph.D. diss., Department of Political Science, University of Rochester.

Kleinman, Daniel. 2003. *Impure Cultures: University Biology at the Millennium.* Madison: University of Wisconsin Press.

Kleinman, Daniel, and Steven Vallas. 2001. "Science, Capitalism, and the Rise of the 'Knowledge Worker': The Changing Structure of Knowledge Production in the United States." *Theory and Society* 30:451–92.

Krimsky, Sheldon. 1982. *Genetic Alchemy: The Social History of the Recombinant DNA Controversy.* Cambridge: MIT Press.

———. 1992. "Regulating Recombinant DNA Research and Its Applications." In *Controversy*, 3rd ed., ed. Dorothy Nelkin, 219–48. Newbury Park, Calif.: Sage.

Leydesdorff, Loet, and Janelle Ward. 2005. "Science Shops: A Kaleidoscope of Science-Society Collaborations." *Public Understanding of Science* 14 (4): 353–72.

Lipson, Mark. 1997. *Searching for the "O-Word": Analyzing the USDA Current Research Information System for Pertinence to Organic Farming; Executive Summary.* Santa Cruz, Calif.: Organic Farming Research Foundation. http://www.ofrf.org.

Marx, Karl, and Friedrich Engels. 1973. *The German Ideology.* Part 1. Ed. C. J. Arthur. New York: International Publishers.

Mol, Arthur, and Gert Spaargaren. 2000. "Ecological Modernisation Theory in Debate: A Review." *Environmental Politics* 9 (1): 17–49.

Moore, Kelly. 2006. "Powered by the People: Scientific Authority in Participatory Science." In *The New Political Sociology of Science*, ed. Scott Frickel and Kelly Moore, 299–350. Madison: University of Wisconsin Press.

———. 2008. *Disrupting Science: Social Movements, American Scientists, and the Politics of the Military, 1945–1975.* Princeton: Princeton University Press.

Moore, Kelly, and Nicole Hala. 2002. "Organizing Identity: The Creation of Science for the People." *Research in the Sociology of Organizations* 19:309–35.

Nowotny, Helga, and Hilary Rose, eds. 1979. *Counter-movements in the Sciences: The Sociology of the Alternatives to Big Science.* Dordrecht: D. Reidel.

Nowotny, Helga, Peter Scott, and Michael Gibbons, eds. 2001. *Re-thinking Science: Knowledge and the Public in an Age of Uncertainty.* Malden, Mass.: Oxford University Press.

Organic Farming Research Foundation. 2003. *State of the States, 2nd Edition: Organic Systems Research at Land Grant Institutions, 2001–2003.* Santa Cruz, Calif.: Organic Farming Research Foundation. http://www.ofrf.org.

Power Engineering International. 2009. "DOE Budget Lifts Renewables, Cuts Nuclear and Coal." http://pepei.pennet.com.

Puckett, John, and Ira Harkavy. 1999. "The Action Research Tradition in the United States." In *Action Research*, ed. Davydd Greenwod, 147–68. Philadelphia: John Benjamins.

Renner, Michael. 2001. "U.S. Contempt for Alternatives." *World Watch* 14 (5): 2.

Sissine, Fred. 1999. "Renewable Energy: Key to Sustainable Energy Supply." *Congressional Research Service Report to Congress 97031.* Washington: Library of Congress. http://www.ncseonline.org.

Slaughter, Sheila, and Larry Leslie. 1997. *Academic Capitalism: Politics, Policies, and the Entrepreneurial University*. Baltimore: Johns Hopkins University Press.

Strand, Kerry, Sam Marullo, Nick Cutforth, Randy Stoecker, and Patrick Donahue. 2003. *Community-Based Research and Higher Education: Principles and Practices*. San Francisco: Jossey-Bass.

Strathern, Marilyn, ed. 2000. *Audit Cultures: Anthropological Studies in Accountability, Ethics, and the Academy*. New York: Routledge.

Traweek, Sharon. 1992. "Border Crossings: Narrative Strategies in Science Studies and among Physicists in Tsukuba Science City, Japan." In *Science as Practice and Culture*, ed. Andrew Pickering, 429–65. Chicago: University of Chicago Press.

U.S. Department of Energy. 2009. "President's Budget Draws Clean Energy Funds from Climate Measure." http://www.eere.energy.gov.

U.S. Environmental Protection Agency. 2004. "Science of Environmental Justice Working Conference." http://www.epa.gov/osp/EJ/ejconf0405.pdf.

Walchelder, Joseph. 2003. "Democratizing Science: Various Routes and Visions of Dutch Science Shops." *Science, Technology, and Human Values* 28 (2): 244–273.

Woodhouse, Edward. 2006. "Nanoscience, Green Chemistry, and the Privileged Position of Science." In *The New Political Sociology of Science*, ed. Scott Frickel and Kelly Moore, 148–81. Madison: University of Wisconsin Press.

Woodhouse, Edward, and Steve Breyman. 2005. "Green Chemistry as Social Movement?" *Science, Technology, and Human Values* 30 (2): 199–222.

KARIN BÄCKSTRAND

## 25. Civic Science for Sustainability:
## Reframing the Role of Experts, Policy-Makers
## and Citizens in Environmental Governance

AT THE JOHANNESBURG WORLD SUMMIT on Sustainable Development, the science and technology communities, along with other non-state actors, were singled out as major partners in the quest for sustainability. This is in line with calls for refashioning scientific expertise into a more transparent, accountable, and democratic enterprise. Participatory, civil, citizen, civic, stakeholder, and democratic science are catchwords that signify the ascendancy of participatory paradigms in science policy. The participatory turn to scientific expert advice can be interpreted as a resistance to the perceived scientization of politics, which implies that political and social issues are better resolved through technical expertise than democratic deliberation. The notion of civic science, which is rather vague and elusive, serves as an umbrella for various attempts to increase public participation in the production and use of scientific knowledge. Civic science alludes to a changing relationship between science, expert knowledge, and citizens in democratic societies. In this perspective, citizens and the public have a stake in the science-politics interface, which can no longer be viewed as an exclusive domain for scientific experts and policy-makers only.

What is the scope for restructuring scientific expertise in a more democratic fashion? Is it possible, or even desirable, to include citizen participation in the production, validation, and application of scientific knowledge? While there is lip service paid to the need for civic science, the question of how it can be realized is largely unresolved. The rhetoric of civic science, which can be conceived as a response to the dangers of Bovine Spongiform Encephalopathy (BSE) disease and the risks of genetically modified food, signifies the heightened public concern about environment issues. Hence, the status of scientific expert knowledge in democratic societies as well as the role of the citizen in the age of experts has been brought to the fore. Public concern and controversy also surround the application of biotechnology and reproductive technology, the storing of toxic and nuclear waste, climate change, and the human genome project.

## Conceptualizing Civic Science

In this chapter, I review the notion of civic science by mapping how the concept is articulated in international relations, science studies, and democratic theory. I also examine the account of civic science underpinning the field of sustainability science that purportedly embraces a more participatory account of scientific expertise. A central proposition is that the promotion of civic science needs to be coupled with a theoretical understanding of the institutional, normative, and epistemological divisions characterizing the term. This chapter begins such an effort by mapping the rationales, justifications, and limitations of civic science. It aims to provide a conceptual grounding for future case studies of civic science in the context of biodiversity, bio-safety, climate change, and desertification.

Climate change, management of natural resources, and bio-safety represent areas where participatory expert knowledge is called for. The rise of global environmental regimes has meant that models for scientific advice on the domestic level now are extended to multilateral scientific assessment (Miller 2001a, 253).This prompts the question of how to find a balance between specialized expert knowledge and public participation in science.

In international relations the science-politics interface has been framed primarily as a matter for scientists and decision-makers. Scientists inform policy-makers and policy-makers turn to science for knowledge and technical assistance. I suggest that the science-politics interface needs to be reframed to include the triangular interaction between scientific experts, policy-makers and citizens. The citizen is not just the recipient of policy but an actor in the science-policy nexus. This is in line with the argument that "[a]ny model of the relationship between scientific expertise and public policy-making should include the public sphere, that is those common spaces in which citizens meet to discuss public matters—In normative terms, the concept of public sphere refers to democratic values, namely public accountability and active citizenry" (Edwards 1999, 169). In this vein, scientific knowledge can be conceived as a global public good in which the citizens have a stake.

The first section reviews how the discipline of international relations has grappled with scientific advice and how the question of civic science is featured in this scholarship. The second section conceptualizes the elusive concept of civic science. Civic science hosts many ambitions, such as enhancing public understanding of science, increasing citizen participation, diversifying representation in, and promoting democratization of science. In the third section I spell out three rationales for civic science mirrored in the literatures of risk society, science studies, and normative democratic theory. The fourth section proceeds by examining the notion of civic science that underpins the evolving field of sustainability science. The concluding section summarizes the institutional, epistemological, and normative challenges connected to civic science.

## International Relations and Civic Science

Civic science is a nascent issue in the discipline of international relations (IR) that primarily has addressed the institutional aspects of advisory science in global environmental politics. As of today, international scientific and technical advisory bodies are central in providing input for international environmental negotiations. The rise of "negotiated science" is a prominent feature in the ongoing diplomatic endeavors associated with climate change, air pollution, ozone depletion, biodiversity, and desertification. Scientific assessment is increasingly organized on a multi-national and multi-disciplinary basis. For example, the negotiation and operation of the long-range transboundary air pollution regime (LRTAP) rests on scientific assessment involving almost two thousand scientific and technical experts from a multitude of countries (Bäckstrand 2001).

There is a lacuna in IR with respect to the relationship between expertise and democratic governance in environmental politics. The normative aspects of scientific expert advice, including the issues of representation, transparency, participation, accountability, and legitimacy are largely absent (Bäckstrand 2003). The legacy of isolating IR from social theory at large precludes a notion of "political" that includes the public. The dichotomy between the orderly and democratic inside of domestic politics and the disorderly anarchic outside of international affairs pervades the discipline (Walker 1993). Consequently, in this perspective, democratic participation in science is primarily confined to the context of domestic policy-making and is more limited in international diplomacy and scientific assessment.

In IR, research has revolved around the links between scientific expert knowledge and processes of global environmental governance. The research agenda has been framed around primarily two sets of issues. Liberal-institutionalism has been preoccupied with the conditions for effective uptake of scientific expert knowledge in international regimes. In contrast, the locus of constructivist IR scholarship has been on the contingent, uncertain, and normative context for scientific expertise. The first issues concern the optimal conditions for making scientific experts influential in the decision-making process and international institutions. Regime-theoretical studies primarily focus on how science effectively can assist in mitigating global environmental risks through diplomacy, regime-building, and multilateral negotiations (Young 1997; Young 1999; Andresen and Skodvin et al. 2000). Knowledge-based explanations of regime formation, such as the epistemic community, signify this approach (Haas 1989; Haas 1992). The central argument is that the mobilization of consensus among transnational networks of scientific experts is instrumental in facilitating international policy coordination and agreement. Another issue is how the organization of scientific expertise can promote utilization of scientific knowledge in international environmental regimes and prevent the politicization of scientific expertise and the exploitation of scientific uncertainties by recalcitrant actors. A precondition for the effective use of scientific knowledge is that there is a shared understanding of the nature of the problem among the authorized experts and that this consensus, in turn, is transmitted to international institutions as well as incorporated into policy. Recent studies move beyond the assumption of shared norms and aim to explain why some global norms—such as the normative compromise of liberal environmentalism—become selected and institutionalized (Bernstein 2001).

The constructivist research agenda revolves around how scientific knowledge and practices are embedded in various cultural and political contexts as well as in societal discourses. Research in this direction adopts insights from a multiplicity of perspectives such as discourse analysis, science and technology studies, and constructivism. Studies of the role of scientific discourses in propelling policy action with regard to stratospheric ozone depletion, climate change, and biological diversity signify this approach (Litfin 1995; Miller and Edwards 2001a). The plethora of literature on global environmental assessment underlines the importance of enhancing saliency, credibility, and legitimacy of scientific assessment (Cash and Clark 2001, 9). However, the question of how, and by what means, to institutionalize credibility and legitimacy of scientific assessment is unanswered. This issue looms large partly because there is a lack of theoretical foundation for coupling democratic citizen participation with scientific assessment.

In the wider post-positivist scholarship there is an ongoing critical revaluation of the status of expert knowledge in modern society. What are the boundaries between scientific and non-scientific knowledge, expert and lay knowledge, global and local knowledge, risk assessment, and risk management? On what basis can these boundaries be maintained? Recent work marrying international relations and science studies starts from an analysis of the co-production of the political order and scientific knowledge (Miller and Edwards 2001a). The production of scientific knowledge is not viewed as external to environmental politics as in the epistemic community approach. The boundaries between institutions of scientific expert advice and policy-making are blurred (Miller 2001a).

An underlying premise is that scientific knowledge and practices operate inside rather than outside of politics. A key question is what counts as credible, authoritative, and legitimate expert knowledge. Instead of taking shared understanding and scientific consensus at face value, the purpose is to unravel the process by which actors come to share common worldviews. Science and politics are in this vein indistinct realms with fluid boundaries subject to negotiation. Research on boundary work (Gieryn 1995) and boundary organizations (Guston 2001) highlight how legitimacy, credibility, and authority of scientific expert knowledge are maintained by establishing borders between the scientific and political spheres. The implication of this analysis is that scientific advisory processes are deeply intertwined with political processes. Without denying the critical importance of scientific knowledge to environmental policy, this perspective highlights the normative and value-laden context for scientific inquiry. Recent studies of climate science and governance illustrate the conflict between a top-down and a bottom-up scientific assessment process (Miller 2001b). This opens up a space for theorizing the tensions between democratic and technocratic governance in environmental affairs. Research in risk society (Beck 1992; Beck and Giddens et al. 1994), environmental sociology, science studies, and democratic theory has addressed the prospect for democratic expertise in policy-making as well as examined the promises and pitfalls of enacting civic science. The next section discusses the contested concept of civic science.

## Civic Science: Participation, Representation, or Democratization?

Civic science has many meanings and aspirations. It is used interchangeably with civil, participatory, citizen, stakeholder, democratic science, and lay knowledge. Civic science has been defined as the efforts by scientists to reach out to the public, communicate scientific results, and contribute to scientific literacy (Clark and Illman 2001). Citizen science, on the other hand, denotes a science that is developed and enacted by the citizens, who are not trained as conventional scientists (Irwin 1995). There is wide disagreement with respect to the question if citizens can, or should be able to deliberate on scientific matters. For instance, should the citizenry be invited to deliberate about the application of science or technology or should they be engaged in scientific problem formulation? In other words, should lay knowledge be limited to the process of risk management or should it also be integrated in risk assessment processes (Kleinmann 2000)?

Civic science harbors many ambitions, such as increasing public participation in science and technology decisions, securing a more adequate representation in science, vitalizing citizen and public deliberation in science, or even installing a democratic governance of science. Representation, participation, and democratization can be conceived as three different but interconnected dimensions. First, civic science as *participation* underlines the importance of increasing public participation by bringing citizens and civil society to the heart of the scientific endeavor and by embarking on participatory practices in the conduct of science. Consensus conferences, participatory technology assessment, citizen juries, and public hearings in science and technology affairs are examples of institutionalized practices that attempt to incorporate citizens in environmental risk management (Weale 2001). Secondly, civic science defined in terms of *representation* aims at reversing the skewed representation in the production of science. The lack of representation of women and indigenous people in the scientific enterprise was highlighted at the World Summit on Sustainable Development (International Council for Science 2002a). Moreover, the poor representation of scientists from developing countries and countries in transition in international scientific assessment processes is recognized as highly problematic both for the quality and legitimacy of scientific knowledge (VanDeveer 1998; Biermann 2002). Which and whose knowledges are represented as true, legitimate, and authoritative? These insights are supported by critical feminist epistemology questioning the universal aspiration of modern science and calling for an inclusion of local, subjugated knowledge in societal and technological decision-making (Haraway 1996; Harding 1998). The representative paradox of science is that a very small group who holds the title of "scientist" can speak on behalf of a universal humanity (Fuller 2000, 8).

Thirdly, civic science as *democratization* challenges the conduct of scientific problem-solving by aspiring to transform the institutions of science to incorporate democratic principles. Proposals to increase representation and participation in science do not necessarily entail a transformation of scientific norms, methods, and practices. However, the aim to democratize science is a more challenging issue that goes beyond the issue of stakeholder representation and participation. Can the rules of modern democracy be readily transferred to the heart of scientific inquiry without compromising scientific quality and politicizing scientific expertise?

Embracing civic science can be conceived as a response to two developments; the emergence of "big" planetary science and the "legitimacy crisis" for modern science. First, civic science can be conceived as a reaction to the expansion of "mega-science" enabled by innovations in global environmental modelling. The international co-ordination, standardization, and harmonization of scientific assessment signify the emerging Earth Systems Science (Jasanoff and Wynne 1998, 58). This is epitomized by the expansion of global models of atmospheric, hydrological, and terrestrial systems in international negotiations, research programs, and international organizations. This emerging global environmental change science has been represented as global and universal knowledge even if the modelling activities are concentrated in a few laboratories in the Northern hemisphere. The top-down model of environmental problem-solving grants power to networks of scientific experts, specialists, and bureaucrats in environmental science. Critics point to a failure to couple global western scientific knowledge with local and indigenous knowledge, agendas, needs, and concerns. A remedy for this is to increase public participation in scientific assessment processes, recognizing the "glocal" level of knowledge production.

Secondly, the call for civic science is a response to the legitimacy crisis of science, which is more pronounced in Europe in the backdrop of food safety scares in the 1990s. The increased reliance of expert advice, negotiated and regulatory science defines issue areas from global warming, toxic waste, and genetically modified organisms (GMOs). However, inflationary use of expert advice has paradoxically produced more uncertainty (Rutgers and Metzel 1999, 148). Science has been called on to provide a firm basis for justifying and making political decisions credible. Scientific knowledge is in many areas provisional, uncertain, and incomplete. Thus, competing expert knowledge has in many instances given rise to a battle between experts and counter-experts. Corporate science has contested environmental advocacy science and vice versa (Jasanoff 1990; Fischer 2000). This politicization of scientific knowledge has paved the way for the erosion of the authority and legitimacy of science as objective knowledge. When the public experiences that science can be both contested and uncertain, the policy-process, which relies on purportedly objective knowledge, loses credibility. The erosion of the legitimating function of science in certain domains has spurred the calls for making science more accountable and democratic. In the next section I explore three rationales for civic science and highlight the normative and epistemological divides surrounding the term.

## Three Rationales for Civic Science

What are the reasons for enhancing public participation in science and making science democratically accountable? First, civic science, if geared toward enhancing public understanding, can potentially mitigate the growing public disenchantment with scientific expertise. Secondly, the sheer complexity of global environmental problems necessitates a reflexive scientific expertise that incorporates a wide array of lay and local knowledge. Thirdly, the primary purpose of civic science is to extend the principles of democracy to the production of scientific knowledge.

### Civic Science as Restoring Public Trust in Science

The first rationale for civic science is to enhance public understanding of science by improved communication, scientific literacy, and outreach. This emerged in the backdrop of the rhetoric of openness that marked the European policy debate on science and technology issues in the 1990s. The rationale was to enhance transparency, civil participation, dialogue, and accountability in science policy (Levidow and Marris 2001, 345). An overarching effort was to bridge the increasing gulf that existed between science and society, which was epitomized by the vehement public reaction to the BSE disease and genetically modified food in Europe and calls for implementing the precautionary principle. A contrasting tale can be found in the United States where GMO food is largely accepted in the public eye and where risk assessment and "sound science" are entrenched practices for assessing health and environmental risks associated with GM crops. The food crisis in Europe reflects a fundamental lack of confidence among citizens toward the scientific and regulatory management of these issues. As a corollary, the public has become more skeptical of both governmental and corporate science while investing more trust in the perceived "independence" of science authorized by nongovernmental organizations such as Greenpeace.

Better communication from the scientists to the public, deeper public understanding of science, and improved scientific literacy have been seen as remedies. In this perspective, the basic root of the declining confidence in expert knowledge is the public misunderstanding of science. The so-called deficit model emerged as a dominant framework for governments' science policy in response to the reactions among the citizens. A central assumption in this model is that the strong reaction of the public is based on irrationality, fear, ignorance, and lack of knowledge. In this vein, the mismatch between scientific and popular risk assessment stems from insufficient and inadequate knowledge among the public. The remedial strategy is information dissemination and "getting the scientific facts right." If citizens were more scientifically literate, the reasoning goes, they would do the same risk assessment as scientific professionals.

The deficit model has been criticized on many accounts and is increasingly rejected for its problematic assumptions (Frewer and Salter 2002). While dressed in the language of transparency, dialogue, and participation, the traditional mode of top-down scientific expert knowledge is still retained. A hierarchy is established between scientists and non-scientists and between enlightened scientific experts and ignorant laymen. Communication is one-way and on unequal terms, from the scientists to the public. The nature of scientific knowledge is not problematized in spite of the growing recognition that scientific knowledge is provisional and uncertain in many regulatory domains. This assumes that scientific knowledge is superior compared to other forms of knowledge. The stewards for sustainability should be scientists and engineers who need to reach out to the public. Needless to say, this model of civic science falls short from a more democratic model of public understanding that seeks to establish dialogue, collaboration, and deliberation between experts and citizens.

## Civic Science and the Complexity of Environmental Problems

The second rationale for civic science is a response to what has been perceived as the accelerating complexity of global environmental problems. In this sense, the condition of indeterminacy prompts the need for a new kind of science: "In terms of nature, the central paradox is that while the scale of control afforded by science and technology continues to increase, so does the domain of uncertainty and risk" (Sarewitz 2000, 91). Civic science is ultimately justified by an epistemological argument. Collective decision-making in the global environmental arena is fraught with uncertainty since scientific knowledge of global environmental risks is inherently limited, provisional, and value-laden. This condition of uncertainty, contingency, and indeterminacy prompts a need for a more pragmatic and open-ended decision process. In this respect, politics is a substitute for certainty (Saward 1993, 77). In light of non-remedial scientific uncertainties, ecological vulnerability and irreversibility, the policy process should be open, transparent, and institutionalize self-reflection.

The gist of the argument is that we are witnessing a transition from normal to post-normal science. The concept of post-normal science captures issues defined by high decision stakes, large system uncertainties, and intense value disputes (Funtowicz and Ravetz 1992, 267). Problems such as climate change, GMOs, or biodiversity, which are fraught with uncertainties, cannot be adequately resolved by resorting to the puzzle–solving exercises of Kuhnian normal science. Established normal scientific practices for problem-solving and risk assessment cannot provide the final answers to post-normal problems. In a situation involving large complexity, radical uncertainty, and high stakes, new scientific practices to ensure quality control have to be established. This encompasses a re-orientation of science toward incorporating multiple stakeholders. Peer review should include "extended peer communities" in order to enhance dialogue between stakeholders such as the NGOs, industry, public, and the media. This is in line with the call for a "democratization of science," i.e., wider participation in scientific assessment beyond a narrow group of scientific elites. However, the proponents for increasing citizenry and public accountability in scientific endeavors are driven not by a general desire for democratization but to make science more effective (ibid., 273). The incorporation of lay knowledge in scientific assessment does not rest on the assumption that lay knowledge is necessarily truer, better, or greener (Wynne 1994). However, due to the uncertainty of future environmental outcomes, possible surprises, and ecological catastrophes, a multiplicity of perspectives can prevent the narrowing of alternatives.

The implications of this paradox of incalculability, uncertainty, and even undecidability of environmental risks (Adam and Loon 2000, 13) have also been addressed in theories of risk society and reflexive modernization (Beck 1992; Beck and Giddens et al. 1994). The transition from industrial society (with its calculable risks) to risk society (with its incalculable mega-hazards) requires a redefinition of the rules, principles, and institutions of decision-making. The reality of the new environmental risk will force the redesign of the basic norms and institutions of societies. This includes the discourses and practices of science, which are at the heart of theories of risk society and reflexive modernization. The de-monopolization and democratization of science imply that authoritative decisions should not be made by a narrow group of experts, but should include a wider spectrum of stakeholders (Beck 1992, 163). NGOs, the public, and business should become active co-producers in the social process of constructing knowledge, revitalizing "sub-politics" as conceived in the risk society thesis. The whole argument rests on the assumption that we face new types of global ecological threats and techno-hazards. Beck's notion of reflexive scientization captures the idea that scientific decision-making on environmental risks should open up for social rationality. A modernization of modernity and science is needed. Hence, the traditional objectivist account of science has to be replaced by a more inclusive science that institutionalizes self-doubt, self-interrogation, and self-reflexivity (Beck 1992).

*Civic Science as the Democratization of Science*

The most far-reaching notion of civic science is found in democratic theory and post-positivist policy studies. Citizen participation and deliberation on issues that have bearing on people's everyday lives are regarded as the normative core of democracy (Cunningham 2002). The realm of science and technology constitutes such an arena. What are the reasons for bringing citizen participation and knowledge(s) to the scientific sphere? The first justification for a broader citizen involvement in science and technology is made by those who favor "strong" democracy (Barber 1984), which encompasses participatory, not only representative, democracy. Secondly, people should be able to deliberate on issues that affect their lives. Basically, those who bear the consequences of decisions should be able to have a say (Harding 2000, 127). Science and technology decisions have in many instances ramifications on the everyday life of citizens. The release of GM food, storing of toxic and nuclear waste, and reproductive technologies constitute such a domain. Thirdly, citizen participation can in many cases contribute significantly to scientific inquiry. Local knowledge has in many cases positively complemented professional scientific expertise. Diversity in expert knowledge is a desirable goal in itself (Harding 1998).

There seems to be an incompatibility between the quest for open-ended deliberation in democracies and the aim of prediction and control in science. "The fact that indeterminacy is not only inevitable, but essential, to democracy—something to be embraced rather than overcome—does not comport well with a scientific worldview whose most legitimating measures of success are predictive certainty and control of nature" (Sarewitz 2000, 92). However, the conflict between these two realms eases if science is viewed as bounded rationality (Miller and Edwards 2001b, 19). This perspective recognizes the contingency of scientific claims and that scientific practices are deeply ingrained in cultural and political processes. The democratization of scientific expertise prompts us to rethink our understanding of scientific knowledge itself. This entails questioning the borders between science and non-science, expert and lay knowledge, universal and local knowledge. A constructivist conception of knowledge paves the way for a more citizen-oriented deliberative approach to risk analysis, where local knowledge can be incorporated into risk assessment (Fischer 2000, 246). The democratic version of civic science argues that the ordinary citizen is capable of more participation than is generally recognized. This echoes discursive or deliberative democracy that has dealt with the scope of citizen participation beyond traditional electoral politics. A basic tenet in this model is to promote public use of reason, argument, and free deliberation. Free deliberation has the potential to transform preferences, enable a new collective will, and render public decisions more legitimate. The model of deliberative democracy can therefore bridge the gap between the expert and citizen. Participatory risk assessment can be conceived as an extension of deliberative democracy. However, can insights from the participatory, deliberative, and communicative model of democracy be applied to the institutions of scientific knowledge production? Most experiments with consulting citizens for technological decisions—such as citizen juries, consensus conferences, and technology assessments—are more situated in the realm of public policy while risk assessment is still regarded as the exclusive domain for scientific experts. The method of integrated assessment focus groups is one exception that aims to incorporate citizen knowledge in scientific problem formulation (Durrenburger and Kastenholz et al. 1999, 342).

Four questions have been raised against a democratic version of civic science. First, is it possible to extend principles of democracy to the heart of science, which has its own internal procedures and mechanisms for the production, verification, and control of authoritative knowledge? An unsettled issue is whether the rules for production of scientific knowledge will have to change in order to enact civic science. Is it possible or even desirable to reform the basic operation of science to incorporate effectively citizens and other stakeholders? Civic science can be conceived as an instrument to dethrone science or to deprive scientific knowledge from its authority and legitimacy conferred by society. Little guidance is provided on how the practices tied to scientific knowledge production, such as peer review, should be redesigned, complemented, or replaced.

Secondly, skeptical voices argue that citizen deliberation in science will be cumbersome, time-consuming, ineffective, and slow. Even an educated citizenry would have problems grasping the complexities of the highly specialized knowledge of environmental science. Elite models of democracy are highly skeptical of lay citizen participation. The ordinary citizen does not only lack time and capacity to understand the complexity of issues, but the public can be outright ignorant and irrational. Citizens do not have the knowledge to rationally calculate the risk of new technologies. They should trust specialized experts as they trust their political representatives.

Thirdly, the advent of global environmental problem-solving may limit the scope for civic science. Scientific assessments are increasingly global in scope relying on multi-disciplinary and multi-national collaborative research networks. The ongoing experiments with citizen and participatory expertise have primarily taken place at the domestic level. Is the strong version of civic science compatible with the effort to manage global environmental risks relying on global modelling and "big science"? How can local expertise be coordinated to provide alternative knowledges in transboundary or global risk management?

Fourthly, deliberative democracy may be insufficient in promoting the democratization of scientific expertise. The application of science and technology may be subject to public deliberation but not necessarily the production of science (Gaffaney 2001, 17). Deliberation does not necessarily change the ground rules for debate and may ignore the way power enters speech itself. The power largely resides in setting the agenda and establishing norms and rules for decision-making. For example, if "sound science" and risk assessment is the dominant framework for public deliberation on environmental risks, this will ultimately exclude alternative discourses and actors. Protest and resistance could change the decision-making framework from the risk assessment paradigm to a precautionary approach. Hence, participatory democracy has been advanced as an alternative model as it represents a more manifest critique of power and makes the exercise of power transparent.

Civic science should not be seen as a magical recipe for all cases and circumstances. Proponents for the democratization of science strongly stress that subjugated, local, and indigenous knowledge should not necessarily be regarded as better or truer than modern scientific knowledge. In the end, to find the appropriate balance between technical and communicative rationality is a pragmatic and context-dependent judgment. Both technical expert knowledge and ethical judgments are needed in science-based decision-making (Barry 1999, 215). In certain cases technocratic strategies may prove to be more adequate in resolving environmental problems and attaining sustainability goals. Vice versa, in post-normal environmental risk areas surrounded by large scientific uncertainties and even ignorance, a model of civic science that includes societal stakeholders may be more effective. Public questioning of science constitutes a healthy feature of democracy, and calls for transparency in science do not automatically represent an anti-scientific position. A democratic model of civic science will enhance active citizenry, public engagement, and scrutiny (Durant 1999, 317). The next section takes stock of theory and practice of sustainability science to examine how this field has grappled with civic science.

## Civic Science in Sustainability Science

How do the current proposals to restructure science toward the goals of sustainable development fare with civic science? Gearing science toward sustainable development means "that sustainability science must be created through the processes of co-production in which scholars and stakeholders interact to define important questions, relevant evidence, and convincing forms of argument" (Kates et al. 2000a, 2). Hence, in the evolving field of sustainability science a more participatory account of scientific expertise is articulated.

The concept of sustainability science articulates a proactive, inter-disciplinary, transparent science that works in tandem with the needs of society (Kates et al. 2000b). A key focus is the dynamic interaction and interdependence between nature and society. In the past decade, national science academies have worked in collaboration with international scientific associations to redefine the functions, mandate, and scope for scientific inquiry. The ensuing self-reflection within the scientific community itself has consolidated a new vision for a science that is harnessed for the goals of sustainable development. An overarching idea is that science needs to turn toward society and even establish a "new contract" with society. The new model for sustainability science was consolidated in preparation for the World Summit for Sustainable Development. Interdisciplinarity, policy-relevancy and holistic perspectives are cornerstones of this new model of science. The overarching goal is to uncover the resilience levels for natural and human systems. Collaboration across disciplinary divides is a crucial component, both within and between natural science, engineering, social science, and humanities.

Stakeholder participation, transparency, partnership, and dialogue are code words for enacting a more inclusive science-policy relationship. This entails participatory procedures involving scientists, stakeholders, advocates, active citizens, and users of knowledge (ibid). Sustainability science has to be accountable beyond peer review and include a variety of actors in assessment processes (International Council for Science 2002b, 7). Scientists have to engage more in communication with the public with regard to scientific results. This also means bridging the knowledge gap and digital divide between North and South and providing developing countries with opportunities to participate in scientific assessment on more equal terms. Scientific capacity-building in the Third World and partnerships between industrialized countries and developing countries are therefore crucial components (ibid., 8). Moreover, the local-global connectivity is a central aspect of sustainability science. Global knowledge about environmental degradation has to be coupled with local knowledge to produce sustainable solutions. In the quest for sustainability, "universal" knowledge must be connected to "place-based" knowledge (ibid., 19). As a corollary, indigenous or traditional knowledge is recognized as a cumulative body of knowledge that can provide alternative, local perspectives. Science and traditional knowledge should be coupled in order to realize a more equitable partnership as well as mutual learning (International Council for Science 2002a, 16).

Nevertheless, the focus is more on participation than on changing the rules and practices of scientific knowledge production, utilization, and communication. Sustainability science envisions an increased transparency and participation in science and technology in order to foster the legitimacy of the scientific endeavor. Science also needs to enhance its communicative skills and outreach to initiate broader public involvement in science and technology. These proposals can be conceived as a step toward the kind of reflexive scientization that Beck calls for (Beck 1992). However, increased participation in scientific assessment does not necessarily have bearing on the practices, norms, and institutions of scientific knowledge production.

Sustainability science does not address how the practices of science have to change to accommodate democratic participation. The implications for scientific knowledge production and practice are left unanswered, namely, how norms, institutions, and procedures in science have to change to enable broader participation (Gallopin and Funtowicz et al. 2001, 2). In this sense, there is a lack of a coherent social science perspective. While raising critical issues on how to make science more transparent and responsive to the needs of society, the field of sustainability science is still an expert-driven inter-disciplinary endeavor.

## Conclusion

The notion of civic science prompts us to rethink the relationship between science, knowledge, democracy, and environmentalism. The implications for the field of international relations are that we need to move beyond instrumentalist and managerial conception of science and bring the normative issues tied to the employment of scientific expert advice to the forefront. Representation, democracy, participation, and legitimacy are crucial issues in facilitating a constructive science-policy dialogue. This means paying attention to the intermediary role of citizens in science and technological decision-making.

Civic science is essentially a contested term, hosting conflicting institutional, normative, and epistemological dimensions. In the wake of the declining public trust in scientific expertise, civic science has been advanced as a solution to reverse the growing public distrust in science. A "thin" conception of civic science starts from the premise that public trust in science and technology can be restored through improved science communication, scientific literacy, and public understanding of science. A stronger account of civic science advocates reorienting science toward greater institutional reflexivity and responsiveness to citizens. Finally, the version of civic science as democratization suggests that scientific norms, institutions, and procedures need to be reformed in accordance with democratic principles.

Civic science has been put into practice through various institutional innovations such as public hearings, consensus conferences, deliberative polls, and participatory technology assessments. However, these experiments with participatory inquiry have taken place primarily in the domestic setting. There are limited experiences of citizen participation in multilateral diplomacy and scientific assessment. Another unsettled issue with regard to civic science is whether the citizenry should be invited to the heart of scientific endeavor, i.e., to participate in production of scientific knowledge or confined to deliberations about the applications of science?

The fault-line between the different proposals for institutionalizing civic science, especially the last one, revolves around the epistemological dimension. What is the nature of scientific knowledge? Is it defensible to privilege scientific knowledge over other knowledge forms? Civic science represents a very different project for the post-positivist view of science compared to the objectivist perspective. The former questions the boundary between scientific expert knowledge and lay knowledge, between global western knowledge and local indigenous knowledge. In this perspective, all expert knowledge is situated in a specific political and cultural context, inherently value-laden and imbued with worldviews. As a corollary, scientific and technological decision-making should rest on participation by and collaboration among scientists, citizens, and civil society. In contrast, an objectivist epistemology emphasizes the unique ness of scientific knowledge epitomized by its systematic features, its transformative effects, and its global impacts. The systematic features of science, in terms of the capacity to observe, explain, describe, and represent the world, reflect an unprecedented accumulation and progress of knowledge. Without denigrating the important contributions of local, indigenous, and everyday knowledge, these knowledge forms do not display the systematic and universal features of modern science. In this vein, the uniqueness of science grants natural scientists and engineers a continued privileged status in the quest for uncovering the scientific aspects for sustainability.

Hence, an unresolved issue is if the stewards of scientific knowledge production should be scientists and engineers or if the conduct of science should be geared toward a participatory, reflexive, and collaborative effort involving societal stakeholders. However, no universal solution can be offered with respect to the balance between democratic and technocratic modes of scientific decision-making. The success of civic science is largely dependent on the context, i.e., the nature of the environmental risk and problem at hand. Finding a balance between traditional scientific inquiry and participatory expertise and between technical and deliberative approaches will be an ongoing endeavor.

## References

Adam, Barbara, and Jost van Loon. 2000. "Introduction: Repositioning Risk: The Challenge for Social Theory." In *The Risk Society and Beyond. Critical Issues for Social Theory*, edited by Barbara Adam, Ulrich Beck, and Jost van Loon. London/Thousand Oaks/New Delhi: Sage Publications.

Andresen, Steinar, Tora Skodvin, Arild Underdal, and Jørgen Wettestad. 2000. *Science and Politics in International Environmental Regimes. Between Integrity and Involvement*. Manchester and New York: Manchester University Press.

Bäckstrand, Karin. 2001. *What Can Nature Withstand? Science, Politics and Discourses in Transboundary Air Pollution Diplomacy*. Lund Political Studies 116, Dissertation. Department of Political Science, Lund University.

———. 2003. "Precaution, Scientisation or Deliberation? Towards a Green Science Policy Interface." In *Liberal Democracy and Environmentalism*, edited by M. Wissenburg and Y. Levy. London and New York: Routledge.

Barber, Benjamin. 1984. *Strong Democracy*. Berkeley: University of California Press.

Barry, John. 1999. *Rethinking Green Politics. Nature, Virtue and Progress*. London: Sage Publications.

Beck, Ulrich. 1992. *Risk Society: Towards a New Modernity*. London: Sage Publications.

Beck, Ulrich, Anthony Giddens, and Scott Lash. 1994. *Reflexive Modernization: Politics, Tradition and Aesthetics in the Modern Social Order*. Oxford: Polity Press.

Bernstein, Steven. 2001. *The Compromise of Liberal Environmentalism*. New York: Columbia University Press.

Biermann, Frank. 2002. "Institutions for Scientific Advice: Global Environmental Assessments and their Influence in Developing Countries." *Global Governance* 8: 195–219.

Cash, David, and William Clark. 2001. "From Science to Policy: Assessing the Assessment Process." Faculty Research Working Paper. Kennedy School of Government, Harvard University, Cambridge.

Clark, Fiona, and Deborah L. Illman. 2001. "Dimensions of Civic Science: Introductory Essay." *Science Communication* 23 (1): 5–27.

Cunningham, Frank. 2002. *Theories of Democracy*. London and New York: Routledge.

Durant, John. 1999. "Participatory Technology Assessment and the Democratic Model of the Public Understanding of Science." *Science and Public Policy* 26 (5): 313–19.

Durrenburger, Gregor, Hans Kastenholz, and Jeanette Bearingar. 1999. "Integrated Assessment Focus Groups: Bridging the Gap between Science and Policy." *Science and Public Policy* 26 (5): 341–49.

Edwards, Arthur. 1999. "Scientific Expertise and Policy-making: The Intermediary Role of the Public Sphere." *Science and Public Policy* 26 (3): 163–70.

Fischer, Frank. 2000. *Citizens, Experts and the Environment. The Politics of Local Knowledge.* Durham and London: Duke University Press.

Frewer, Lynn, and Brian Salter. 2002. "Public Attitudes, Scientific Advice and the Politics of Regulatory Policy: The Case of BSE." *Science and Public Policy* 29 (2): 137–45.

Fuller, Steve. 2000. *The Governance of Science.* Buckingham and Philadelphia: Open University Press.

Funtowicz, Silvio O., and Jerome R. Ravetz. 1992. "Three Types of Risk Assessment and the Emergence of Post-Normal Science." *Social Theories of Risk*, edited by Sheldon Krimsky and Daniel Golding. London: Praeger.

Gaffaney, Timothy J. 2001. "Philosopher Citizen and Scientific Experts." Paper presented at the Annual Meeting of American Political Science Association, San Francisco, August 30–September 2.

Gallopin, Gilberto, Silvio O. Funtowicz, Martin O'Connor, and Jerome Ravetz. 2001. "Science for the Twenty-First Century: From Social Contract to Scientific Core." *International Journal for Social Science* 168: 219–29.

Gieryn, Thomas. 1995. "Boundaries of Science." In *Handbook of Science and Technology Studies*, edited by Sheila Jasanoff, Gerald E. Markle, James C. Petersen, and Trevor Pinch. Thousand Oaks/London/ New Delhi: Sage Publications.

Guston, David H. 2001. "Boundary Organizations in Environmental Policy and Science. An Introduction." *Science, Technology and Human Values* 26: 399–408.

Haas, Peter. 1989. *Saving the Mediterranean.* New York: Columbia University Press.

———. 1992. "Introduction: Epistemic Communities and International Policy Coordination." *International Organization* 46 (1): 1–35.

Haraway, Donna. 1996. "Situated Knowledges: The Science Question in Feminism and the Privilege of a Partial Perspective." In *Feminism and Science*, edited by Evelyn Fox Keller and Helen Longino. Oxford and New York: Oxford University Press.

Harding, Sandra. 1998. *Is Science Multi-Cultural? Postcolonialisms, Feminisms, and Epistemologies.* Bloomington and Indianapolis: Indiana University Press.

———. 2000. "Should Philosophies of Science Encode Democratic Ideals?" In *Science, Technology and Democracy*, edited by D. L. Kleinmann. Albany: State University of New York Press.

International Council for Science. 2002a. ICSU Series on Science for Sustainable Development No. 4: Science, Traditional Knowledge and Sustainable Development. ICSU.

———. 2002b. ICSU Series on Science and Technology for Sustainable Development No. 9: Science and Technology for Sustainable Development. ICSU.

Irwin, Allan. 1995. *Citizen Science. A Study of People, Expertise and Sustainable Development*. London and New York: Routledge.

Jasanoff, Sheila. 1990. *The Fifth Branch: Science Advisers as Policymakers*. Cambridge: Harvard University Press.

Jasanoff, Sheila, and Brian Wynne. 1998. "Science and Decisionmaking." In *Human Choice and Climate Change: Volume One*, edited by Steve Rayner and Elisabeth L. Malone, 1–87. Columbus, Ohio: Batelle Press.

Kates, Robert, et al. 2000a. Sustainability Science. Research and Assessment System for Sustainability Program Discussion Paper 2000-33. Belfer Center for Science and International Affairs, Kennedy School of Government, Cambridge.

———, et al. 2000b. Sustainability Science. *Science* 292: 641–45.

Kleinmann, Daniel, ed. 2000. *Science, Technology and Democracy*. Albany: State University of New York.

Levidow, Les, and Claire Marris. 2001. "Science and Governance in Europe: Lessons from the Case of Agricultural Biotechnology." *Science and Public Policy* 28 (5): 345–60.

Litfin, Karen. 1995. "Framing Science: Precautionary Discourse and the Ozone Treaties." *Millennium* 24 (2): 251–77.

Miller, Clark A. 2001a. "Hybrid Management: Boundary Organizations, Science Policy, and Environmental Governance in the Climate Regime." *Science, Technology and Human Values* 26: 478–500.

———. 2001b. "Challenges in the Application of Science to Global Affairs: Contingency, Trust and Moral Order." In *Changing the Atmosphere. Expert Knowledge and Environmental Governance*, edited by Clark A. Miller and Paul N. Edwards. Cambridge: MIT Press.

Miller, Clark A., and Paul N. Edwards. 2001a. *Changing the Atmosphere. Expert Knowledge and Environmental Governance*. Cambridge: MIT Press.

———. 2001b. "Introduction: The Globalization of Climate Science and Climate Politics." In *Changing the Atmosphere: Expert Knowledge and Environmental Governance*, edited by Clark A. Miller and Paul N. Edwards. Cambridge: MIT Press.

Rutgers, M. R., and M. Metzel. 1999. "Scientific Expertise and Public Policy: Resolving Paradoxes?" *Science and Public Policy* 26 (3): 146–50.

Sarewitz, Daniel. 2000. "Human Well-being and Federal Science. What's the Connection?" In *Science, Technology and Democracy*, edited by Daniel Kleinmann. Albany: State University of New York Press.

Saward, Michael. 1993. "Green Democracy?" In *The Politics of Nature. Explorations in Green Political Theory*, edited by Andrew Dobson and Paul Lucardie. London and New York: Routledge.

VanDeveer, Stacy D. 1998. "European Politics with a Scientific Face: Transition Countries, International Environmental Assessment, and Long-Range Transboundary Air Pollution." ENEP Discussion Paper E-98-9, Kennedy School of Government, Harvard University.

Walker, Robert B. J. 1993. *Inside/Outside. International Relations as Political Theory.* Cambridge: Cambridge University Press.

Weale, Albert. 2001. "Scientific Advice, Democratic Responsiveness and Public Policy." *Science and Public Policy* 28 (6): 413–22.

Wynne, Brian. 1994. "Scientific Knowledge and the Global Environment." In *Social Theory and the Global Environment*, edited by Michael Redclift and Ted Benton. London and New York: Routledge.

Young, Oran, ed. 1997. *Global Governance: Drawing Insights from the Environmental Experience.* Cambridge : MIT Press.

———. 1999. *The Effectiveness of International Environmental Regimes. Causal Connections and Behavioral Mechanisms.* Cambridge: MIT Press.

# Copyright Acknowledgments

JOHN M. HOBSON, "Discovering the Oriental West," in *The Eastern Origins of Western Civilisation* (New York: Cambridge University Press, 2004). Copyright 2004 John M. Hobson. Published by Cambridge University Press, reproduced with permission.

STEVEN J. HARRIS, "Long-Distance Corporations, Big Sciences, and the Geography of Knowledge," *Configurations* 6, no. 2 (1998): 269–304. Copyright 1998 The Johns Hopkins University Press and The Society for Literature and Science. Reprinted with permission of The Johns Hopkins University Press.

MARY TERRALL, "Heroic Narratives of Quest and Discovery," *Configurations* 6, no. 2 (1998): 223–42. Copyright 1998 The Johns Hopkins University Press and The Society for Literature and Science. Reprinted with permission of The Johns Hopkins University Press.

ELLA REITSMA, "Maria Sibylla Merian: A Woman of Art and Science," in *Maria Sibylla Merian and Daughters: Women of Art and Science* (Zwolle: Waanders Publishers). Copyright the Rembrandt House Museum, Amsterdam; The J. Paul Getty Museum, Los Angeles; and Waanders Publishers, Zwolle. Reproduced with the kind permission of Ella Reitsma and Waanders Publishers.

LONDA SCHIEBINGER, "Prospecting for Drugs: European Naturalists in the West Indies," in *Colonial Botany: Science, Commerce and Politics in the Early Modern World*, ed. Londa Schiebinger and Claudia Swan (Philadelphia: University of Pennsylvania Press, 2005), 137–54. Reprinted with permission of University of Pennsylvania Press.

LUCILE H. BROCKWAY, "Science and Colonial Expansion: The Role of the British Royal Botanic Gardens," excerpted from *Science and Colonial Expansion: The Role of the British Royal Botanic Gardens*, by Lucile H. Brockway (New Haven: Yale University Press, 2002), 6–8, 187–96. Reproduced with permission of Yale University Press.

JUDITH CARNEY, "Out of Africa: Colonial Rice History in the Black Atlantic," in *Colonial Botany: Science, Commerce, and Politics in the Early Modern World*, ed. Londa Schiebinger and Claudia Swan (Philadelphia: University of Pennsylvania Press, 2005), 204–20. Reprinted with permission of University of Pennsylvania Press.

WARD H. GOODENOUGH, "Navigation in the Western Carolines: A Traditional Science," in *Naked Science: Anthropological Inquiry into Boundaries, Power, and Knowledge*, ed. Laura Nader (New York: Routledge, 1996). Reproduced with permission of Taylor and Francis Group and Ward H. Goodenough.

COLIN SCOTT, "Science for the West, Myth for the Rest?" in *Naked Science: Anthropological Inquiry into Boundaries, Power, and Knowledge*, ed. Laura Nader (New York: Routledge, 1996). Reproduced with permission of Taylor and Francis Group.

PETER MÜHLHÄUSLER, "Ecolinguistics, Linguistic Diversity, Ecological Diversity," in *On Biocultural Diversity: Linking Language, Knowledge, and the Environment*, ed. Luisa Maffi (Washington: Smithsonian Institution Press, 2001). Copyright Luisa Maffi 2010.

HELEN APPLETON, Maria E. Fernandez, Catherine L. M. Hill, and Consuelo Quiroz, "Gender and Indigenous Knowledge," in *Missing Links*, ed. Gender Working Group, United Nations Commission on Science and Technology for Development (Ottawa: International Development Research Centre, 1995). This material is reproduced here with the permission of Canada's International Development Research Centre, http://www.idrc.ca.

STEPHEN B. BRUSH, "Whose Knowledge, Whose Genes, Whose Rights?," in *Valuing Local Knowledge: Indigenous People and Intellectual Property Rights*, ed. Stephen B. Brush and Doreen Stabinsky (Washington: Island Press, 1996). Reproduced by permission of the Island Press.

D. MICHAEL WARREN, "The Role of the Global Network of Indigenous Knowledge Resource Centers in the Conservation of Cultural and Biological Diversity," in *On Biocultural Diversity: Linking Language, Knowledge, and the Environment*, ed. Luisa Maffi (Washington: Smithsonian Institution Press, 2001). Copyright Luisa Maffi 2010.

ARTURO ESCOBAR, "Development and the Anthropology of Modernity," in *Encountering Development: The Making and Unmaking of the Third World*, by Arturo Escobar (Princeton: Princeton University Press, 1995). Reprinted by permission of Princeton University Press.

CATHERINE V. SCOTT, "Tradition and Gender in Modernization Theory," in *Gender and Development: Rethinking Modernization and Dependency Theory*, by Catherine V. Scott (Boulder: Lynne Rienner, 1995). Copyright 1995 by Lynne Rienner Publishers, Inc. Used with permission of the publisher.

BETSY HARTMANN, "Security and Survival: Why Do Poor People Have Many Children?," in *Reproductive Rights and Wrongs: The Global Politics of Population Control*, by Betsy Hartmann (Boston: South End Press, 1995). Reproduced with permission of South End Press.

COMMITTEE ON WOMEN, Population and the Environment, "Call for a New Approach," in *Reproductive Rights and Wrongs: The Global Politics of Population*

*Control*, by Betsy Hartmann (Boston: South End Press, 1997). Reproduced with permission of South End Press.

JENNY REARDON, "The Human Genome Diversity Project: What Went Wrong?," in *Race to the Finish: Identity and Governance in an Age of Genomics*, by Jenny Reardon (Princeton: Princeton University Press, 2005). Reprinted by permission of Princeton University Press.

CORI HAYDEN, "Bioprospecting's Representational Dilemma," *Science as Culture* 14, no. 2 (2005). Reproduced with permission of Taylor and Francis Group Ltd., http://www.informaworld.com, and Cori Hayden.

ZIAUDDIN SARDAR, "Islamic Science: The Contemporary Debate," in *Encyclopedia of the History of Science, Technology, and Medicine in Non-Western Cultures*, ed. Helaine Selin (Dordrecht: Kluwer, 1997). Reproduced with kind permission from Springer Science and Business Media and Ziauddin Sardar.

SUSANTHA GOONATILAKE, "Mining Civilizational Knowledge," from *Toward a Global Science: Mining Civilizational Knowledge*, by Susantha Goonatilake (Bloomington: Indiana University Press, 1998). Reproduced with permission of Indiana University Press.

CATHERINE A. ODORA HOPPERS, "Towards the Integration of Knowledge Systems: Challenges to Thought and Practice," from the introduction to *Indigenous Knowledge and the Integration of Knowledge Systems: Toward a Philosophy of Articulation*, ed. Catherine A. Odora Hoppers (Claremont, South Africa: New Africa Books, 2002). Reproduced with permission of Catherine A. Odora Hoppers and New Africa Books.

DANIEL SAREWITZ, "Human Well-Being and Federal Science: What's the Connection?," in *Science, Technology, and Democracy*, ed. Daniel Lee Kleinman (Albany: State University of New York Press, 2000). Reproduced with permission of State University of New York Press.

DAVID J. HESS, "Science in an Era of Globalization: Alternative Pathways," in *Alternative Pathways in Science and Industry: Activism, Innovation, and the Environment in an Era of Globalization* (Cambridge: MIT Press, 2007). The material that appears here is a seven-thousand-word condensation of chapter 2. Copyright 2007 Massachusetts Institute of Technology, reprinted by permission of the MIT Press and David Hess.

KARIN BÄCKSTRAND, "Civic Science for Sustainability: Reframing the Role of Experts, Policy-Makers, and Citizens in Environmental Governance," *Global Environmental Politics* 3, no. 4 (November 2003). Reproduced with the kind permission of MIT Press.

# Index

Abuelos de Plaza de Mayo, 322
Academia dei Lincei, 103
Academy of Sciences, France, 87, 99
Actor-network theory (ANT), 345–47,
    354
Adams, Vincanne, 365–66
Adanson, Michel, 121
Affirmative action, 325, 333–34
Africanism and developmentalism,
    273, 277
African Renaissance, 388
Agency: Eastern, 42, 49; identity
    and, 57
Agency for Healthcare Research and
    Quality, 425
Agricultural technologies, 214, 429
Ahmad ibn-Mājid, 54
Aid agencies, 5, 7, 248. See also
    United Nations (UN)
AIDS, 409–10
Aldrovandi, Ulisse, 104
Aligarh school, 376–78
American Cyanamid, 349
American Home Products, 314
Amerindians, 112
Anderson, Warwick, 7, 365–66
Animal/human reciprocators, 179,
    182–83, 189–93
ANT (actor–network theory), 345–47,
    354
Anthropology and development,
    280–83
Anti-Eurocentrism, 40–41, 53
Antimilitarism, 4
Arawaks, 111, 113, 118
Argonauts (mythological figures),
    89–90
Asad, Talal, 280
Asian achievements. See under East
Assaliya brewing, 218
Atomic Energy Commission, 404
Averill, Mary Beth, 302

Bacon, Francis, 61, 79, 86
Badajos Junta, 67
Bailly, Jean-Sylvain, 94, 96, 99
Bancroft, Edward, 117
Banks, Joseph, 121, 130–31
Barbot, James B., 144
Barbot, Jean, 143
Barrère, Pierre, 113–14
Beauvoir, Simone de, 13
Beck, Ulrich, 448, 453
Benedict, Ruth, 39
Bernal, Martin, 40
Bhabha, Homi, 275, 277, 283
"Bioactive Agents from Dryland
    Biodiversity of Latin America,"
    348–50, 358
Biocontact zones, 116–19
Biodiversity: anti-biopiracy and,
    368; compensation and, 234–38;
    conservation of, 155–56, 230–31,
    234, 242, 253; crop germplasm and,
    231–32, 240, 242; cultural diversity
    and, 155–56, 247; cultural knowl-
    edge and, 226–28; gender and,
    213–14, 219–21, 250; indigenous ap-
    proaches to, 249–50, 389; landraces
    and, 231–32, 235, 240; linguistic
    diversity and, 200–202; promo-
    tion of, 219–21; prospecting and,
    110, 215, 241–42, 267, 335, 343–45,
    347–49, 357–60; resource centers
    for, 251–53; rights and, 234; sustain-
    able ethics and, 394; threats to, 225,
    227–28. See also Ethnobotany
Biomedical research, 409
Biopower, 325
Bioprospecting. See Biodiversity:
    prospecting and
Birth rates. See Population expansion
Blaut, J. M., 34
Bligh, William, 131
Bodard, Pierre-Henri-Hippolyte, 123

Botanicals, 71–76, 110–15, 118–23, 232–33; colonial expansion and, 123–24, 127–38; compensation and, 234–38. *See also* Ethnobotany; specific plants

Bottle-feeding, 213–14

Bouguer, Pierre, 89, 92, 98

Bourgeois, Nicolas-Louis, 110–13, 119, 123

Bouvet, Joachim, 75

Bovine Spongiform Encephalopathy (BSE), 439, 446

Breast-feeding, 313–14

British East India Company, 130–31

Brown, Phil, 424, 426–27

BSE. *See* Bovine Spongiform Encephalopathy (BSE)

Bucaille, Maurice, 375

Bucaillism, 375

Caldwell, John, 311

Cano, Sebastien del, 67

Capitalism: modern, 39, 44, 46, 48, 50, 52, 54, 300; warfare and, 51

Caribs, 111–13, 118

Carolinean navigation. *See* Western Caroline Islands, navigation system

Cartography. *See* maps; *specific projections*

Casa de la Contratación de las Indias, 65–69

Cassini, Jacques, 87–88

Cassini, Jean-Dominique, 95

Caterpillars, 103–8

Cavilli-Sforza, Luca, 322, 331

CBD. *See under* United Nations

Celsius, Anders, 88

Center for Studies on Science, India, 374

Cesi, Federico, 103

CGIAR (Consultative Groups on International Agricultural Research), 240

Chakrabarty, Dipesh, 346

Cheselden, William, 122

Chiflet, Jean-Jacques, 74

Childbirth. *See* Population expansion

Chinese medicine, 221

Chodorow, Nancy, 297

Christianity: on native peoples, 57

Cinchona bark, 73–76, 110, 114, 123, 128, 132–33. *See also* Quinine

Civic science: democratization and, 444–45, 448–51, 453; description of, 439–41, 443–44, 453–55; developments leading to, 445; environmental problems and, 447–48, 450, 453; international relations and, 441–43; public trust and, 446–47; representation and, 444; in sustainability science, 451–53

Clairaut, Alexis, 88

Class: conflict and, 304; gender and, 13, 16, 269; middle, 130, 265, 293, 315; race and, 13; science and, 386; struggle and, 48–49, 327

Cleyer, Andreas, 72

Climate change, 411, 440

Cold War: arms race and, 2; modern science and, 403–6, 413, 415; politics of, 4, 7; Third World and, 3; weaponry and, 404

Coleman, James C., 301, 306

Colonialism: botany and expansion and, 123–24, 127–38; on contributions of non-Euro-Americans, 248; genome project and, 321; modernity and, 33, 266; opposition to, 1; power and, 277, 280; racial/cultural hierarchies in, 275; research and, 392; sciences and, 3, 5–6, 33, 127–38, 366, 390, 396. *See also* Imperialism; Neocolonialism

Columbus, Christopher, 57, 111, 123, 151

Commelin, Caspar, 72

Commelin, Joannis, 72

Commission of Pilots, 67

Committee on Women, Population and the Environment, 265, 318–20

*Communist Manifesto, The* (Marx), 47

Compagnie des Indes, 121

COMSTECH. *See* Islamic Conference Standing Committee on Scientific and Technological Cooperation (COMSTECH)
Condorcet, Caritat de, 85, 93–94, 98–99
Consejo Real y Supremo de las Indias, 65, 67–69
Consultative Groups on International Agricultural Research (CGIAR), 240
Consumption patterns, 318
Contraception, 314–15, 319
*Contra yerva*, 119–20
Convention on Biological Diversity, 235. *See also under* United Nations
Cook, James, 131, 206
Cook-Deegan, Robert, 322
Co-production, 326–30, 452
Copyright, 216, 398
Corporate travel. *See* Travel
Corporations, long-distance. *See* Long-distance corporations
Council of the Indias, 65, 67–69
Cranes, Diana, 130
Creation theories, 156
Cree of James Bay knowledge construction: animal/human reciprocators and, 179, 182–83, 189–93; eating/sexuality metaphors and, 182–83; in hunting, 178, 183–89; literal/figurative knowledge and, 178, 180–81, 183, 187, 194; reciprocity in, 176, 179, 181–83, 189–90, 192–94; relating/differentiating in, 180–82; root metaphors, 175–78, 181, 193–94; signs in epistemology, 178–80
*Crisis in Modern Science* (Third World Network), 8
Crop germplasm. *See* Germplasm
Crusades, 57
Cultural anthropology, 172–73
Cultural capital, 248
Cultural diversity, 10, 156, 227–28, 242, 247, 282

Cultural identity, 140, 147, 153, 229, 373
Curtin, Philip, 3

d'Alembert, Jean Le Rond, 84–85, 99, 110
Darwin, Charles, 37, 131, 135
Dead reckoning, 164, 171–72
Debt coercion, 133
*Declaration on Science and the Science Agenda Framework*, 390
Deep Blue (computer), 406
Democracy and indeterminacy, 407–8
Department of Agriculture, U.S., 430
Department of Defense, U.S. (DOD), 403–4, 415
Department of Energy, U.S. (DOE), 321, 330–31, 405, 430
Dependency theory, 280, 305
Despotism, Eastern, 44–45, 48–49, 51–52
Development discourse, 271–78; anthropology and, 280–83; deconstructing, 278–80; dependency theory and, 280, 305; development syndrome and, 301–2; discipline in, 304; political development and, 303–5; tradition in, 307
Development policies, 264–66, 270, 281
Díaz de Solis, Juan, 67
Diderot, Denis, 99, 110
Differentiation, 180–82, 290, 297, 301
Diversity. *See* Biodiversity; Cultural diversity; Linguistics
Dobzhansky, Theodosius, 331–32
DOD (Department of Defense), 403–4, 415
DOE (Department of Energy), 321, 330–31, 405, 430
*Do It Herself* (study), 217–19
Drug discovery/prospecting, 111–15, 120–23, 343–45, 348–49, 351–54, 357–59. *See also* Ethnobotany
Du Bois, W. E. B., 41–42, 47

DuGuay-Trouin, René, 93
Duhem, Pierre, 423
Dutch East India Company, 65; medical/botanical interests and, 70–72, 120; mortality rates and, 70–71; opium trade and, 120; Society of Jesus and, 70–76

Earth measurements, 87–93, 95–96
Easlea, Brian, 99
East: achievements and, 34, 54–56; agency of, 42, 49; appropriation of resources by West and, 56–58; as despotic, 44–45, 48–49, 51–52; as inferior, 43–47, 49, 53; Marx on, 47–49; as Other, 43–44; patriarchal construction of, 44–45; Peter Pan theory, 44, 50; resource portfolios of, 54–55; rise of West and, 40; as separate from West, 39, 51; stagnation of, 44–45, 48, 52, 56; subjugation of, 45, 57–58; as unchangeable, 48; world development by, 54. *See also* Orientalism
Eating/sexuality metaphors, 182–83
Ecolinguistics, 198–99
Ecology: anthropological, 179; cultural, 198; destruction of, 264, 318; human, 129; industrial, 414–15; movements, 4; preservation, 137; scientific, 194
Economic botany, 36
Economic development: academic research and, 425; of East, 42, 53; European colonialism and, 57; obstacles to, 299–300; political development and, 304; role of ideas and, 280; theories, 271; UN on, 2, 270
Empathy, 296–97
Empirical research, 103
Empirical thought, 176
*Encyclopedie* (Diderot and d'Alembert), 99, 110
Enga tree names, 200–202
Enlightenment: Chinese ideas and, 55; dichotomies, 294, 296, 306, 382;

discovery narratives, 85–86, 90, 96, 98; modernization theory and, 2–3, 46; program for science, 406–13, 415; scientific rationality and, 2, 84, 100, 327; standpoint theory and, 20; universal knowledge and, 114. *See also* Scientific Revolution
Environmental degradation, 203, 208, 238, 243, 318–20, 452
Environmental Protection Agency (EPA), 425, 429
Epistemic modernization, 419–21, 432; community research in, 423–25; public communication models in, 425–27; scientific reform in, 428–31; social diversification in, 421–23
Epstein, Steven, 427
Ethnobotany: bioactivity and, 268, 352–53, 359; disappearance of knowledge, 354–55; drug discovery and, 343–45, 348–49, 351–54, 357–59; entitlement and, 344–45; epistemological advocacy and, 347–48; ethnobotanical knowledge, 344–45, 351–57; fact vs. content in, 352–53; as high-risk science, 358–59; legitimacy of, 349–50; literature of, 233; patents and, 355–56; project code in, 351–52; proof of knowledge and, 343, 347; representation and, 356–57; screening and, 353–54; secrecy in, 351, 356; sponsors of, 348–51; testing and, 121. *See also* Biodiversity; Botanicals
Ethnography, 172–73, 176, 178, 282
Eugenics, 331–32, 334
Eurocentrism: achievements and, 33–35; vs. anti-Eurocentrism, 40–41, 53; assumptions of, 41; critique of, 1; definition of, 40; as false view, 40; in feminist science/technology studies, 16, 22; foundations of, in theories, 43–47; ignorance of, 34; illusion of, 53–58; as linear prog-

ress, 49; in map projections, 42–43; modernity and, 53–54; Orientalist clause, 55–56. *See also* Orientalism

Exceptionalism, 6–7, 9, 21, 56, 58, 152, 368–69

Fanon, Frantz, 3

FAO (Food and Agricultural Organization), 235

Feminism: agendas and, 12; on development policies, 265; on gender bias, 4–5; perceived as Western, 17–18; postcolonial science theory and, 12, 16–18, 21; on standpoint methodology, 19; on Third World women, 274–75; use of term, 24 n. 18

Ferguson, James, 278–79

Fermin, Philippe, 119

Fernández-Armesto, Felipe, 41

Fertility decline, 311–12

Figurative signification. *See* Literal/figurative knowledge

Flax, Jane, 305–6

Folbre, Nancy, 312

Fontenelle, Bernard de, 89, 91

Food and Agricultural Organization (FAO), 235

Fortune, Robert, 131

Foucault, Michel, 325

Freedom: masculinist conception of, 293, 296; women's, 318–19

French Academy of Sciences, 87, 99

Fundamentalisms, 156; scientific/mystical, 373–75

Galileo Galilei, 103

Gama, Vasco de, 57; myth of, 54

Gamble, Samuel, 144–45

GATT (General Agreement on Tariffs and Trade), 235

GEAR (Growth, Equity and Reconstruction), 388

Gender: bias and feminism, 4–5; biodiversity and, 213–14, 250; class and, 13, 16, 269; definition

of, 13; European expansion and, 35; hierarchies and, 11; indigenous knowledge and, 211–13, 217–23; inequalities, 320; intellectual property rights and, 214–16; modernization theory on, 267, 291, 299, 307; postcolonial science theory and, 11–21; race and, 13; relations, 10, 13, 15–16, 35; science studies and, 14–15; in Scientific Revolution, 99; studies, 10–11, 18–21, 24 n. 18. *See also* Women

Gendzier, Irene, 280

General Agreement on Tariffs and Trade (GATT), 235

Genetically modified food, 439

Genetic engineering, 136–37

Geography: of knowledge, 61–64, 76, 80; of movement, 62; questionnaires and, 68–69

*German Ideology, The* (Marx), 48–49

Germplasm, 231–32, 240, 242

Global Diversity Strategy, 247

Globalization: of cultural production, 276; indigenous knowledge and, 368, 388–89; neutrality of term, 23 n. 1; oriental, 40; race and, 326; scientific research and, 419, 431–33; theory, 4, 7

Goedaert, Johannes, 104–5

Goldstene, Paul, 135

Goose hunting knowledge, Cree, 183–89, 191–92

Greece, Ancient, 46, 48–49

Green chemistry, 429–31

Green Revolution, 410

Gross, Michael, 302

Growth, Equity and Reconstruction (GEAR), 388

Guerra, Francisco, 120

Gupta, K. A, 212–13, 393

HapMap, 323

Haraway, Donna, 18, 151–52, 306–7, 422

Hartmann, Betsy, 265

Harvard University, School of Public Health, 424
Hegel, G. W. F., 49
Herbal remedies. *See* Botanicals; Ethnobotany
*Herbarium Ambionense* (Rumpf), 71
Hermann, Paul, 71
Herolt, Jakob Hendrik, 106
*Hevea*, 132. *See also* Rubber plantations
Hirschmann, Nancy, 296, 305
*Histoire des mathématiques* (Montucla), 94–95
Hobo-Dyer world map projection, 43
Hodgson, Marshall, 43
Holling, C. S., 407
Hooker, Joseph, 131, 135
Hortus Medicus (Amsterdam), 115
Hountondji, P., 396
House of Trade. *See* Casa de la Contratación de las Indias
Hubbard, Ruth, 263
HUGO (Human Genome Organization), 321
Hull, David, 413
Human Genome Diversity Project, 267; colonialism and, 321; co-production and, 326–30; debates, 321–24; human rights and, 322; identity and, 335; indigenous gene sampling in, 325; race and, 330–36; use of term, 337 n. 2; vision for, 327
Human Genome Organization (HUGO), 321
Humanity and science, 406–13
Humans and animals. *See* Animal/human reciprocators
Humboldt, Alexander von, 117–18
Hunting knowledge, Cree, 178, 183–89, 191–92
Huntington, Samuel, 303–5
Huygens, Christiaan, 88

ICBG (International Cooperative Biodiversity Groups), 348, 353–54, 357–59

Identity: agency and, 57; crises, 301; cultural, 140, 147, 153, 229, 373; genetics and, 335, 338 n. 11; masculine, 297; politics, 333–34; Western, 44, 46
Ijmalis, 374, 376–78
IK. *See* Indigenous knowledge (IK)
ILO (International Labour Organisation), 222
IMF (International Monetary Fund), 307
Imperialism: appropriation of economic resources, 40; gender and, 12–13; modernity and, 33, 266; opposition to, 1; research and, 392; science and, 135–36, 138, 366; subjugation of East and, 45, 57–58; technology and, 33. *See also* Colonialism; Neoimperialism
Indian House, 65–69
Indigenous knowledge (IK): analytical framework for, 395; biodiversity and, approaches to, 249–50, 389; bioprospecting and, 110, 215, 241–42, 267, 335, 343–45, 347–49, 357–60; case studies/resources and, 249, 251–53; challenges to systems and, 395–98; colonialism and, 396; at community level, 393–95; compensation for, 234–38; as cultural capital, 248–49; cycle of, 250; definition of, 395; ecological, 175, 177, 193, 214; empowerment strategy and, 399–400; erosion of, 393; feminism and, 17; gender and, 211–13, 217–23; globalization and, 368, 388–89; industrial use of, 233; as intellectual property, 226; loss of, 248–49; programs, 216–17; protection of, 215–16; representation and, 345–48; resource centers, 251–53; science development and, 390–400; systems, 10, 17, 20, 151, 249, 388–400; technical, 247; use of term, 247; value addition to, 394.

*See also specific groups and systems*

Indigenous people: decline of, 112–13, 155; epistemology and, 151, 392; genetics study of, 321, 322–23, 334–35; representation and, 444; sovereignty rights of, 335, 349; use of term, 229. *See also* Indigenous knowledge (IK)

Individualism, 87, 296, 301

INE (National Institute of Ecology), 349

Inkeles, Alex, 290, 293–94

Intellectual property rights (IPRS), 214–16, 225–26, 238–41, 248, 343, 349–50, 389, 398, 419; as conservation tool, 239–41

International Cooperative Biodiversity Groups (ICBG), 348, 353–54, 357–59

International Federation of Institutes of Advanced Studies, 373

International Haplotype Map, 323

International Labour Organisation (ILO), 222

International Monetary Fund (IMF), 307

*Invention of Africa* (Mudimbe), 273

IPRS. *See* Intellectual property rights (IPRS)

Irwin, Howard S., 129

Islam: achievements, 34; Bucaillism, 375; Europe vs., 57; fundamental concepts of, 377; mysticism and, 375; science debates about, 373–79

"Islamabad Declaration," 376

Islamic Conference Standing Committee on Scientific and Technological Cooperation (COMSTECH), 376

Jardin du Roi (Amsterdam), 115

Jardin Royal des Plantes (Paris), 115

Jason (mythological figure), 89–90

Jesuits. *See* Society of Jesus (SJ)

Jesuits' bark. *See* Cinchona bark

Jordanova, L. J., 294

K'ang Hsi, 75

Kasparov, Garry, 406

Keller, Evelyn Fox, 99

Kellogg Foundation, 425

*Kew Bulletin*, 132

Kew Gardens, 115, 127–28, 131–35

Kinematics, 62–64

King, Mary-Claire, 322

King, Philip Gidley, 206

Kipling, Rudyard, 45

Knowledge systems: benefits of studying, 152–55; challenges of studying, 155–57; cultural identity and, 153; internationalization of, 132; local vs. distributed, 81. *See also* Indigenous knowledge (IK); *specific groups and systems*

Koerner, Lisbet, 122

Kuhn, Thomas, 78–80, 448

Labadie, Jean de, 105

La Condamine, Charles-Marie de, 89, 93, 98, 114, 117–19

Landes, David, 40–41, 53–54, 58

Landraces, 231–32, 235, 240

Languages: Enga, 200–202; extinction of, 228–29, 248–49; local innovation and, 393–94; Mota, 207–8; past experience and, 209; Pitkern, 203–5; West Indian, 116–18

Lapland expedition, 87–88, 96

LaPolombara, Joseph, 302

Latin America ICBG. *See* "Bioactive Agents from Dryland Biodiversity of Latin America"

Latour, Bruno, 65, 115, 346

Laveaga, Gabriela Soto, 358–59

Law, John, 65

Leibnizian mathematics, 87

Leiden Botanical Garden, 71–72

Leite, Cerquiera, 137

Leopold Wilhelm, Archduke, 74

Lerner, Daniel, 296

Levins, Richard, 263

Lévi-Strauss, Claude, 176

Lewontin, Richard, 263

Linguistics: biodiversity and, 200–202; diversity and, 198–200; ecolinguistics, 198–99; independence hypothesis, 199; mobility and, 202

Linneaus, Carl, 108, 121

Literal/figurative knowledge, 178, 180–81, 183, 187, 194

Long, Edward, 115

Long-distance corporations, 62, 65–66. *See also* specific corporations

Long-range transboundary air pollution regime (LRTAP), 441

López de Velasco, Juan, 68

Louis XIV, 75

LRTAP (Long-range transboundary air pollution regime), 441

MAAS *Journal of Islamic Science*, 373

Macpherson, C. B., 302–3

Maffie, James, 151–53

Male supremacy, 12–13, 25 n. 27

Manhattan Project, 2, 385

Manzo, Kate, 280

Maps: by Commission of Pilots, 67; Eurocentrism of, in projections, 42–43; pattern of, 66–67, 69; schematic of, 165–67; thread of, 62–64, 76. *See also* specific projections

Marginalization, 41–42, 263, 279, 322, 325, 347, 357

Marks, Jonathan, 156

Marxism, 40–41, 47–49, 272, 327, 346

Matthew effect, 137

Matthews, John, 144

Maupertuis, Pierre-Louis Moreau de, 89–92, 96, 98–99, 110

Maurepas, Comte de, 88

McClellan, James E., 33

McClelland, David, 293, 297–98

Medea (mythological figure), 90

Medicine, Chinese, 221

Medicines, plant-based. *See* Botanicals; Ethnobotany

Mercator world map projection, 42–43, 64

Merchant, Carolyn, 99

Merck Pharmaceutical, 241

Merian, Dorothea Maria, 105

Merian, Johanna Helena, 105–6

Merian, Maria Sibylla, 36, 103–9, 116; caterpillar study by, 103–5, 107–8; Suriname insect book by, 106–7

Merino sheep, 138 n. 2

Merton, Robert, 137

Microscopes, 103

Mignolo, Walter, 33

Military and science, 405–6

*Millennium* (Fernández-Armesto), 41

Mitchell, Timothy, 273–74

Mobility. *See* Travel

Modernity/modernization: anthropology of, 277–78; capitalist, 39, 44, 46, 48, 50, 52, 54; colonialism and, 33; development and, 279–80, 300–301; dichotomies and, 294; empire and, 33; epistemic, 420–21, 432; Eurocentrism and, 53–54; on gender differences, 291; individualism and, 296, 301; male citizenship and, 293–99; masculine, 291, 296–97, 302, 306; order and truth in, 274; reflexive, 448; science/technology and, 2; single science and, 6; take-off in, 298; tradition and, 290–91, 294–96, 303; traits of, 294; Western vs. local, 9

Modernization theory: criticism of, 303, 305; gender and, 267, 291, 299, 307; male citizenship and, 293–99; model of, 266; psychocultural, 293, 296; scientific rationality and, 2; sexism and, 291–93; structural-functional, 293, 301–2

Mohanty, Chandra, 274, 277, 279, 282, 293

Montucla, J. F., 94–95, 99

Moore, Francis, 143

Morandé, Pedro, 280

Moreiras, Alberto, 346–47, 357, 360 n. 1, 360 n. 3

Morgan, Henry, 113
Mortality rates, 70, 310, 312–14
Mota language, 207–8
Mudimbe, V. Y., 151, 273
Muslim Association for the Advance-
 ment of Science (MAAS), 373
Mystical fundamentalism, 373–75
Mystical vs. empirical thought, 176

NAFTA (North American Free Trade
 Agreement), 349
Nandy, Ashis, 365
Narayan, Uma, 20
NASA, 404–5
Nassy, David de Isaac Cohen, 114
National Autonomous University
 (UNAM), 349–51, 353, 355, 359
National Cancer Institute, U.S., 241,
 343, 348
National Human Genome Research
 Institute (NIIGRI), 323
National Innovations Fund, 399
National Institute of Ecology (INE),
 349
National Institute of General Medical
 Sciences (NIGMS), 321
National Institutes of Health, U.S.
 (NIH), 137, 330–31, 348, 350, 358,
 405, 414, 427
National Medical Institute, Mexico,
 343
National Renewable Energy Labora-
 tory, 430
National Research Council, U.S., 247
National Science Foundation (NSF),
 2, 321, 405
National System of Innovation, 388,
 399
Native peoples. *See* Indigenous
 people
Neocolonialism, 12, 16, 21, 280
Neoimperialism, 12, 16
Nestlé Company, 314
New science, 61, 84, 86, 99, 385
Newton, John, 144
Newtonian physics, 87, 92, 95, 97–98

NGOs (non-governmental organiza-
 tions), 217–18, 420–21, 446, 448
NHGRI (National Human Genome
 Research Institute), 323
NIGMS (National Institute of General
 Medical Sciences), 321
NIH (National Institutes of Health),
 137, 330–31, 348, 350, 358, 405, 414,
 427
Non-governmental organizations
 (NGOs), 217–18, 420–21, 446, 448
Norfolk Island, 206–8; life form nam-
 ing, 207–8
North American Free Trade Agree-
 ment (NAFTA), 349
NSF (National Science Foundation),
 2, 321, 405

Object relations theory, 297, 306
Office de la Recherche Scientifique
 et Technique d'Outre-Mer
 (ORSTOM), 8
Oral traditions, navigational, 170–71
Organization of Islamic Conference,
 373–74
Orientalism: capitalism and, 52;
 developmentalism and, 272, 277; as
 discredited, 22 n. 1; foundations in
 theories, 43–47; Orientalist clause
 and, 55–56; superiority and, 1; West
 and, 39–43. *See also* East; Eurocen-
 trism
*Orientalism* (Said), 1, 43, 272–73, *277*
*Origin of Species* (Darwin), 135
ORSTOM, 8
Other/otherness, 43–44, 283, 390, 392
Outhier, Réginald, 97

*Padron Real* (pattern map), 66–67,
 69, 76
Paris Observatory, 87
Parsons, Talcott, 299–300, 307
*Passing of Traditional Society, The*
 (Lerner), 296
Patents, 216
Paternalism, 292

Patriarchy, 44–45, 305–6, 315, 386
Patrimonialism, 303
Peace and prosperity, 263–64
Peru expedition, 87–93
Peruvian bark. *See* Cinchona bark
Peters, Arno, 42–43
Peters-Gall world map projection, 43
Phillips, Thomas, 144
Physicians for Human Rights, 322
Piailug, 171–72
Pigg, Stacy Leigh, 281
Pijl, L., 72
Pinckard, George, 145–46
Pitcairn Island, 202–6; life form nam-
    ing, 204–5; Norfolk Island and, 206
Plantation system, 132–33. *See also*
    *specific plants*
*Political Order in Changing Societies*
    (Huntington), 303–5
Population expansion: breast-feeding
    and, 313–14; contraception and,
    314–15, 319; environmental condi-
    tions and, 319; security and, 312;
    statistics of, 310; survival and, 311;
    women and, 165
Population genetics, 327, 331–32, 334–35
Positivism, 20–21, 58, 347, 374
Postcolonial, use of term, 23 n. 3
Postcolonial science/technology the-
    ory, 3–6; analysis and, 345; cause
    of disinterest in, 3–4; dissonance
    in, 15–18; feminism and, 12, 16–18,
    21; gender and, 11–21; on indig-
    enous knowledge systems, 17;
    issues/conferences and, 7–9; on
    multiple scientific traditions, 5–6,
    9; new approaches to, 21–22
Postmodernism, 283, 381–82
Post-positivist scholarship, 442, 449,
    454
Pouppé-Desportes, Jean-Baptiste-
    René, 112, 114, 117
Poverty: belief systems and, 2; biodi-
    versity and, 242–43; development

policies and, 264–65; feminism
    and, 18
Prain, David, 130
Prakash, Gyan, 33
Pratt, Mary Louise, 116
Project Camelot, 136
Provincialism, 39, 42, 47
Public health, 120, 310, 324, 334,
    409–10, 425
Public/private distinction, 50, 291
Pye, Lucian, 301

Quinine, 3, 110, 114, 123, 128, 132–33.
    *See also* Cinchona bark

Rabinow, Paul, 334
Race: categorization by, 324–25, 327,
    329–30, 332–33, 336; difference and,
    333; gender and, 13; hierarchies of,
    in colonialism, 275; purity and,
    324–25; relations, 12–13, 133, 267,
    275, 326; science and, 324, 327–36
Racism: in biology, 321; categoriza-
    tion and, 324–25, 327, 329–30, 332–
    33, 336; Christian, 57; contributions
    of non-Euro-Americans and, 248;
    science and, 324, 327–30
Randall, Vicky, 293
Rationality. *See* Scientific rationality
RDP (Reconstruction and Develop-
    ment Programme), 388
Reality, colonization of, 271
Reardon, Jenny, 367
Reciprocity, Cree concept of, 176, 179,
    181–83, 189–90, 192–94
Reconstruction and Development
    Programme (RDP), 388
Redi, Francesco, 104
Reede tot Drakenstein, H. A. van, 72
*Relaciones Geográficas*, 68
Relating/differentiating, 180–82
Renewable energy, 429–30
Representation: in ethnobotany, 356–
    57; indigenous knowledge and,
    345–48; indigenous people and,

444; Third World and, 272; of
Third World women, 274–75, 299
Ribeiro, Diogo, 67
Rice cultivation, 140–47; accessions,
232; African, 141–42, 147; in South
Carolina, 143–47
Richer, Jean, 95–96
Rochefort, Charles de, 117
Rooney, Phyllis, 294
Root metaphors, 175–78, 181, 194
Rosenthal, Joshua, 358–59
Rostow, W. W., 290, 292, 298–99
Royal Botanic Gardens, Kew (London), 115, 127–28, 131–35
Royle, Forbes, 130–31
Rubber plantations, 128, 133–34
Rudolph, Lloyd I., 303
Rudolph, Susan Hoeber, 303
Rugeley, Luke, 120
Rumpf, Georg Eberhard, 71

Said, Edward: on anthropological
literature, 291; on imaginative geographies, 276; on Orientalism, 1,
43, 272–73, 277; on role of Western
Sciences, 6–7
Salley, A. S., 144
Sameness discourses, 333
Scale of practice, 76–81
Science(s): autonomy of, 2, 366,
425, 432–33; big vs. small, 76–80;
Bucaillism, 375; civic, 439–55; as
class-based, 386; Cold War roots
of, 403–6, 413, 415; colonialism
and, 3, 5–6, 33, 366, 390, 396;
community-oriented research in,
423–25; connection of, to society,
21; constructivist research agenda
of, 442; counterstories to history
of, 3; epistemic modernization in,
419–21, 432; epistemic primitive accumulation of, 36–37, 420; epistemologies for, 10; ethics of practice,
394, 411; European expansion and,
35–37; exceptionalism of, 6; female

admirers and, 96–97; fundamentalism and, 373–75; funding for
research of, 430–32; gender studies
and, 14–15; government oversight
of, 2–3; heroism and, 16, 35–36, 85–
86, 89–100; humanity in, 406–13;
ideals of, 408; imperialism and,
135–36, 138, 366; indeterminacy
and, 447, 449; indigenous knowledge systems and, 390–400; Islamic debates and, 373–79; long-distance travel and, 61; masculinity
of, 93; multiple, 9–10, 17, 18, 154;
nature of, 263; Nazi, 331–32; new,
61, 84, 86, 99, 385; politicization
of, 442, 445; positivist view and,
20–21, 58, 347, 374; post-positivist
view and, 442, 449, 454; power
and, 323, 325; public communication models in, 425–27, 446–47;
pure, 130; race and, 324, 327–36;
radical movements and, 4; reform
movements in, 428–31; regulation
of, 136–37; representation and, 346;
role of scientists, 411–12; scale of
practice and, 76–81; scientific revolution, 34; social diversification
and, 421–23; social studies of, 3, 7,
365–67, 413; S&T programs, 216–17;
sustainability, 214, 393, 407, 414–15,
429, 439–41, 451–53; Third World
and, 7–8, 23 n. 6, 368, 373–79; value
neutrality of, 2, 4–5, 154, 366–67;
Western priorities and, 369–70. *See
also* Postcolonial science/technology theory
*Science and Empires*, 8
Scientific rationality, 17, 34, 153, 264,
348; Enlightenment and, 2, 84, 100,
327; Orientalist view of, 51
Scientific Revolution: capitalist
modernity and, 46; cartography
and, 62; founding fathers of, 84–85;
gender in, 99; heroic narratives of,

Scientific Revolution (*cont.*) 89–93, 97–100; Kuhn's model and, 78–80, 448; vs. new science and, 385; non-European contributions to, 34, 37, 55–56; small/big sciences and, 77–80; travel and, 61. *See also* Enlightenment

Serpent herb, 114

Sexism and modernization theory, 291–93, 305

Sexuality metaphors, 182–83

Shaman Pharmaceuticals, 241, 343

Shiva, Vandana, 18, 151

Shockley, William, 322

Sibling assistance chains, 311

Sidereal compass, 160, *161*, 162, 164, 172

Signs, epistemological, 178–80

Sikkink, Kathryn, 280

Sisal industry, 128, 132

SJ. *See* Society of Jesus (sj)

Slaves: as doctors, 113–14; rice cultivation and, 140–41, 143–46; scientific contributions of, 36

Sloane, Hans, 106, 111, 113–14, 120

Smeathman, Henry, 146

Smith, Linda, 392

Soap-making, 222–23

Social balance of power, 50

Social Science Research Council (ssrc), 291, 301–2, 306

Social studies of science, 3, 7, 365–67, 413

Society of Jesus (sj), 65; academic network of, 82 n. 16; cinchona production, 64, 73–76; as corporation, 35, 65, 72–76, 80; Dutch East India Company and, 70; medical interests, 72–73; missionary work, 16, 64, 73

Son preference, 312–13

South African indigenous knowledge systems, 388–89; analytical framework of, 395; challenges to, 395–98; community-level innovation and,

393–95; development imperatives and, 390–92; empowerment strategy for, 399–400; relation to Western knowledge systems and, 398–99; research challenges and, 392–93

Soviet Union scientists, 405–6

Spary, Emma, 115

Spivak, Gayatri, 346–47, 365

Spontaneous generation, 36

Sputnik, 403

ssrc (Social Science Research Council), 291, 301–2, 306

*Stages of Growth* (Rostow), 298

Standpoint methodology, 18–21

Star structure navigation system, 159–69, *160–61*, *163*, *165–67*; description of, 159–62; drags and, 164–65, *165*, 169, 171; living seamarks and, 163, 170; oral traditions and, 170–71; as practical science, 171–72; in practice, 168–69; predicting weather and, 167–68; protective rituals and, 169–70; sailing direction exercises and, 162–63, *163*; schematic mapping and, 165–67, *166–67*

"Statement on Scientific Knowledge seen from Islamic Perspective," 376

Stel, Simon van der, 70–72

Stephens, Joanna, 122

Strathern, Marilyn, 350, 355

Stratification, 300

Structuralism, 176

Subaltern studies, 346–47, 357, 360 n. 3

Sufism, 375

Sugar tree, 114–15

Sustainability science, 214, 393, 407, 414–15, 429, 439–41, 451–53. *See also* Civic science

Sustainable agriculture, 429–31

Sustainable Agriculture Research and Education, 429

Sydenham, Thomas, 121

*Systema Naturae* (Linneaus), 108

Tainos, 111, 113, 118

Talbor, Robert, 120

Technology: imperialism and, 33; natural resources and, 318; role of, in world power, 137; social studies of, 7; S&T programs and, 216–17; universal design and, 423; wealth distribution and, 410, 431. *See also* Postcolonial science/technology theory; Science(s)

Telescopes, 103

Theobald, Robin, 293

Third World: Cold War and, 3; debt crises and, 264; development discourse on, 272–78; development policies for, 264–66, 270; First World construct and, 22 n. 1; media images of, 278; nation-building in, 293; representations of, 272; science and, 7–8, 23 n. 6, 368, 373–79; theory and, 6; women and, representation of, 274–75, 299

Third World Network, 8, 322, 365, 368

Thouin, André, 115

Thread maps, 62–64

Thunberg, Charles, 120

Timmermann, Barbara, 353

TNCS (Trans-national corporations), 389

Todorov, Tzvetan, 151

Tradition-modernity dualism, 290–91, 294–96, 303

Trans-national corporations (TNCS), 389

Travel: increased, 64–65; long-distance companies and, 62, 65–66; regulation of, 68–69; Scientific Revolution and, 61

Traweek, Sharon, 263, 422

Tree names, Enga, 200–202

Trigger fish schematic mapping, 165–67, 166

Triumphalism, 6, 9, 41, 46–47, 53, 152, 368, 369

Tropical medicine. *See* Botanicals; Ethnobotany

Truman doctrine, 269–70

Turnbull, David, 365

UANM (National Autonomous University), 349–51, 353, 355, 359

UN. *See* United Nations (UN)

UNCED (Conference on Environment and Development), 247, 389

UNDP (UN Development Programme), 389

UNEP (UN Environmental Programme), 389

UNESCO, 7–8, 331–32, 339, 390

Union for the Protection of Plant Varieties (UPOV), 233

Union of Concerned Scientists, 136

United Nations (UN): Conference on Environment and Development (UNCED), 247, 389; Convention on Biological Diversity (CBD), 344, 348, 349; Convention on the Elimination of All Forms of Discrimination Against Women, 320; Department of Social and Economic Affairs, 269; Development Programme (UNDP), 389; on economic development, 2, 270; Environmental Programme (UNEP), 389; National Research Council, 247–48

United Nations Educational, Scientific and Cultural Organization (UNESCO), 7–8, 331–32, 339, 390

United States Agency for International Development (USAID), 182, 264, 281

Universal humanism, 333

Universal truths, 327

University of Pennsylvania, 424

UPOV (Union for the Protection of Plant Varieties), 233

Urbanization, 269, 311, 318

USAID (United States Agency for International Development), 182, 264, 281

U.S. Global Climate Change Research Program (USGCRP), 405

Verenigde Oost-indische Compagnie. *See* Dutch East India Company
Vespucci, Amerigo, 81 n. 7
Vespucci, Juan, 67
Virtue, 39, 121, 376
VOC. *See* Dutch East India Company
Voltaire, 89–92

Wagner, Roy, 180
Ward's pills, 122
Warmaking and environmental degradation, 318
Watson-Verran, Helen, 365
*Wealth and Poverty of Nations, The* (Landes), 40–41
Weber, Max: on capitalism, 50, 52; Orientalist foundations of Weberianism, 40, 49–53
*We Have Never Been Modern* (Latour), 346
West: appropriation of Eastern resources and, 56–58; balance of power in, 51–52; as inferior, 54; patriarchal construction of, 44–45; pristine myth of, 39–40, 46, 50, 58; as separate from East, 39, 51; as superior to East, 43–47. *See also* Eurocentrism
Western Caroline Islands, navigation system, 159–69, *160–61, 163*; living seamarks and, 164, 170; oral traditions and, 170–71; predicting weather and, 167–68; schematic mapping and, 165–67, *166–67*
West Indies: biocontact zones and, 116–19; drug prospecting in, 111–15,
120–21, 123–24; transcultural exchange in, 118
White, Lynn, 54
WHO (World Health Organization), 314, 340
WID. *See under* Women
Wild pig balm, 114–15
Willis formula, 47
Wolf, Eric, 39, 46
Women: agricultural technologies and, 214; biological resources and, 250; in development (WID), 279; freedom of, 318–19; as healers, 121–22; indigenous knowledge and, 121–22, 211–13, 217–23, 250; modernity and, 292–93; population expansion and, 165; promotion of diversity by, 219–21; rice cultivation and, 142; rights of, 15, 293, 320; standpoint methodology and, 19–21; sterilization of, 319, 322; in S&T programs, 216–17; subordination of, 291, 300, 305–7, 312, 315; in Third World, representation of, 274–75, 299. *See also* Feminism; Gender
Wood, Peter, 141
World Bank, 264, 278, 281, 307
World Congress of Indigenous Peoples, 322
World Health Organization (WHO), 314, 340
World peace, 1–2, 263, 270
World Summit for Sustainable Development, 452
World Trade Organization (WTO), 349, 389

Young, Robert J., 1

SANDRA HARDING is professor of education and women's studies at the University of California, Los Angeles. She is a philosopher, and she co-edited *Signs: Journal of Women in Culture and Society* from 2000 to 2005. She has authored or edited fifteen books, including *Sciences from Below: Feminisms, Postcolonialities, and Modernities* (Duke University Press, 2008), *The Feminist Standpoint Theory Reader: Intellectual and Political Controversies* (Routledge, 2004), and *Is Science Multicultural? Postcolonialisms, Feminisms, and Epistemologies* (Indiana University Press, 1998).

Library of Congress Cataloging-in-Publication Data

The postcolonial science and technology studies reader /
edited by Sandra Harding.
p. cm.
Includes bibliographical references and index.
ISBN 978-0-8223-4936-5 (cloth : alk. paper)
ISBN 978-0-8223-4957-0 (pbk. : alk. paper)
1. Science—Social aspects. 2. Feminism and science.
3. Science and civilization. 4. Postcolonialism.
I. Harding, Sandra G.
Q175.5.P678 2011
303.48'3—dc22
2011006409